AF510402

IET ENERGY ENGINEERING 153

# Signal Processing for Fault Detection and Diagnosis in Electric Machines and Systems

**Other volumes in this series:**

| | |
|---|---|
| Volume 1 | **Power Circuit Breaker Theory and Design** C.H. Flurscheim (Editor) |
| Volume 4 | **Industrial Microwave Heating** A.C. Metaxas and R.J. Meredith |
| Volume 7 | **Insulators for High Voltages** J.S.T. Looms |
| Volume 8 | **Variable Frequency AC Motor Drive Systems** D. Finney |
| Volume 10 | **$SF_6$ Switchgear** H.M. Ryan and G.R. Jones |
| Volume 11 | **Conduction and Induction Heating** E.J. Davies |
| Volume 13 | **Statistical Techniques for HighVoltage Engineering** W. Hauschild and W. Mosch |
| Volume 14 | **Uninterruptible Power Supplies** J. Platts and J.D. St Aubyn (Editors) |
| Volume 15 | **Digital Protection for Power Systems** A.T. Johns and S.K. Salman |
| Volume 16 | **Electricity Economics and Planning** T.W. Berrie |
| Volume 18 | **Vacuum Switchgear** A. Greenwood |
| Volume 19 | **Electrical Safety: A guide to causes and prevention of hazards** J. Maxwell Adams |
| Volume 21 | **Electricity Distribution Network Design, 2nd Edition** E. Lakervi and E.J. Holmes |
| Volume 22 | **Artificial Intelligence Techniques in Power Systems** K. Warwick, A.O. Ekwue and R. Aggarwal (Editors) |
| Volume 24 | **Power System Commissioning and Maintenance Practice** K. Harker |
| Volume 25 | **Engineers' Handbook of Industrial Microwave Heating** R.J. Meredith |
| Volume 26 | **Small Electric Motors** H. Moczala *et al.* |
| Volume 27 | **AC-DC Power System Analysis** J. Arrillaga and B.C. Smith |
| Volume 29 | **High Voltage Direct Current Transmission, 2nd Edition** J. Arrillaga |
| Volume 30 | **Flexible AC Transmission Systems (FACTS)** Y-H. Song (Editor) |
| Volume 31 | **Embedded Generation** N. Jenkins *et al.* |
| Volume 32 | **High Voltage Engineering and Testing, 2nd Edition** H.M. Ryan (Editor) |
| Volume 33 | **Overvoltage Protection of Low-Voltage Systems, Revised Edition** P. Hasse |
| Volume 36 | **Voltage Quality in Electrical Power Systems** J. Schlabbach *et al.* |
| Volume 37 | **Electrical Steels for Rotating Machines** P. Beckley |
| Volume 38 | **The Electric Car: Development and future of battery, hybrid and fuel-cell cars** M. Westbrook |
| Volume 39 | **Power Systems Electromagnetic Transients Simulation** J. Arrillaga and N. Watson |
| Volume 40 | **Advances in High Voltage Engineering** M. Haddad and D. Warne |
| Volume 41 | **Electrical Operation of Electrostatic Precipitators** K. Parker |
| Volume 43 | **Thermal Power Plant Simulation and Control** D. Flynn |
| Volume 44 | **Economic Evaluation of Projects in the Electricity Supply Industry** H. Khatib |
| Volume 45 | **Propulsion Systems for Hybrid Vehicles** J. Miller |
| Volume 46 | **Distribution Switchgear** S. Stewart |
| Volume 47 | **Protection of Electricity Distribution Networks, 2nd Edition** J. Gers and E. Holmes |
| Volume 48 | **Wood Pole Overhead Lines** B. Wareing |
| Volume 49 | **Electric Fuses, 3rd Edition** A. Wright and G. Newbery |
| Volume 50 | **Wind Power Integration: Connection and system operational aspects** B. Fox *et al.* |
| Volume 51 | **Short Circuit Currents** J. Schlabbach |
| Volume 52 | **Nuclear Power** J. Wood |
| Volume 53 | **Condition Assessment of High Voltage Insulation in Power System Equipment** R.E. James and Q. Su |
| Volume 55 | **Local Energy: Distributed generation of heat and power** J. Wood |
| Volume 56 | **Condition Monitoring of Rotating Electrical Machines** P. Tavner, L. Ran, J. Penman and H. Sedding |
| Volume 57 | **The Control Techniques Drives and Controls Handbook, 2nd Edition** B. Drury |
| Volume 58 | **Lightning Protection** V. Cooray (Editor) |
| Volume 59 | **Ultracapacitor Applications** J.M. Miller |
| Volume 62 | **Lightning Electromagnetics** V. Cooray |
| Volume 63 | **Energy Storage for Power Systems, 2nd Edition** A. Ter-Gazarian |
| Volume 65 | **Protection of Electricity Distribution Networks, 3rd Edition** J. Gers |
| Volume 66 | **High Voltage Engineering Testing, 3rd Edition** H. Ryan (Editor) |

| | |
|---|---|
| Volume 67 | **Multicore Simulation of Power System Transients** F.M. Uriate |
| Volume 68 | **Distribution System Analysis and Automation** J. Gers |
| Volume 69 | **The Lightening Flash, 2nd Edition** V. Cooray (Editor) |
| Volume 70 | **Economic Evaluation of Projects in the Electricity Supply Industry, 3rd Edition** H. Khatib |
| Volume 72 | **Control Circuits in Power Electronics: Practical issues in design and implementation** M. Castilla (Editor) |
| Volume 73 | **Wide Area Monitoring, Protection and Control Systems: The enabler for smarter grids** A. Vaccaro and A. Zobaa (Editors) |
| Volume 74 | **Power Electronic Converters and Systems: Frontiers and applications** A.M. Trzynadlowski (Editor) |
| Volume 75 | **Power Distribution Automation** B. Das (Editor) |
| Volume 76 | **Power System Stability: Modelling, analysis and control** A.A. Sallam and Om P. Malik |
| Volume 78 | **Numerical Analysis of Power System Transients and Dynamics** A. Ametani (Editor) |
| Volume 79 | **Vehicle-to-Grid: Linking electric vehicles to the smart grid** J. Lu and J. Hossain (Editors) |
| Volume 81 | **Cyber-PhysicalSocial Systems and Constructs in Electric Power Engineering** S. Suryanarayanan, R. Roche and T.M. Hansen (Editors) |
| Volume 82 | **Periodic Control of Power Electronic Converters** F. Blaabjerg, K. Zhou, D. Wang and Y. Yang |
| Volume 86 | **Advances in Power System Modelling, Control and Stability Analysis** F. Milano (Editor) |
| Volume 87 | **Cogeneration: Technologies, optimisation and implementation** C.A. Frangopoulos (Editor) |
| Volume 88 | **Smarter Energy: From smart metering to the smart grid** H. Sun, N. Hatziargyriou, H.V. Poor, L. Carpanini and M.A. Sánchez Fornié (Editors) |
| Volume 89 | **Hydrogen Production, Separation and Purification for Energy** A. Basile, F. Dalena, J. Tong and T.N. Veziroðlu (Editors) |
| Volume 90 | **Clean Energy Microgrids** S. Obara and J. Morel (Editors) |
| Volume 91 | **Fuzzy Logic Control in Energy Systems with Design Applications in MATLAB®/Simulink®** İ.H. Altaş |
| Volume 92 | **Power Quality in Future Electrical Power Systems** A.F. Zobaa and S.H.E.A. Aleem (Editors) |
| Volume 93 | **Cogeneration and District Energy Systems: Modelling, analysis and optimization** M.A. Rosen and S. Koohi-Fayegh |
| Volume 94 | **Introduction to the Smart Grid: Concepts, technologies and evolution** S.K. Salman |
| Volume 95 | **Communication, Control and Security Challenges for the Smart Grid** S.M. Muyeen and S. Rahman (Editors) |
| Volume 96 | **Industrial Power Systems with Distributed and Embedded Generation** R. Belu |
| Volume 97 | **Synchronized Phasor Measurements for Smart Grids** M.J.B. Reddy and D.K. Mohanta (Editors) |
| Volume 98 | **Large Scale Grid Integration of Renewable Energy Sources** A. Moreno-Munoz (Editor) |
| Volume 100 | **Modeling and Dynamic Behaviour of Hydropower Plants** N. Kishor and J. Fraile-Ardanuy (Editors) |
| Volume 101 | **Methane and Hydrogen for Energy Storage** R. Carriveau and D.S-K. Ting |
| Volume 104 | **Power Transformer Condition Monitoring and Diagnosis** A. Abu-Siada (Editor) |
| Volume 106 | **Surface Passivation of Industrial Crystalline Silicon Solar Cells** J. John (Editor) |
| Volume 107 | **Bifacial Photovoltaics: Technology, applications and economics** J. Libal and R. Kopecek (Editors) |
| Volume 108 | **Fault Diagnosis of Induction Motors** J. Faiz, V. Ghorbanian and G. Joksimović |

| | |
|---|---|
| Volume 110 | **High Voltage Power Network Construction** K. Harker |
| Volume 111 | **Energy Storage at Different Voltage Levels: Technology, integration, and market aspects** A.F. Zobaa, P.F. Ribeiro, S.H.A. Aleem and S.N. Afifi (Editors) |
| Volume 112 | **Wireless Power Transfer: Theory, technology and application** N. Shinohara |
| Volume 114 | **Lightning-induced Effects in Electrical and Telecommunication Systems** Y. Baba and V.A. Rakov |
| Volume 115 | **DC Distribution Systems and Microgrids** T. Dragičević, F. Blaabjerg and P. Wheeler |
| Volume 117 | **Structural Control and Fault Detection of Wind Turbine Systems** H.R. Karimi |
| Volume 119 | **Thermal Power Plant Control and Instrumentation: The control of boilers and HRSGs, 2nd Edition** D. Lindsley, J. Grist and D. Parker |
| Volume 120 | **Fault Diagnosis for Robust Inverter Power Drives** A. Ginart (Editor) |
| Volume 121 | Monitoring and Control using Synchrophasors in Power Systems with Renewables I. Kamwa and C. Lu (Editors) |
| Volume 123 | **Power Systems Electromagnetic Transients Simulation, 2nd Edition** N. Watson and J. Arrillaga |
| Volume 124 | **Power Market Transformation** B. Murray |
| Volume 125 | **Wind Energy Modeling and Simulation Volume 1: Atmosphere and plant** P. Veers (Editor) |
| Volume 126 | **Diagnosis and Fault Tolerance of Electrical Machines, Power Electronics and Drives** A.J.M. Cardoso |
| Volume 128 | **Characterization of Wide Bandgap Power Semiconductor Devices** F. Wang, Z. Zhang and E.A. Jones |
| Volume 129 | **Renewable Energy from the Oceans: From wave, tidal and gradient systems to offshore wind and solar** D. Coiro and T. Sant (Editors) |
| Volume 130 | **Wind and Solar Based Energy Systems for Communities** R. Carriveau and D.S-K. Ting (Editors) |
| Volume 131 | **Metaheuristic Optimization in Power Engineering** J. Radosavljević |
| Volume 132 | **Power Line Communication Systems for Smart Grids** I.R.S Casella and A. Anpalagan |
| Volume 139 | **Variability, Scalability and Stability of Microgrids** S.M. Muyeen, S.M. Islam and F. Blaabjerg (Editors) |
| Volume 145 | **Condition Monitoring of Rotating Electrical Machines** P. Tavner, L. Ran and C. Crabtree |
| Volume 146 | **Energy Storage for Power Systems, 3rd Edition** A.G. Ter-Gazarian |
| Volume 147 | **Distribution Systems Analysis and Automation 2nd Edition** J. Gers |
| Volume 152 | **Power Electronic Devices: Applications, failure mechanisms and reliability** F Iannuzzo (Editor) |
| Volume 155 | **Energy Generation and Efficiency Technologies for Green Residential Buildings** D. Ting and R. Carriveau (Editors) |
| Volume 157 | **Electrical Steels, 2 Volumes** A. Moses, K. Jenkins, Philip Anderson and H. Stanbury |
| Volume 158 | **Advanced Dielectric Materials for Electrostatic Capacitors** Q Li (Editor) |
| Volume 159 | **Transforming the Grid towards Fully Renewable Energy** O. Probst, S. Castellanos and R. Palacios (Editors) |
| Volume160 | **Microgrids for Rural Areas: Research and case studies** R.K. Chauhan, K. Chauhan and S.N. Singh (Editors) |
| Volume 166 | **Advanced Characterization of Thin Film Solar Cells** N. Haegel and M. Al-Jassim (Editors) |
| Volume 167 | **Power Grids with Renewable Energy Storage, Integration and Digitalization** A.S. Sallam and O.P. Malik |
| Volume 172 | **Lighting Interaction with Power Systems, 2 volumes** A. Piantini (Editor) |
| Volume 905 | **Power System Protection, 4 Volumes** |

# Signal Processing for Fault Detection and Diagnosis in Electric Machines and Systems

Edited by
Mohamed Benbouzid

The Institution of Engineering and Technology

Published by The Institution of Engineering and Technology, London, United Kingdom

The Institution of Engineering and Technology is registered as a Charity in England & Wales (no. 211014) and Scotland (no. SC038698).

© The Institution of Engineering and Technology 2021

First published 2020

This publication is copyright under the Berne Convention and the Universal Copyright Convention. All rights reserved. Apart from any fair dealing for the purposes of research or private study, or criticism or review, as permitted under the Copyright, Designs and Patents Act 1988, this publication may be reproduced, stored or transmitted, in any form or by any means, only with the prior permission in writing of the publishers, or in the case of reprographic reproduction in accordance with the terms of licences issued by the Copyright Licensing Agency. Enquiries concerning reproduction outside those terms should be sent to the publisher at the undermentioned address:

The Institution of Engineering and Technology
Michael Faraday House
Six Hills Way, Stevenage
Herts, SG1 2AY, United Kingdom

www.theiet.org

While the authors and publisher believe that the information and guidance given in this work are correct, all parties must rely upon their own skill and judgement when making use of them. Neither the authors nor publisher assumes any liability to anyone for any loss or damage caused by any error or omission in the work, whether such an error or omission is the result of negligence or any other cause. Any and all such liability is disclaimed.

The moral rights of the authors to be identified as authors of this work have been asserted by them in accordance with the Copyright, Designs and Patents Act 1988.

**British Library Cataloguing in Publication Data**
A catalogue record for this product is available from the British Library

**ISBN 978-1-78561-957-1 (Hardback)**
**ISBN 978-1-78561-958-8 (PDF)**

Typeset in India by MPS Limited

# Contents

# About the editors

**Mohamed Benbouzid** is a Full Professor of Electrical Engineering at the University of Brest, France. He is a Distinguished Professor and 1000 Talent Expert at the Shanghai Maritime University, Shanghai, China. He is an IEEE and IET Fellow. He is the Editor-in-Chief of the International Journal on Energy Conversion and the Applied Sciences Section on Electrical, Electronics and Communications Engineering. He is a Subject Editor for the IET Renewable Power Generation received the Ph.D. degree in electrical and computer engineering from the National Polytechnic Institute of Grenoble, Grenoble, France, in 1994, and the Habilitation à Diriger des Recherches degree from the University of Amiens, France, in 2000.

After receiving the Ph.D. degree, he joined the University of Amiens where he was an Associate Professor of electrical and computer engineering. Since September 2004, he has been with the University of Brest, Brest, France, where he is a Full Professor of electrical engineering. Prof. Benbouzid is also a Distinguished Professor and a 1000 Talent Expert at the Shanghai Maritime University, Shanghai, China. His main research interests and experience include analysis, design, and control of electric machines, variable-speed drives for traction, propulsion, and renewable energy applications, and fault diagnosis of electric machines.

Professor Benbouzid has been elevated as an IEEE Fellow for his contributions to diagnosis and fault-tolerant control of electric machines and drives. He is also a Fellow of the IET. He is the Editor-in-Chief of the *International Journal on Energy Conversion and the Applied Sciences* (MDPI) Section on Electrical, Electronics and Communications Engineering. He is a Subject Editor for the IET *Renewable Power Generation*. He is also an Associate Editor of the *IEEE Transactions on Energy Conversion*.

## About the contributing authors

**Yassine Amirat** is an Associate Professor at ISEN Yncréa Ouest, Brest, France, and an affiliated member of the Institut de Recherche Dupuy de Lôme (UMR CNRS 6027). He is an IEEE Senior Member, an Associate Editor for the Springer *Electrical Engineering Journal*, and an Editor of the MDPI *Journal of Marine Science and Engineering*. He is also interested in renewable energy applications such as wind turbines, marine current turbines and hybrid generation systems.

**Claude Delpha** is currently Associate Professor, in Université Paris Saclay, France. He is graduated in Electrical and Signal Processing Engineering and obtained his PhD in the field of Instrumentation, Measurements and Signal Processing with smart sensors applications. Since 2001. He is with the Laboratoire des Signaux et Systèmes (CNRS, CentraleSupelec) and works on signal processing solutions for complex systems security. His main areas of interests are fault diagnosis and prognosis (modelling, detection, estimation) and data hiding.

**Demba Diallo** is a Full Professor at the Université Paris-Saclay. He received the MSc and PhD degrees in Electrical and Computer Engineering, from the National Polytechnic Institute of Grenoble, France, in 1990 and 1993, respectively. He is currently with the Group of Electrical Engineering Paris. His current area of research includes fault detection and diagnosis, fault tolerant control and energy management. The applications are related to more electrified transportation systems (EV and HEV) and microgrids.

**Elhoussin Elbouchikhi** is an Associate Professor at ISEN Yncréa Ouest, LABISEN, Brest, France, and an affiliated member of the Institut de Recherche Dupuy de Lôme (UMR CNRS 6027). He is an IEEE Senior Member, an Associate Editor for IET Generation, Transmission & Distribution and a Member of the Editorial Board of the MDPI Energies journal. His main current research interests include diagnosis, fault-tolerant control, energy management systems in microgrids, power electronics, and renewable energy applications.

**Lotfi Saidi** received the PhD degree in electrical engineering from the Université de Tunis, Tunisia, in 2014. He is an Associate Professor and head of electronics and computer engineering department at the University of Sousse, Tunisia. Dr Saidi is an IEEE Senior Member and his research interests include the application of advanced signal processing tools for electrical machines condition monitoring; prognosis and health management (PHM) of power converters; PHM of batteries; and electrical systems' remaining useful life prediction.

**Tianzhen Wang** is a Full Professor and Doctoral Supervisor with the Department of Electrical and Automation, Shanghai Maritime University. Prof. Tianzhen is an IEEE Senior Member. Her awards and honours include a Committee Member of fault diagnosis and safety on the Technical Process Specialized Committee, and the Cognitive Computing and System Specialized Committee China Automation Society. Her research interests include fault diagnosis, and fault-tolerant control and their application in inverters and renewable energy conversions systems.

# Introduction

*Mohamed Benbouzid*[1] *and Demba Diallo*[2]

The search for competitiveness and growth gains has contributed for three decades now to the evolution of maintenance policies. Indeed, the industry has moved from passive maintenance to active maintenance intending to improve productivity. This active maintenance requires continuous monitoring of industrial systems in order to increase reliability and availability rates, and guarantee the safety of people and property. Up to now, for electromechanical systems condition monitoring, vibration sensors are still preferred. However, electrical signals (e.g. currents flowing in machine windings), usually already measured for control purposes, are also becoming popular, as it does not require additional cost. Moreover, the current analysis has several advantages since it is a non-invasive technique.

Indeed, electrical current processing-based fault detection and diagnosis of electromechanical systems have received intense research interest for several decades. Moreover, the International Standard 'ISO FDIS 20958' dealing with 'Condition monitoring and diagnostics of machine systems – Electrical signature analysis of three-phase induction motors' sets out guidelines for the online techniques recommended for condition monitoring and diagnostics of machines, based on the electrical signature analysis. Hence, many studies have shown that relevant fault features could be extracted from the time-series or spectrum of the currents flowing in the machine windings. In the time domain, several characteristics or features can be processed using statistical tools, residuals from observers or machine learning techniques. In the frequency domain, most of the used fault detection and diagnosis techniques perform spectral analysis, such as Fourier or MUltiple SIgnal Classification (MUSIC) techniques. Although these techniques exhibit good results, they have several drawbacks. Because systems are becoming more complex, accurate analytical models necessary for observer-based methods are tedious to obtain. Moreover, the systems have variable operating points and are highly integrated, creating complex interactions difficult to cope within physics-based models. As a consequence, these methods may fail because they cannot handle non-stationary conditions, closely spaced frequencies, incipient faults and harsh environments (noise). In this context, it is then

[1]Institut de Recherche Dupuy de Lôme, CNRS, University of Brest, Brest, France
[2]Group of Electrical Engineering Paris, CNRS, CentraleSupelec, University of Paris-Saclay, Gif/Yvette, France

obvious that several challenges have to be addressed for fault detection and diagnosis in such applications using specific signal processing tools.

In this challenging context, this book, mainly research-oriented, identifies opportunities of advanced signal processing techniques for electromechanical systems' fault detection and diagnosis. It provides methodologies and algorithms with several illustrative examples and practical case studies, and includes extensive application features not found in academic textbooks. This book is primarily intended for researchers and postgraduates in the field of fault detection and diagnosis.

Chapter 1 deals with the use of parametric spectral estimation techniques and detection theory. This approach is used for fault characteristics estimation. The generalized likelihood ratio test is used for automatic decision-making. The proposed fault detection approach uses fault frequency signature bins and amplitude estimators, and a fault decision module based on statistical tools. Maximum likelihood estimator is used for fault characteristics computation. Then, composite hypothesis testing is used as a decision module. The main objective is to discriminate a healthy system from a faulty one. Fault severity measurement criterion is also proposed.

Chapter 2 deals with the use of demodulation techniques for fault detection. Indeed, most of the electrical machine faults lead to current modulation (amplitude and or phase). In this context, fault detection and diagnosis will rely on the extraction of the instantaneous amplitude and or the instantaneous frequency. It is, therefore, sufficient to demodulate the current signal for fault detection and diagnosis purposes. However, demodulation techniques depend on the signal type and dimension. This chapter will specifically highlight the use of demodulation techniques for mono- and multi-dimensional signals, and for mono- and multicomponent signals.

Chapter 3 deals with Kullback–Leibler divergence-based methodology for early fault detection and diagnosis. A four-step methodology is proposed, including modelling, preprocessing, feature extraction and feature analysis. After the definition of incipient fault based on the levels of fault, signal and environmental nuisances, a paradigm is drawn between information-hiding domain and fault detection and diagnosis. The chapter will show that the dissimilarity measure of the probability density function used for data hiding is efficient for incipient fault detection.

Chapter 4 deals with the fault detection and diagnosis applicability issue of higher-order spectra. Indeed, a significant problem with this kind of signal processing tool is the interpretation of the obtained results because much uncertainty still exists about the relation between higher-order spectra contribution compared to second-order statistics. In this context, various higher-order spectra-based algorithms and their challenging problems are discussed.

Chapter 5 deals with another fault detection and diagnosis approach. Indeed, principal component analysis (PCA) and mainly its improved versions are explored – here, the relative and normalized PCAs, which address the PCA limitations for fault detection and diagnosis in complex systems.

*Chapter 1*

# Parametric signal processing approach

*Elhoussin Elbouchikhi[1] and Mohamed Benbouzid[2]*

Induction machines are characterized by their ruggedness, reliability, efficiency, easiness of control, and attractive cost. Moreover, advances in power converters have significantly enhanced the performance of electrical machines and drives, and have extended their use to variable speed drives in both industrial applications and energy production systems. However, their useful life has been decreased due to failures and performance reduction. In fact, electrical drives are subject to various failures that may affect power supply devices, electrical machines, control devices, cables, connections, and protective devices. Moreover, mechanical loads can fail due to bearings wearing or breakage, shaft misalignment, gearbox defects, etc.

These failures have increased the need for predictive maintenance to increase electrical drives resilience and lifetime. Predictive maintenance can be performed using condition-based monitoring. It offers the possibility to schedule the maintenance activities depending on the operating conditions. Moreover, it allows detecting incipient faults and preventing breakdowns. Consequently, electrical drives components maximize their lifespan leading to reduction of downtime and maintenance costs. These condition monitoring techniques require additional hardware and software, which increases the system complexity and cost. However, electrical drives overall maintenance cost is significantly decreased.

Condition monitoring can be performed based on some physical signals processing such as vibration, temperature, and voltage/currents analysis. Motor current signature analysis (MCSA) is a cost-effective solution since it does not require additional sensors and can be easily implemented [1,2]. Moreover, stator currents are often measured on the drive system for control and protection purposes. Various approaches have been proposed to perform the stator current processing such as fast Fourier transform (FFT), MUSIC (MUltiple SIgnal Characterization), ESPRIT (Estimation of Signal Parameters via Rotational Invariance Techniques) [2], and MLE (maximum likelihood estimator) [3,4]. For decision-making, sophisticated techniques have been investigated such support vector machines (SVMs) [5], artificial neural networks (ANNs) [6], and fuzzy logic [7].

[1]ISEN Yncréa Ouest, LABISEN, Brest, France
[2]Institut de Recherche Dupuy de Lôme, CNRS, University of Brest, Brest, France

This chapter addresses the issue of condition monitoring based on MCSA using parametric spectral estimation techniques and detection theory. This approach is used for fault characteristics estimation. Then, generalized likelihood ratio test (GLRT) is used for automatic decision-making. The proposed fault detection approach uses fault frequency signature bins and amplitude estimators, and a fault decision module based on statistical tools. MLE is used for fault characteristics computation. Then, composite hypothesis testing is used as a decision module. The main objective is to discriminate the healthy induction motor from a faulty one. Finally, a fault severity measurement criterion is proposed and demonstrated for several induction motor fault detection.

## 1.1 Fault effects on intrinsic parameters of electromechanical systems

### *1.1.1 Main failures and occurrence frequency*

Electrical machines and drives can be affected by several failures that can be combined. These failures include stator faults such as open or short-circuit stator phase windings, broken rotor bar or managed rotor end rings, permanent magnets demagnetization, static and/or dynamic air-gap eccentricity, bent shaft and misalignment, bearings and gears breakage, and power electronic components failure of the drive system such as switching devices. These faults can be classified into two categories, namely electrical and mechanical faults.

Electrical faults include stator and rotor faults. In general, stator faults are much more frequent than rotor faults [8]. Nonetheless, broken rotor bars and magnets demagnetization are critical since they can lead to catastrophic failure. This can be justified by the following reasons:

- Rotor faults are hard to detect since rotor electrical quantities are not accessible for measurement.
- Stator electrical faults have been reduced by recent advances in stator winding design, manufacturing processes, and insulation high performance.

Stator winding failures include open-phase fault and short circuit of few turns of phase windings. Turns short-circuit fault can lead to catastrophic failure if not detected at an early stage. Regarding rotor faults, squirrel-cage rotors are subject to two main failures, which are bars and end-ring segments damage. Broken end ring appears especially in case of fabricated squirrel cage, in contrary to cast cage that are more rugged. Moreover, permanent magnets on the rotor of permanent magnet synchronous machine (PMSM) can experience demagnetization as their magnetic force can weaken locally or uniformly.

Mechanical faults enclose bearing faults, air-gap irregularities, bent shaft, and misalignment. Bearing fault causes include inherent eccentricity, vibration, internal stresses, and bearing currents due to power electronics. Air-gap eccentricity includes static, dynamic, and mixed (or combined) eccentricities [9]. Eccentricity faults can be due to bad bearing positioning during motor assembly, worn bearing, bent rotor shaft, or operation under a critical speed and excessive load [10]. These mechanical faults cause excessive mechanical stress on the machine and increase the bearing wear and induce torque oscillations. Moreover, eccentricity can lead to radial magnetic force, which may expose the stator windings to harmful vibrations. The second significant effect of mechanical faults is load torque oscillations. This failure is characterized by load torque periodic variations which lead to mechanical speed oscillations. This fault can be also due to load unbalance, shaft misalignment, and bearing and gearbox faults.

The distribution of the aforementioned failures within electrical machines and power electronics subassemblies is reported in several reliability surveys [8,11] and are summarized in Figure 1.1. It is worth to notice that bearings and insulation are the Achilles heels in the electromechanical systems. Moreover, as the power rating of the electrical drives increases, the reliability of the power electronic becomes more critical and maintenance cost higher. The failure rate distribution in power electronics is shown in Figure 1.1(b).

### *1.1.2 Origins and consequences*

Electrical drives failures are due to several causes, which are related to design, manufacturing, or employment processes. These causes include internal and external causes as shown in Figure 1.2(a) and (b), respectively. Electromechanical systems' fault origins are manifold and can be summarized as follows:

- Electrical origins: copper wear, voltage stresses owing to power electronics usage, high power variations in inductive circuits, windings voltage inhomogeneous distribution, deterioration of insulating materials, and common mode voltages and currents caused by capacitive and inductive coupling of the circuit composed of the rotor, the shaft, and the two bearings;
- Mechanical causes: tangential and radial forces due to the presence of the magnetic field, air-gap irregularities, mechanical vibration, and frictional wear for bearings;
- Thermal stresses: mechanical overload, unbalanced power supply, large number of consecutive starts, insulation aging, and poor cooling;
- Environmental reasons: high ambient temperature, contaminated environment due to dust, humidity, and air acidity.

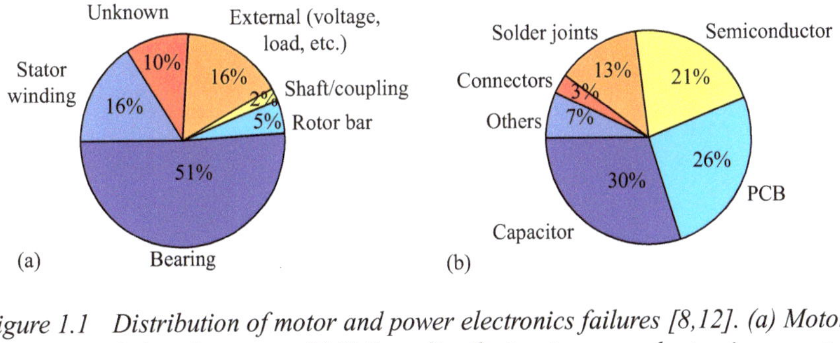


*Figure 1.1 Distribution of motor and power electronics failures [8,12]. (a) Motor failure frequency. (b) Failure distribution in power electronic converters*

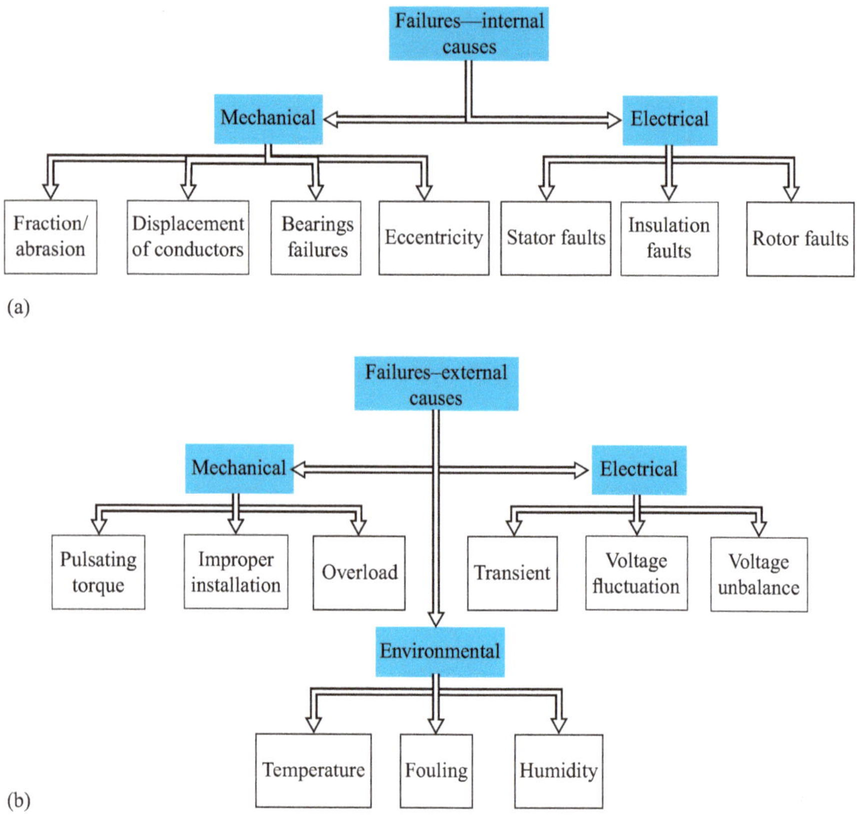


*Figure 1.2 Main causes of electrical machines failures [13]. (a) Internal causes. (b) External causes*

These faults can have several consequences such as magnetic field distortion, overheating phenomena, risks of electric arcs, vibration effects, abnormally high or destructive currents, electromechanical torque oscillation, noise, problem of additional torque, and risk of stator damages. These faults can lead to catastrophic failures if undetected at an incipient stage. Consequently, it is mandatory to develop a maintenance approach that ensures electrical machines reliability, availability, and safety, at minimum cost.

### *1.1.3 Condition-based maintenance*

Maintenance is mandatory in various industrial applications. It can be classified as either corrective, preventive, or predictive maintenance. Corrective maintenance is intended to repair the system after a failure. It suffers from several disadvantages such as production loss, reliability decrease, and risk of catastrophic failures. In contrary, preventive maintenance is required to reduce failure probability and includes planned, predictive, and condition-based maintenances. Unlike planned maintenance, which is carried out at predetermined intervals, condition-based maintenance allows finding the optimum time for performing the required maintenance actions by supervising the current state of the system components. Furthermore, it allows minimizing downtime and repair costs. This kind of maintenance is still under investigation and several approaches have been proved to be efficient based on several transducers and various physical quantities measurement.

#### 1.1.3.1 Fault detection methods

Condition-based maintenance of electrical machines is based on performance and parameters monitoring. According to the sensor measurement used, most methods for condition monitoring could be classified into several categories: vibration monitoring, torque monitoring, temperature monitoring, oil/debris analysis, acoustic emission monitoring, optical fiber monitoring, and current/power monitoring [14,15]. The most common used techniques can be classified into two major classes:

- Model-based method: It is used to measure the deviation between the model output and the actual machine output and then predict a potential failure signature. This approach is depicted in Figure 1.3 [16,17]. The goal is to generate several symptoms indicating the difference between nominal behaviour and abnormal operating conditions.
- Signal-based approach: Any kind of fault modifies the symmetrical properties of electrical machines. Therefore, characteristic fault frequencies appear in some

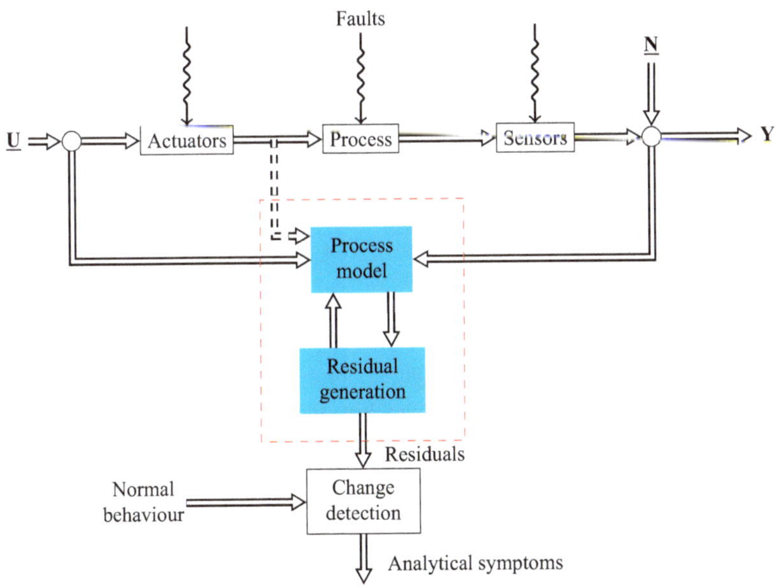


*Figure 1.3 Model-based approaches for diagnosis [16]*

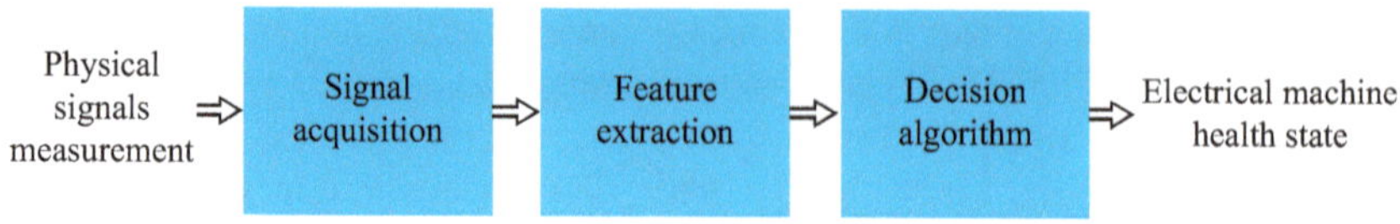


*Figure 1.4 Signal-based approaches for diagnosis*

physical signals issued from sensors. The analysis of these signals allows to enhance the knowledge about a specific fault, its impact on intrinsic parameters of the machine, and its frequency signature. Signal analysis is performed using suitable signal conditioning and processing techniques for fault features extraction. Then, a fault decision algorithm performed for distinguishing faulty cases from healthy ones and classification purposes. This approach principle is illustrated in Figure 1.4.

Compared to model-based approaches, the signal-based methods do not require any knowledge about the machine parameters. Moreover, the fault detection procedure may be performed without any knowledge about the operating conditions of the machine. A promising technique relies on current/power monitoring. It is based on current and/or voltage measurements that are already available for control and protection purposes. Nevertheless, the challenge in using current and/or voltage signals for condition monitoring is to propose signal processing techniques allowing to extract a fault detection and diagnosis criteria in stationary and non-stationary environment (variable speed drives) and smart diagnosis scheme able to classify faults and foresee a potential failure.

#### 1.1.3.2 Fault effects on stator currents

Electrical machines are a highly symmetrical electromagnetic systems. Hence, any fault can cause a certain degree of asymmetry. These failures lead to various effects on intrinsic parameters and physical quantities of the electrical machines, which can be classified in three major categories [18]:

- Faults leading to eccentricity between stator and rotor: bearing defects, shaft misalignment, and centring defect;
- Failures introducing torque oscillations: mechanical load defect and bearing faults;
- Defects leading to disturbance in magnetomotive forces (MMFs): stator short-circuit defects, broken electrical connections in the stator, and magnets failure.

These fault effects on stator currents have been widely investigated and can be broadly classified into three major contributions:

- Introduction of additional frequencies on the stator current power spectral density (PSD) depending on the type of fault and machine dimensions [1,2]. For instance, Broken rotor bar induces the increase in resistance of the broken bar, which leads to asymmetry of the resistance in rotor phases. Consequently, the broken bar induces asymmetry of the rotating electromagnetic field in the air gap, which introduces additional frequencies in the stator current.

- Phase and/or amplitude modulation of stator currents due to presence of specific fault [18,19].
- Impact over the negative-sequence component [20].

### *1.1.4 Motor current signature analysis*

Healthy electrical machine contains a great number of spectral components due to its supply voltage, rotor slotting, and possible iron saturation as it can be seen from Table 1.1. Here, $\omega_r$ denotes the rotational frequency, $\omega_c$ corresponds to fault frequency introduced by modified rotor MMF, and $N_r$ is the number of rotor bars or rotor slots.

In various works, numerical machine models and analytical development accounting for faults have been developed allowing to understand the effect of some phenomena on stator currents.

#### 1.1.4.1 Fault frequency signatures

Bearing defects have been typically categorized as distributed or local. Local defects cause periodic impulses in vibration signals. Amplitude and frequency of such impulses are determined by shaft rotational speed, fault location, and bearing dimensions (Figure 1.5). The frequencies of these impulses are given by (1.1)

$$\begin{cases} \omega_c = \frac{\omega_r}{2}\left(1 - \frac{d}{D}\cos(\alpha)\right) \\ \omega_{bd} = \frac{D}{d}\omega_r\left(1 - \frac{d^2}{D^2}\cos^2(\alpha)\right) \\ \omega_{id} = \frac{n_r}{2}\omega_r\left(1 + \frac{d}{D}\cos(\alpha)\right) \\ \omega_{od} = \frac{n_r}{2}\omega_r\left(1 - \frac{d}{D}\cos(\alpha)\right) \end{cases} \tag{1.1}$$

where $\omega_{cd}$ corresponds to fundamental cage frequency, $\omega_{bd}$ is ball defect frequency, $\omega_{id}$ is inner race defect frequency, and $\omega_{od}$ corresponds to outer race defect frequency. $\omega_r$ refers to shaft rotation frequency, $n_r$ is the number of rollers, $d_r$ is the roller diameter, $D_r$ is the pitch diameter of the bearing, and $\alpha$ is the contact angle (Figure 1.5).

*Table 1.1 Synopsis of stator current frequency components under healthy conditions*

| **Stator current harmonics** | **Frequency ($l \in \mathbb{N}$)** | **Origins** |
|---|---|---|
| Fundamental angular frequency | $\omega_s$ | Supply voltage |
| Time harmonics | $l \times \omega_s$ | Harmonics of supply voltage<br>PWM inverters |
| Rotor slot harmonics | $kN_r\omega_r \pm l\omega_s$ | Modified air gap |
| Saturation harmonics | $\omega_s \pm 2k\omega_s$ | Deformation of flux density |
| Torque/speed oscillation | $l\omega_s \pm \omega_r$ | Modified air gap |
| Eccentricity | $l\omega_s \pm \omega_r$ | Torque oscillation |

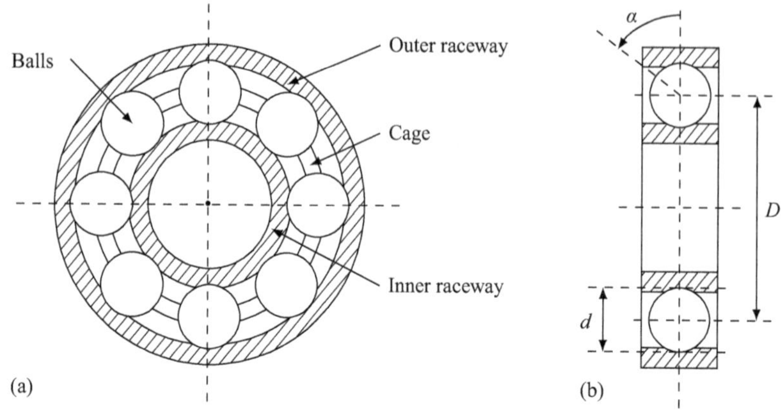


*Figure 1.5 Ball bearing structure and main characteristics*

In [21,22], it has been demonstrated that the characteristic bearing fault frequencies in vibration can be reflected on stator currents. Since ball bearings support the rotor, any bearing defect will produce a radial motion between the rotor and the stator of the machine (air-gap eccentricity), which may lead to anomalies in the air-gap flux density. As the stator current for a given phase is linked to flux density, the stator current is affected as well by the bearing defect. The relationship between vibration frequencies and current frequencies for bearing faults can be described by (1.2)

$$\omega_{bng} = |\omega_s \pm k\omega_d| \tag{1.2}$$

where $\omega_s$ is the supply fundamental frequency, $\omega_d$ is one of the characteristic vibration frequencies given above, and $k \in \mathbb{N}^*$.

Eccentricity fault effect has been studied to model the fault impact on stator current [23,24]. It has been proved that under eccentricity faults, the stator currents contain the frequencies given by (1.63). It worth noting that when other mechanical problems exist, torque oscillation characteristic frequencies may be hidden:

$$\omega_{ecc} = \omega_s \left| 1 \pm k \left( \frac{1-s}{N_p} \right) \right| \tag{1.3}$$

Broken rotor bar induces a bar resistance increase, which leads to asymmetry of the resistance in rotor equivalent phases. Consequently, broken rotor bars fault induces asymmetry of the rotating electromagnetic field in the air gap. Since stator currents are linked to the air-gap electromagnetic field, any broken rotor bar may have an effect over the stator current waveform [25]. This effect is modelled by adding some frequency components on the stator current PSD [26,27], which are located at

$$\omega_{brb} = (1 + 2ks)\omega_s \tag{1.4}$$

A summary of induction machine stator current faults-related frequencies is presented in Table 1.2. Here $N_p$ is the pole pair number, $s$ is the per unit slip, $\omega_s$ is the supply fundamental frequency, and $\omega_x$ is the fault characteristic frequency. These frequencies are used to monitor the induction machines using the stator currents. When a fault occurs, the amplitude at these frequencies increases and reveals abnormal operating conditions.

#### 1.1.4.2 Stator current AM/FM modulation

In [19,28], the authors have presented an analytical approach for the modelling of mechanical and bearing faults based on traditional MMF and permeance wave approach for computation of the air-gap magnetic flux density [29]. These studies have demonstrated two main results:

- Mechanical faults lead to eccentricity and load oscillation faults. The eccentricity fault is responsible for the amplitude modulation of the stator currents and the load oscillation leads to frequency modulation of the stator currents. The modulation frequency depends on the operating conditions of the machine and the fault severity. Therefore, the fault severity is proportional to the modulation index.
- Depending on the defective components of the bearings, the stator currents are modulated (amplitude and/or frequency modulation). Table 1.3 gives a summary of bearing-related frequencies in the stator current spectrum.

In general way, in presence of fault, the current is sinusoidally frequency or amplitude modulated. Based on this signal modelling approach, it seems that the most appropriate tool to extract fault indicator is demodulation techniques. However, this modelling approach suffers from many restrictive assumptions and lack of generality since it cannot be applied for all type of faults.

*Table 1.2 Faults characteristic frequencies [1,2]*

| Induction machine fault | Fault-related frequency $k \in \mathbb{N}$ |
|---|---|
| Bearing damage | $\lvert\omega_s \pm k\omega_d\rvert$ |
| Broken rotor bars | $\omega_s(1 \pm 2ks)$ |
| Air-gap eccentricity | $\omega_s \left\lvert 1 \pm k\left(\frac{1-s}{N_p}\right)\right\rvert$ |
| Load oscillation | $\omega_s \left[1 \pm k\left(\frac{1-s}{N_p}\right)\right]$ |

*Table 1.3 Comparison of two studies on bearing fault-related frequencies*

| Faulty bearing components | According to Schoen [30] | According to Blodt [28] | |
|---|---|---|---|
| | | Eccentricity | Torque oscillations |
| Outer raceway | $\omega_s \pm k\omega_{od}$ | $\omega_s \pm k\omega_{od}$ | $\omega_s \pm k\omega_{od}$ |
| Inner raceway | $\omega_s \pm k\omega_{id}$ | $\omega_s \pm \omega_r \pm k\omega_{id}$ | $\omega_s \pm k\omega_{id}$ |
| Ball defect | $\omega_s \pm k\omega_{bd}$ | $\omega_s \pm \omega_{cage} \pm k\omega_{bd}$ | $\omega_s \pm k\omega_{bd}$ |

The two approaches seem to be different. However, we can consider reasonably that the two approaches are equivalent since both of them assume introduction of frequency components in the stator currents due to faults. Moreover, the modulation approach is more restrictive than the first one since it assumes the upper sidebands and lower sidebands have the same amplitude. Furthermore, in case of phase modulation, the amplitude of sidebands is governed by Bessel functions.

## 1.2 Fault features extraction techniques

An automatic fault detection and diagnosis in electrical machines is generally performed in two stages: fault features computation and decision-making. Signal processing techniques such as PSD estimation and demodulation techniques are used for fault features computation. The PSD estimation techniques allow to estimate fault frequency signature, while the demodulation techniques highlight the AM or FM modulation introduced by a specific fault. In this section, we introduce a parametric spectral estimation approach for fault characteristics estimation.

### *1.2.1 Introduction*

In steady-state conditions, techniques based on conventional PSD estimators are required. These techniques can be categorized into two classes: the conventional periodogram and its extensions and the high resolution techniques [31]. In non-stationary environment, the time-frequency and time-scale techniques are performed to highlight the fault signature. These methods allow tracking the fault-related frequencies in the time-frequency plane. These representations allow to monitor fault characteristics and severity evolution over time. Both PSD estimation techniques and time-frequency approaches are summarized in Figure 1.6 with some relevant references.

Demodulation techniques are also used to reveal the presence of mechanical and electrical faults in electrical machines. These techniques estimate the instantaneous amplitude and frequency of the stator currents. The computation of the modulation index allows monitoring the fault and even distinguish the fault type. Moreover, demodulated signals are generally further processed in order to measure failure severity. Demodulation techniques are classified as depicted in Figure 1.7.

### *1.2.2 Stator current model under fault conditions*

#### 1.2.2.1 Model assumptions

The stator current signal model presented herein is based on the following assumptions:

- $\mathscr{H}_1$: The measured stator currents are modelled as a sum of $2 \times L + 1$ sine waves embedded in noise. Here, $2 \times L$ corresponds to the number of the sidebands introduced by the fault. Their amplitude allows detecting the fault.

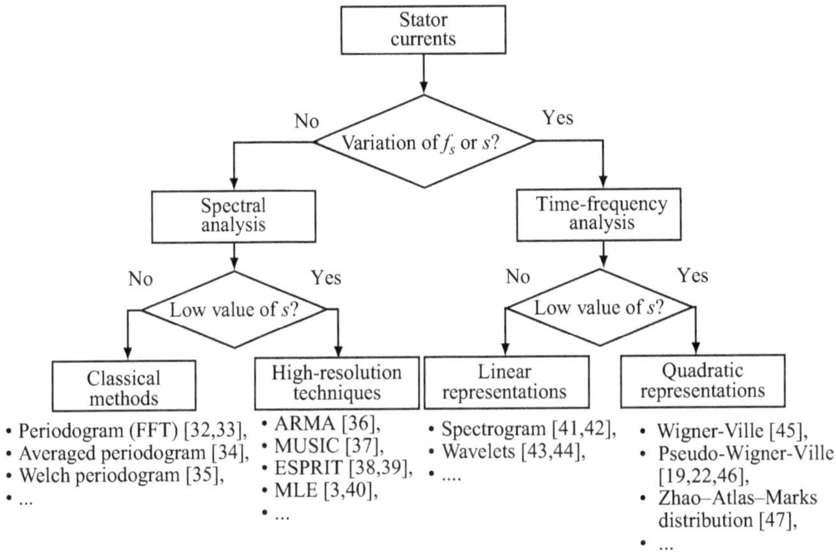


*Figure 1.6 Spectral analysis and time-frequency analysis*

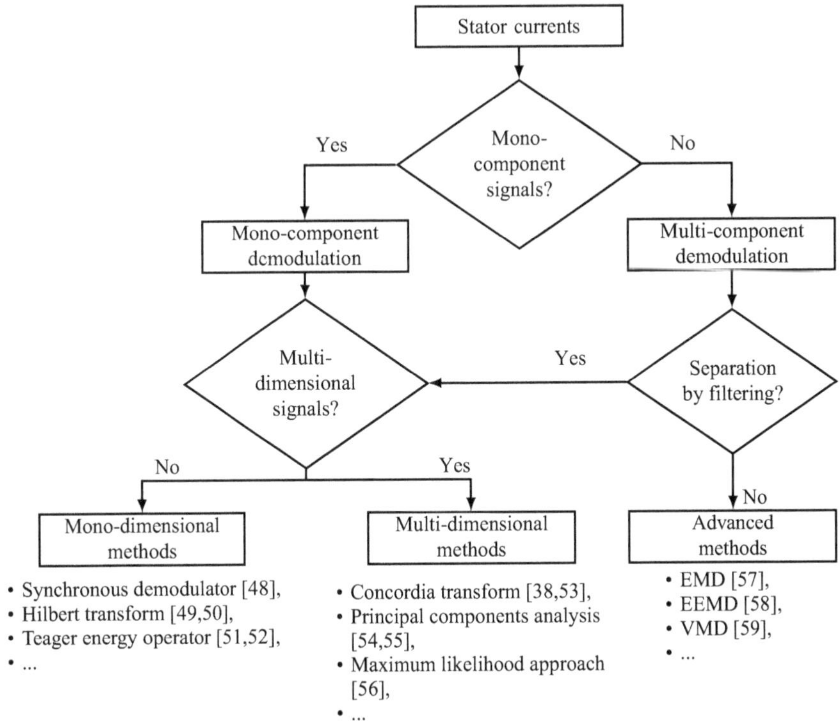


*Figure 1.7 Demodulation techniques classification*

- $\mathcal{H}_2$: The noise is assumed to be white Gaussian with zero-mean and variance $\sigma^2$. The Gaussian assumption is motivated by the following:
  - The central limit theorem which establishes, given certain conditions, that the sum of a sufficiently large number of independent and identically random variables are approximately Gaussian distributed even if the original variables are not normally distributed [60,61].
  - The Gaussian noise assumption leads to minimize the worst-case asymptotic Cramér–Rao bound (CRB) [62,63].
  - The minimum-variance unbiased estimator is equal to mean least square estimator assuming the noise to be white Gaussian [64].
- $\mathcal{H}_3$: The signal spectrum contains the frequency bins given by Table 1.2.
- $\mathcal{H}_4$: The sinusoids phases are independent and uniformly distributed in $[-\pi, \pi[$.

In practice, it should be pointed out that the assumption $\mathcal{H}_1$ requires the knowledge of $L$. In the present chapter, a technique to estimate $L$ based on information criterion rules [65] and the knowledge of the stator current samples $x[n]$. Regarding the assumption $\mathcal{H}_2$, it is not particularly restrictive since the noise can be whitened by an appropriate choice of the sampling frequency [31]. Moreover, if the noise process is not white and has unknown spectrum, then accurate frequency estimates can be computed by estimating the sinusoids using the non-linear least squares (NLS) estimator [31, Chapter 4, Introduction].

#### 1.2.2.2 Stator current modelling

Under assumptions $\mathcal{H}_1$–$\mathcal{H}_4$, the stator current samples $x[n]$ can be modelled as follows:

$$x[n] = \sum_{k=-L}^{L} a_k \cos\left(\omega_k(\Omega) \times \left(\frac{n}{F_s}\right) + \phi_k\right) + b[n] \tag{1.5}$$

where

- Parameters $\omega_k(\Omega)$, $a_k$ and $\phi_k$ correspond to the angular frequency, the amplitude, and the phase of the $k$th frequency component, respectively. $F_s$ corresponds to the sampling rate and $b[n]$ stands for noise samples.
- $\Omega$ is a set of parameters that need to be estimated. For instance, in case of broken rotor bars, the fault characteristic frequency $\omega_{brb}$ is given by (1.65) and the corresponding parameters to be estimated are $\Omega = \{\omega_s, s\}$. In the general way, the faults signature studied within this chapter are given by $\omega_d = \omega_s \pm k\omega_d$ with $k \in \mathbb{N}^*$, which gives a set of parameters to be estimated as $\Omega = \{\omega_s, \omega_c\}$.

PSD is defined as the discrete-time Fourier transform of the covariance function of $x[n]$ [31]. Under assumptions $\mathcal{H}_2$ and $\mathcal{H}_4$, the theoretical PSD of $x[n]$ is given in Figure 1.8. In practice, the PSD is unknown, and must be estimated from $N$

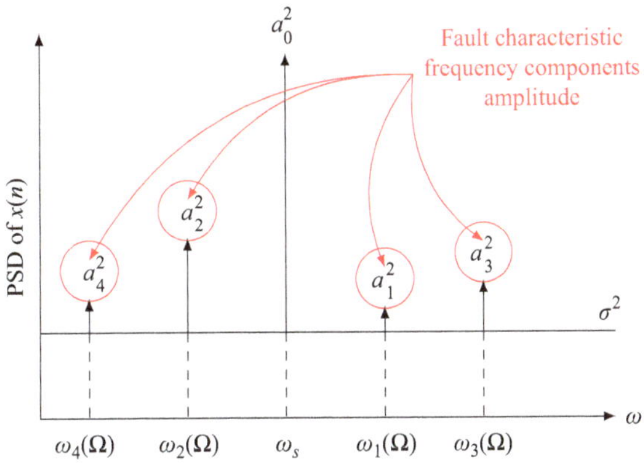


*Figure 1.8 Theoretical PSD for L = 2*

measured signal samples. Based on stator current model in (1.5), signal samples can be expressed as follows:

$$x[0] = \sum_{k=-L}^{L} a_k \cos\phi_k + b[0] \tag{1.6}$$

$$x[1] = \sum_{k=-L}^{L} a_k \cos\left(\omega_k(\Omega) \times \frac{1}{F_s} + \phi_k\right) + b[1] \tag{1.7}$$

$$\vdots$$

$$x[N-1] = \sum_{k=-L}^{L} a_k \cos\left(\omega_k(\Omega) \times \frac{N-1}{F_s} + \phi_k\right) + b[N-1] \tag{1.8}$$

Consequently, using a matrix notation, signal samples $x[n]$ $(n = 0, \ldots, N-1)$ can be expressed as follows:

$$\mathbf{x} = \mathbf{H}(\Omega)\theta + \mathbf{b} \tag{1.9}$$

where

- $\mathbf{x} = [x[0], \ldots, x[N-1]]^T$ is a $N \times 1$ column vector of stator current samples;
- $\mathbf{b} = [b[0], \ldots, b[N-1]]^T$ is a $N \times 1$ column vector of noise samples;
- $\theta$ is a $2 \times (2 \times L + 1) \times 1$ column vector aggregating the amplitudes and phases of fault characteristic frequency components. This vector is expressed as follows:

$$\theta = \begin{bmatrix} a_{-L}\cos(\phi_{-L}) & \cdots & a_L\cos(\phi_L) & -a_{-L}\sin(\phi_{-L}) & \cdots & -a_L\sin(\phi_L) \end{bmatrix}^T \tag{1.10}$$

- $\mathbf{H}(\Omega)$ is a $N \times 2(2L+1)$ matrix that has a rank of $2 \times L + 1$ and is given by

$$\mathbf{H}(\Omega) = \begin{bmatrix} \mathbf{z}_{-L} & \cdots & \mathbf{z}_L & \mathbf{y}_{-L} & \cdots & \mathbf{y}_L \end{bmatrix} \tag{1.11}$$

with

$$\mathbf{z}_k = \left[1 \quad \cos\left(\omega_k(\Omega) \times \tfrac{1}{F_s}\right) \quad \cdots \quad \cos\left(\omega_k(\Omega) \times \tfrac{N-1}{F_s}\right)\right]^T$$
$$\mathbf{y}_k = \left[0 \quad \sin\left(\omega_k(\Omega) \times \tfrac{1}{F_s}\right) \quad \cdots \quad \sin\left(\omega_k(\Omega) \times \tfrac{N-1}{F_s}\right)\right]^T \tag{1.12}$$

PSD estimators can be broadly classified into two categories: non-parametric and parametric PSD estimators. Non-parametric estimators estimate the PSD from the stator current samples $\mathbf{x}$ without need for any *a priori* knowledge about the signal and include the periodogram and its extensions. Unlike non-parametric methods, parametric ones take profit from the knowledge about signal characteristics to enhance estimates accuracy. These approaches include MUSIC and ESPRIT algorithms as well as MLE. Departing from this second approach, the remaining parts of this section propose a parametric spectral estimator that exploits the signal model in (1.9). In this regard, PSD estimation based on stator current samples $\mathbf{x}$ is considered as a statistical estimation problem.

### *1.2.3 Non-parametric spectral estimation techniques*

The periodogram is an estimate of the spectral density of a signal. It is the most common tool for computing the amplitude versus frequency characteristics of a signal. The periodogram of a complex discrete-time stationary signal $x[n]$ is defined as follows:

$$\widehat{P}_x(\omega) = \frac{1}{N}\left|\sum_{n=0}^{N-1} x[n]e^{\frac{-j\omega n}{F_s}}\right|^2 \tag{1.13}$$

The frequency resolution, which is defined as the ability of the periodogram to distinguish too close frequency components, is equal to the inverse of the signal acquisition duration. The periodogram is usually implemented using FFT algorithm since it efficiently computes discrete Fourier transform (DFT). It should be stressed that the periodogram is biased (the distance between the average of the estimates, and the single parameter being estimated is not zero) and inconsistent estimator (the variance does not decrease to zero when the data record length goes to infinity) of the PSD [66]. This can be overcome if several realizations $x_m[n]$ of the same random process $x[n]$ are available. This is performed using Welch periodogram; the signal is split up into overlapping segments, and the periodogram of each segment multiplied by a time window is computed and then averaged. Mathematically speaking, the Welch periodogram is defined as

$$\widehat{P}_w(\omega) = \frac{1}{L}\sum_{k=1}^{k=L} \widehat{P}_{xw}^{(k)}(\omega) \tag{1.14}$$

where $\widehat{P}_{xw}^{(k)}$ corresponds to the periodogram of the windowed signal $x[n]w[n - \tau k]$, where $w[\cdot]$ refers to a time window (Hanning, Hamming, etc.) and $\tau$ is a time delay.

For illustration purpose, Figure 1.9(a) and (b) depicts the stator current periodogram and the Welch periodogram, respectively. The periodogram was computed

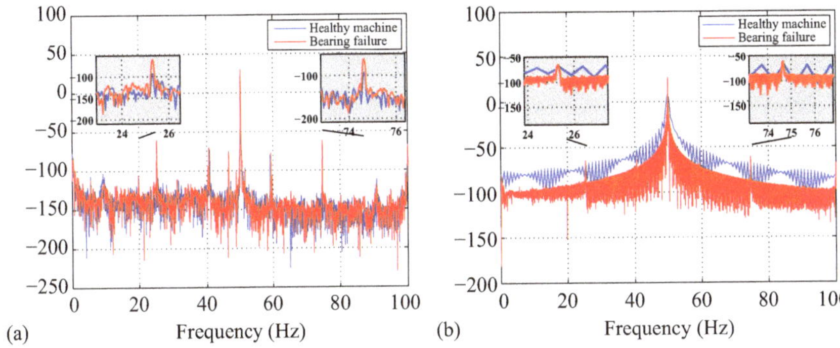


*Figure 1.9 PSD of stator current with bearing fault versus healthy case. (a) Periodogram. (b) Welch periodogram*

using a signal length of 10 s, a sampling frequency of 1 kHz, and the Hamming window. Regarding the Welch method, the signal is split up to eight overlapping segments, each with a 50% overlap. Then, the modified periodograms using Hamming window that is the same length as of the overlapping segments, is computed. Finally, the resulting periodograms are averaged to produce the PSD estimate.

The Welch periodogram enhances estimation statistical performance (decreases the variance of the estimate as compared to a single periodogram estimate of the entire signal). Unfortunately, it decreases the spectral precision and resolution due to segmentation. In summary, there is a trade-off between variance reduction and resolution. In order to increase the frequency resolution, the signal acquisition time must be increased. However, signal may be no longer stationary for long-term acquisition duration.

## *1.2.4 Subspace spectral estimation techniques*

Parametric methods can be of interest to enhance the frequency resolution and statistical performance, in case *a priori* signal proprieties are known. Indeed, parametric methods can yield higher resolution than periodogram-based methods in case of signal short acquisition duration. These techniques have the ability to resolve spectral lines separated by less than $\frac{1}{N}$ cycles per sampling period, which is the resolution limit for the classical non-parametric methods [67]. These approaches are generally called high-resolution methods and include three sub-classes: linear prediction methods, subspace techniques, and maximum likelihood (ML) estimation. The focus is made herein on subspace techniques.

The subspace category includes MUSIC and ESPRIT approaches [67]. To derive these PSD estimates, the noise is assumed to be a white Gaussian noise with zero mean and variance $\sigma^2$. These methods are based on the singular value decomposition of the samples covariance matrix which allows separating signal and noise subspaces.

---

**Algorithm 1:** MUSIC algorithm.

---

**Require:** Signal samples $\mathbf{x}[n]$.

Compute the covariance matrix estimate $\widehat{\mathbf{R}}_x = \frac{1}{G}\sum_{n=0}^{G-1}\mathbf{x}[n]\mathbf{x}[n]^H$, where $(\cdot)^H$ refers to Hermitian matrix transpose. Since $\mathbf{x}[n]$ has length $M$ and we have $N$ observations of $x[n]$, we can thus construct a set of $G = N - M + 1$ different subvectors $\{\mathbf{x}(n)\}_{n=0}^{G-1}$.

Compute the covariance matrix $\widehat{\mathbf{R}}_{xx}$ EVD

$$\widehat{\mathbf{R}}_{xx} = \mathbf{U}\boldsymbol{\Lambda}\mathbf{U}^H \tag{1.15}$$

where $\mathbf{U}$ is composed of the $M$ orthonormal eigenvalues of $\widehat{\mathbf{R}}_{xx}$, and $\boldsymbol{\Lambda}$ is a diagonal matrix of the corresponding eigenvalues $\lambda_k$ listed in decreasing order.

Estimate the model order $P$ using information criteria rules [65].

Evaluate

$$\mathscr{J}(\omega) = \frac{1}{\left\|\mathbf{a}^H(\omega)\widehat{\mathbf{G}}\right\|_F^2} \tag{1.16}$$

where $\|\cdot\|_F$ stands for Frobenius norm and the column vector $\mathbf{a}(\omega)$ is given by

$$\mathbf{a}(\omega)^H = \left[1, \mathrm{e}^{\frac{j\omega}{F_s}}, \mathrm{e}^{\frac{2j\omega}{F_s}}, \ldots, \mathrm{e}^{\frac{(M-1)j\omega}{F_s}}\right] \tag{1.17}$$

Matrix $\widehat{\mathbf{G}}$ is diagonal matrix formed by the $M - P$ less significant eigenvalues $\lambda_k$ spanning the noise subspace [68].

Find the $P$ largest peaks of $\mathscr{J}(\omega)$ to obtain angular frequency estimates $\widehat{\omega}_i$.

**return** Angular frequency estimates $\widehat{\omega}_i$.

---

MUSIC algorithm is based on the eigenvalue decomposition (EVD) of the covariance matrix $\mathbf{R}_{xx}$ of signal samples $\mathbf{x}[n] = [x[n], x[n+1], \ldots, x[n+M-1]]^T$. This decomposition allows computing the eigenvalues and the associated eigenvectors of $\mathbf{R}_{xx}$. MUSIC method for PSD estimation is presented by Algorithm 1.

MUSIC algorithm allows computing the pseudo-spectrum since it gives the frequency bins location, but does not allows computing their magnitude. To overcome this issue, RootMUSIC algorithm has been proposed in the literature. It computes discrete frequency spectrum estimates, along with the corresponding power estimates. Unfortunately, even though MUSIC shows high resolution capability, the PSD is obtained at highly computational cost (searching over parameters space) and excessive data storage. Moreover, its performance depends on covariance matrix estimator and signal-to-noise ratio (SNR). Figure 1.10(a) and (b) depicts the PSD of the stator current using MUSIC and RootMUSIC algorithms.

ESPRIT algorithm has been proposed in the literature to reduce the computational burden of the spectral estimation based on MUSIC. Indeed, unlike MUSIC, this technique relies on covariance matrix eigenvalues computation and signal subspace determination, which allows to extract directly the frequency content, rather than leading to a cost function to be optimized.

ESPRIT method is described by Algorithm 2. Figure 1.10(c) gives the PSD estimation based on ESPRIT. Since ESPRIT method computes only frequency estimates,

**Algorithm 2:** ESPRIT algorithm.

**Require:** Signal samples $\mathbf{x}[n]$.

1: Compute the covariance matrix estimate $\widehat{\mathbf{R}}_x = \frac{1}{G}\sum_{n=0}^{G-1}\mathbf{x}[n]\mathbf{x}[n]^H$.

2: Compute the covariance matrix $\widehat{\mathbf{R}}_{xx}$ EVD

$$\widehat{\mathbf{R}}_{xx} = \mathbf{U}\boldsymbol{\Lambda}\mathbf{U}^H \tag{1.18}$$

3: Estimate the model order $P$ using information criteria rules [65].

4: Determine signal subspace estimate $\widehat{S}$, which is composed of eigenvectors associated with the $P$ greatest eigenvalues. Then, compute $\widehat{S}_1$ and $\widehat{S}_2$ as follows:

$$\begin{aligned}\widehat{S}_1 &= [I_{M-1}\ \ 0]\widehat{S}\\ \widehat{S}_2 &= [0\ \ I_{M-1}]\widehat{S}\end{aligned} \tag{1.19}$$

5: Compute the EVD of $\widehat{S}_{12} = \left[\widehat{S}_1\quad \widehat{S}_2\right]$

$$\widehat{S}_{12} = E\Lambda E^H \tag{1.20}$$

and partition $E$ into $P \times P$ sub-matrices

$$E = \begin{bmatrix} E_{11} & E_{12}\\ E_{21} & E_{22}\end{bmatrix} \tag{1.21}$$

6: Determine $\Phi$ the eigenvalues of $\psi = -E_{12}E_{22}{}^{-1}$.

7: Get angular frequency estimates as follows:

$$\widehat{\omega}_i = \angle(\Phi_i) \tag{1.22}$$

where $\angle(\cdot)$ corresponds to the phase.

8: **return** Angular frequency estimates $\widehat{\omega}_i$.

---

least squares have been used for frequency bins amplitude estimation [69]. Unfortunately, the performance of such techniques significantly decrease for high level of noise.

### *1.2.5 ML-based approach*

This section presents the MLEs for stator currents' parameter estimation. Moreover, a model order estimator is proposed based on information criteria rules.

#### 1.2.5.1 Exact ML estimates

For a fixed set of data measurements and based on statistical model, ML approach determines the set of values for model parameters that maximize the likelihood function. Intuitively, MLE maximizes the agreement of the selected model with the observed data. Indeed, for discrete random variables, it maximizes the observed data probability under chosen statistical distribution. The MLE is used in order to accurately estimate $\theta$ and $\omega_k(\Omega)$ of the signal model given by (1.9). The ML estimates of $\Omega$ and $\theta$ are obtained by maximizing the probability density function (PDF) of the signal

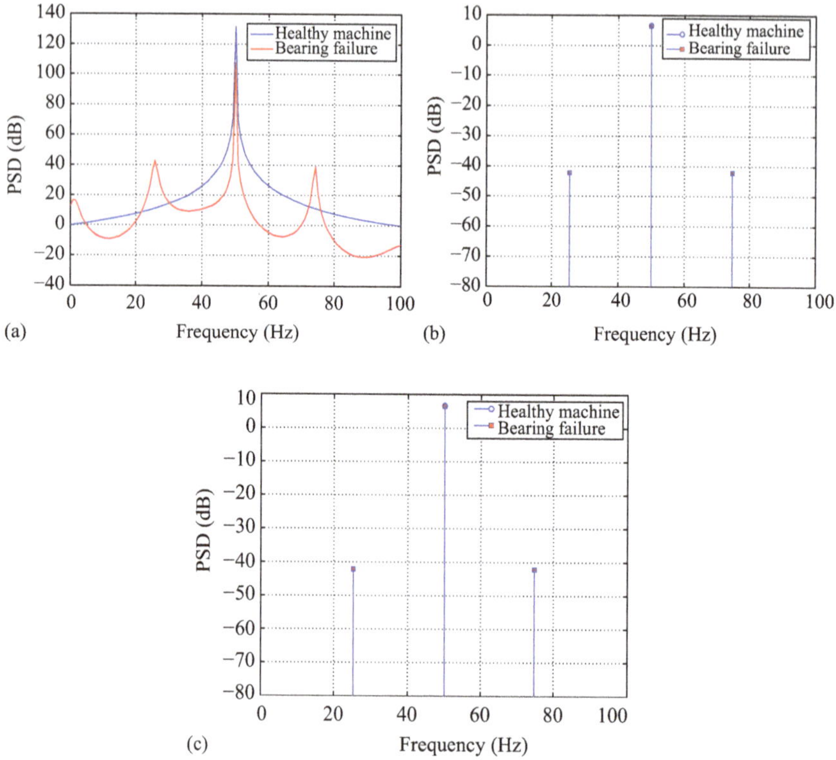


*Figure 1.10 High-resolution techniques-based PSD for bearing fault versus healthy case. (a) MUSIC-based PSD estimation. (b) RootMUSIC-based PSD estimation. (c) ESPRIT-MLE-based PSD estimation*

samples with respect to the unknown parameters. Mathematically, ML estimates are given by

$$\{\widehat{\theta}, \widehat{\Omega}\} = \arg\max_{\theta,\Omega} \log(p(\mathbf{x}; \theta, \Omega)) \tag{1.23}$$

where $p(\mathbf{x}; \theta, \Omega)$ is the PDF of $\mathbf{x}$. Assuming that assumption $\mathscr{H}_2$ holds , i.e. $b[n] \sim \mathscr{N}(0, \sigma^2)$, the PDF of the measurement data $\mathbf{x}$ is given by

$$p(\mathbf{x}; \theta, \Omega) = \frac{1}{(2\pi\sigma^2)^{\frac{N}{2}}} \times \exp\left[-\frac{1}{2\sigma^2}(\mathbf{x} - \mathbf{H}(\Omega)\theta)^T(\mathbf{x} - \mathbf{H}(\Omega)\theta)\right] \tag{1.24}$$

where $(\cdot)^T$ denotes matrix transpose. For the sake of illustration, one-dimensional Gaussian density function is given in Figure 1.11.

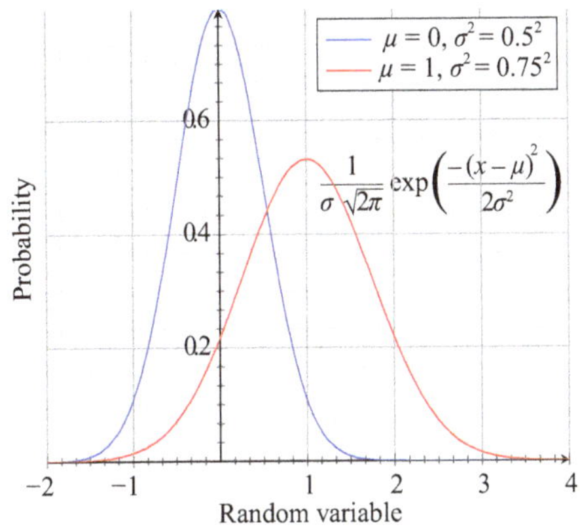


*Figure 1.11 Gaussian distribution PDF $N(\mu, \sigma^2)$*

Based on [69], the maximization in (1.23) is equivalent to the minimization of the following cost function:

$$\mathcal{L}(\mathbf{x}; \theta, \Omega) = (\mathbf{x} - \mathbf{H}(\Omega)\theta)^T (\mathbf{x} - \mathbf{H}(\Omega)\theta) \tag{1.25}$$

Differentiating $\mathcal{L}(\mathbf{x}; \theta, \Omega)$ with respect to $\theta$ and setting the derivative equal to 0 gives the ML estimate of $\theta^*$. By expanding the cost function, we obtain

$$\begin{aligned}\mathcal{L}(\mathbf{x}; \theta, \Omega) &= (\mathbf{x} - \mathbf{H}(\Omega)\theta)^T (\mathbf{x} - \mathbf{H}(\Omega)\theta)\\ &= (\mathbf{x}^T - \theta^T \mathbf{H}^T(\Omega))\,(\mathbf{x} - \mathbf{H}(\Omega)\theta)\\ &= \mathbf{x}^T\mathbf{x} - \theta^T\mathbf{H}^T(\Omega)\mathbf{x} - \mathbf{x}^T\mathbf{H}(\Omega)\theta + \theta^T\mathbf{H}^T(\Omega)\mathbf{H}(\Omega)\theta \end{aligned} \tag{1.26}$$

The derivative of $\mathcal{L}(\mathbf{x}; \theta, \Omega)$ with respect to $\theta$ is equal to (see reference [70]) the following:

$$\begin{aligned}\frac{\partial \mathcal{L}(\mathbf{x}; \theta, \Omega)}{\partial \theta} &= 0 - \mathbf{H}^T(\Omega)\mathbf{x} - \mathbf{H}^T(\Omega)\mathbf{x} + (\mathbf{H}^T(\Omega)\mathbf{H}(\Omega) + (\mathbf{H}^T(\Omega)\mathbf{H}(\Omega))^T)\theta\\ &= -2\mathbf{H}^T(\Omega)\mathbf{x} + 2\mathbf{H}^T(\Omega)\mathbf{H}(\Omega)\theta\end{aligned} \tag{1.27}$$

Setting this derivative to 0, the MLE of $\theta$ is obtained. This estimate is denoted $\widehat{\theta}$ and is given by

$$\left.\frac{\partial \mathcal{L}(\mathbf{x}; \theta, \Omega)}{\partial \theta}\right|_{\theta=\widehat{\theta}} = 0 \Rightarrow \mathbf{x} = \mathbf{H}(\Omega)\widehat{\theta} \tag{1.28}$$

*This is valid for unknown noise distribution. In this regard, the minimization in (1.25) leads to least squares estimator of $\theta$ and $\Omega$.

Finally, the MLE of $\theta$ is expressed as follows:

$$\widehat{\theta} = \mathbf{H}^{\dagger}(\Omega)\mathbf{x} \tag{1.29}$$

where $\mathbf{H}^{\dagger}(\Omega)$ is the pseudo-inverse of $\mathbf{H}(\Omega)$, i.e.

$$\mathbf{H}^{\dagger}(\Omega) = (\mathbf{H}^{T}(\Omega)\mathbf{H}(\Omega))^{-1}\mathbf{H}^{T}(\Omega) \tag{1.30}$$

where $(\cdot)^{-1}$ stands for the matrix inverse.

It must be stressed that the ML estimate of $\theta$ depends on the unknown parameters $\Omega$ which must be estimated. Consequently, the ML estimate of $\omega_k(\Omega)$ is obtained by minimizing $\mathscr{L}(\mathbf{x}; \widehat{\theta}, \Omega)$ with respect to $\Omega$. By replacing $\theta$ by $\widehat{\theta}$ in (1.25), we obtain

$$\begin{aligned}
\mathscr{L}(\mathbf{x}; \widehat{\theta}, \Omega) &= (\mathbf{x} - \mathbf{H}(\Omega)\widehat{\theta})^{T}(\mathbf{x} - \mathbf{H}(\Omega)\widehat{\theta}) \\
&= (\mathbf{x} - \mathbf{H}(\Omega)\mathbf{H}^{\dagger}(\Omega)\mathbf{x})^{T}(\mathbf{x} - \mathbf{H}(\Omega)\mathbf{H}^{\dagger}(\Omega)\mathbf{x}) \\
&= \mathbf{x}^{T}(\mathbf{I}_N - \mathbf{H}(\Omega)\mathbf{H}^{\dagger}(\Omega))\mathbf{x}
\end{aligned} \tag{1.31}$$

Neglecting the terms that do not depend on $\Omega$, it can be shown that the ML estimate of $\Omega$ is expressed as follows:

$$\{\widehat{\Omega}\} = \arg\max_{\Omega} \mathscr{J}(\Omega) = \arg\max_{\Omega} \mathbf{x}^{T}\mathbf{H}(\Omega)\mathbf{H}^{\dagger}(\Omega)\mathbf{x} \tag{1.32}$$

Once the estimation of the set $\Omega$ is performed, the fault-related frequency components can be computed based on the faults characteristics presented in Table 1.2. Finally, the PSD estimate of the stator current for fault feature extraction can be decomposed into two steps:

- Compute the estimate of $\Omega$ and consequently $\omega_k(\Omega)$ based on (1.32).
- Vector $\theta$ containing amplitudes and phases of the fault characteristic frequencies is estimated by replacing $\Omega$ with its estimates in (1.29).

Thanks to its statistical properties, the ML estimation remains the most accurate approach for PSD computation even in case of coloured noise [67]. Specifically, this estimator outperforms the frequency resolution limitation of the periodogram. Moreover, as opposed to other PSD estimation methods discussed earlier, the proposed approach is specifically aimed to deal with the use of faults characteristic frequencies knowledge to improve the accuracy of the PSD estimation. Moreover, the ML estimation is an alternative to the minimum-variance unbiased (MVU) estimator, which does not always exist. Cramér–Rao lower bounds (CRLBs) are a benchmark against which we can compare the performance of any unbiased estimator. The CRLB provides an inferior bound for any unbiased estimator variance. The ML estimation variance approaches asymptotically these bounds [64]. Moreover, owing to the proposed signal model, the optimization problem has been reduced from $2 \times L + 1$-dimensional problem to two-dimensional (2D) optimization problem. Finally, the estimation of the frequency bins and their amplitudes have been decoupled.

As the maximum cannot be found analytically for the optimization problem in (1.32), numerical methods should be used to estimate $\Omega$ and afterwards $\omega_k(\Omega)$. In our context, the cost function depends only on two parameters, which implies a maximization in a 2D space. Moreover, the search space is relatively limited

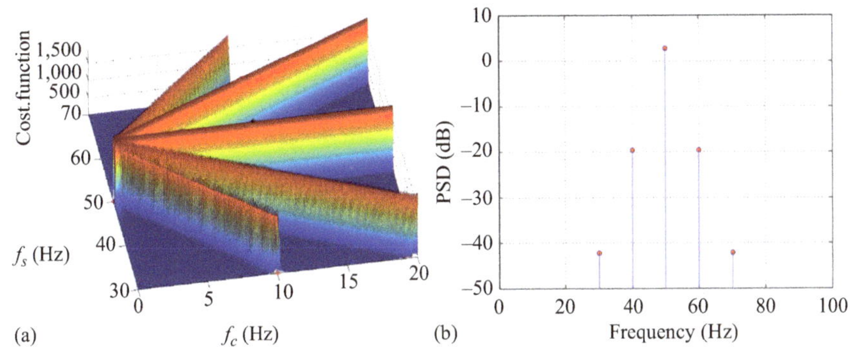


*Figure 1.12 Cost function and ML-based PSD estimation ($\omega_s = 2\pi\,50$ rad/s, $\omega_c = 2\pi\,10$ rad/s and $L = 2$). (a) Exact cost function $\mathscr{J}(\Omega)$. (b) PSD estimation*

since the variation range of the fundamental frequency is approximately known for grid-supplied induction machines. Taking into account these considerations, the maximization in (1.32) can be performed using grid-search algorithm. This optimization procedure evaluates the cost function at the vertices of a rectangular grid, and chooses the vertex with the highest value. Figure 1.12 illustrates the cost function for a synthetic signal with $\omega_k(\Omega) = \omega_s \pm k\omega_c$, $\omega_s = 2\pi\,50$ rad/s, $\omega_c = 2\pi\,10$ rad/s, $L = 2$, and $SNR = 50$ dB. It can be observed that the maximum is reached at the true values of the fundamental frequency and fault-related frequency. It should be highlighted herein that the maximization procedure may be computationally demanding as it requires the inversion of a large matrix for each vertex of the grid.

#### 1.2.5.2 Approximate ML estimates

The computational burden of the PSD estimation can be significantly reduced if the number of signal samples, $N$, is sufficiently large. It must be stressed that an approximate MLE can be obtained if $\omega_k(\Omega)/F_s$ is not close to 0 and $\frac{2\pi}{2}$. By using the following limit (see, e.g., [64]).

$$\lim_{N\to\infty} \frac{2}{N}(\mathbf{H}^T(\Omega)\mathbf{H}(\Omega)) = \mathbf{I}_N \tag{1.33}$$

where $\mathbf{I}_N$ corresponds to the $N \times N$ identity matrix; the cost function for frequencies estimation can be approximated as follows:

$$\begin{aligned}\mathscr{J}_a(\Omega) &= \lim_{N\to\infty} \mathscr{J}(\Omega)\\ &= \frac{2}{N}\mathbf{x}^T\mathbf{H}(\Omega)\mathbf{H}^T(\Omega)\mathbf{x}\\ &= \frac{2}{N}\left\|\mathbf{H}^T(\Omega)\mathbf{x}\right\|_F^2\end{aligned} \tag{1.34}$$

where $\|\cdot\|$ denotes to the Frobenius norm. Based on the structure of $\mathbf{H}(\Omega)$, we obtain

$$\mathcal{J}_a(\Omega) = \frac{2}{N}\sum_{k=-L}^{L}\left(\sum_{n=0}^{N-1} x[n]\cos\left(\omega_k(\Omega)\times\frac{n}{F_s}\right)\right)^2$$

$$+\frac{2}{N}\sum_{k=-L}^{L}\left(\sum_{n=0}^{N-1} x[n]\sin\left(\omega(\Omega)\times\frac{n}{F_s}\right)\right)^2 \tag{1.35}$$

$$= 2\sum_{k=-L}^{L}\left|\frac{1}{\sqrt{N}}\sum_{n=0}^{N-1} x[n]e^{-j\left(\frac{\omega_k(\Omega)}{F_s}\right)n}\right|^2 \tag{1.36}$$

where the last equality is due to the fact that current samples are real valued, i.e. $x[n] \in \mathbb{R}$. This equation can be expressed according to the DFT of $x[n]$ as follows:

$$\mathcal{J}_a(\Omega) = 2\sum_{k=-L}^{L}|DFT_x[\omega_k(\Omega)/F_s]|^2 = 2\sum_{k=-L}^{L}\widehat{P}_x\left(\omega_k(\Omega)/F_s\right) \tag{1.37}$$

where $\widehat{P}_x(\cdot)$ corresponds to the periodogram defined in (1.14) and $DFT_x[\omega]$ is the DFT computed at the angular frequency $\omega$, i.e. (see [64])

$$DFT_x[\omega] = \frac{1}{\sqrt{N}}\sum_{n=0}^{N-1} x[n]e^{-j\omega n} \tag{1.38}$$

Finally the approximate ML estimate of $\Omega$ is simply derived by replacing $\mathcal{J}(\Omega)$ with $\mathcal{J}_a(\Omega)$ in (1.32), i.e.

$$\{\widehat{\Omega}\} = \arg\max_{\Omega} \mathcal{J}_a(\Omega) \tag{1.39}$$

Similarly, the approximate ML estimate of $\theta$ is obtained using

$$\widehat{\theta} = \frac{2}{N}\mathbf{H}^T(\Omega)\mathbf{x} \tag{1.40}$$

Equations (1.36) and (1.40) show that the approximate cost function is reduced to a sum of DFT bins. This makes the approximate approach attractive for the following reasons:

- Most digital signal processor (DSP) boards include functions for DFT computation,
- DFT can be efficiently computed using the FFT algorithm.

Unfortunately, it should be highlighted that the accuracy of the approximation highly depends on the signal length $N$. Specifically, the approximation in (1.33) is no longer valid for short data length. In this case, the DFT of the stator current contains side lobes, which reduces the frequency resolution. These side lobes can hide components close to each other in the frequency domain and then lead to misleading results. Moreover, the side lobes may be interpreted as fault characteristic frequencies and then lead to false alarm.

To summarize the previously discussed aspects, the approximated method is limited by the DFT algorithm resolution: the parameters are estimated correctly as long as the observed signal length $N$ is large enough compared to the inverse of the

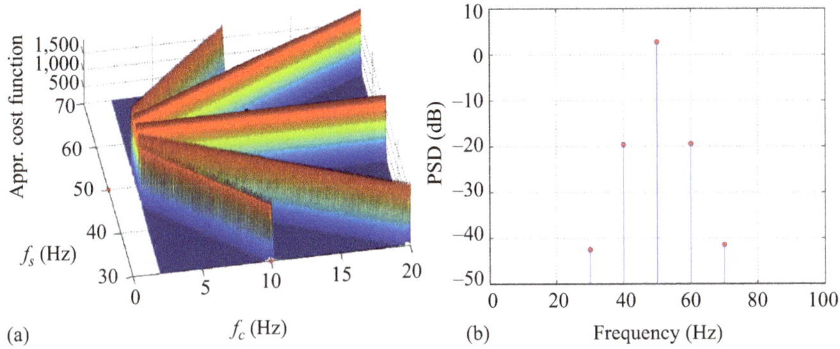


*Figure 1.13 Approximate cost function $\mathcal{J}_a(\Omega)$ and signal PSD ($\omega_s = 2\pi\,50$ rad/s, $\omega_c = 2\pi\,10$ rad/s, and $L = 2$). (a) Approximate cost function $\mathcal{J}_a(\Omega)$. (b) PSD estimation based on approximate MLE*

smallest frequency difference between two neighbouring frequencies of the signal $\omega_{k_2}$ and $\omega_{k_1}$, i.e. (1.41):

$$N \gg \frac{2\pi}{\min_{k_1 \neq k_2} |\omega_{k_2} - \omega_{k_1}|} \times F_s \tag{1.41}$$

Figure 1.13 gives the approximate cost function and the approximate PSD estimate. One can notice that the exact cost function in Figure 1.12 and its approximation in Figure 1.13 have roughly the same shape and differ only in low frequencies due to frequency resolution and windowing. In fact, these two shapes differ if the signal model is composed of closely spaced frequencies. Particularly, the approximate cost function shows a spurious peak located at $\omega_c = 0$ rad/s. Consequently, these peaks should be removed from the approximate cost function to obtain accurate estimate of $\omega_c$. This can be performed by excluding small values of $\omega_c$ from the grid search (see Figure 1.13). Despite the frequency resolution limitation, the approximate approach is of great interest as it leads to a drastic computational cost decrease. For instance, the evaluation of the approximate cost function in Figure 1.13 on a HP ProBook PC at 2.2GHz, using MATLAB® and Simulink®, requires only 4.2 s, while the evaluation of the exact one in Figure 1.12 requires 26.7 s.

### 1.2.5.3 Model order selection

Parametric spectral estimation methods require not only the estimation of a vector of real-valued parameters but also the selection of the number of sinusoidal components for the specification of a data model. This issue is known in the signal processing community as model order selection. In the previously presented results, both exact and approximate ML estimates of the fault characteristics assume the knowledge of the number of sidebands introduced by the fault. However, this is not the case in the fault detection applications. Besides, to efficiently implement the MLEs, an accurate knowledge of $L$ is required. The sidebands number ($2 \times L$) estimation is of great interest in fault detection and diagnosis since it allows distinguishing healthy motor from faulty one. Moreover, if model order is not accurately selected, the fault characteristic frequency or the fundamental frequency may erroneously be estimated as it is illustrated in Figures 1.14 and 1.15.

In this section, we propose to combine the ML estimation with an order-dependent penalty term based on the information criteria rules. The information-theoretic criteria rules include minimum description length (MDL) principle, Akaike information criterion (AIC), Bayesian information criterion (BIC), and generalized information criterion (GIC) [65,71]. In the following, the estimation of $L$ can be performed by maximizing the penalized ML estimate of $\Omega$ [65] as follows:

$$\{\widehat{\Omega}, \widehat{L}\} = \arg\min_{\Omega,L} (-2\log p(\mathbf{x}, \widehat{\theta}, \widehat{\sigma}^2, \Omega, L) + c(g, N)) \tag{1.42}$$

where $c(g,N)$ is a penalty function, which depends on the number of free parameters $g$ and the number of data samples $N$. Several penalty functions have been proposed in

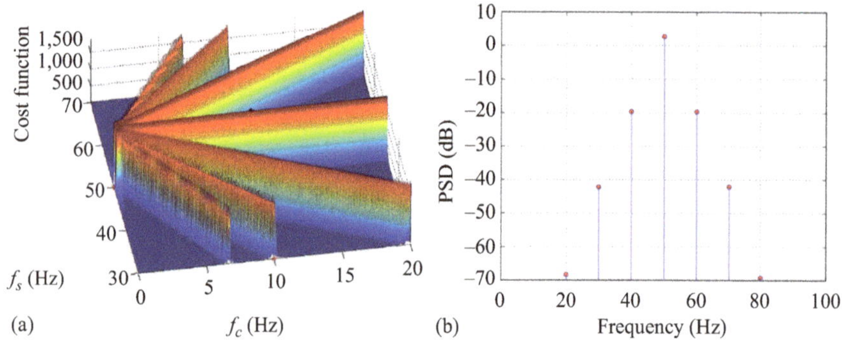


*Figure 1.14 Exact cost function $\mathcal{J}(\Omega)$ and signal PSD ($\omega_s = 2\pi\,50$ rad/s, $\omega_c = 2\pi\,10$ rad/s, and $L = 2$) for a wrong value of model order. (a) Exact cost function. (b) Exact PSD estimation*

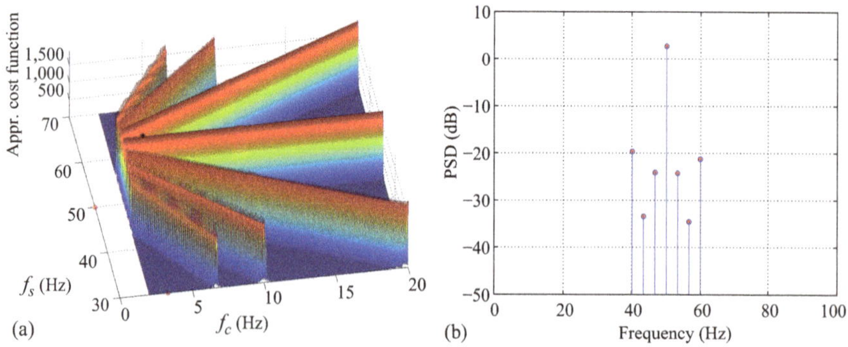


*Figure 1.15 Approximate cost function $\mathcal{J}_a(\Omega)$ and signal PSD ($\omega_s = 2\pi\,50$ rad/s, $\omega_c = 2\pi\,10$ rad/s, and $L = 2$) for a wrong value of model order. (a) Approximate cost function. (b) Approximate PSD estimation*

the literature as discussed in reference [65]. In this study, MDL principle is used since it minimizes the complexity of the model and maximizes the fitness [72]. Under the assumption that the number of frequency components is $2 \times L + 1$, the number of free parameters $g$ is given by $g = 4 \times L + 5$. Consequently, the MDL-based penalty function is given by the following expression:

$$c(g, N) = g \log(N) \tag{1.43}$$

As the exponential function is a monotonic increasing function, a straightforward computation allows simplifying the cost function in (1.42) as follows:

$$\{\widehat{\Omega}, \widehat{L}\} = \arg\max_{\Omega, L} \; -(\mathbf{x}^T\mathbf{x} - \mathscr{J}(\Omega)) \times \exp\left(\frac{g\log(N)}{N}\right) \tag{1.44}$$

Similarly to the exact estimates, the approximate ML estimation can be extended to include the model order selection as follows:

$$\{\widehat{\Omega}, \widehat{L}\} = \arg\max_{\Omega, L} \; -(\mathbf{x}^T\mathbf{x} - \mathscr{J}_a(\Omega)) \times \exp\left(\frac{g\log(N)}{N}\right) \tag{1.45}$$

Fundamental frequency can be assumed to be known for grid connected induction machines. Consequently, the optimization problem in (1.44) reduces to 2D problem. Figure 1.16 illustrates the cost function for a synthetic signal with $\omega_k(\Omega) = \omega_s \pm k\omega_c$, where $\omega_s = 2\pi 50$ rad/s, $\omega_c = 2\pi 10$ rad/s, $L = 2$, and $SNR = 50$ dB. The acquisition time and the sampling frequency are equal to 1 s and $F_s = 1$ kHz, respectively. The grid search algorithm evaluates the cost function for $\omega_c$ ranging from 0 to $2\pi 40$ rad/s with a step size of $2\pi 0.01$ rad/s and $L$ varying from 0 to 5. Figure 1.16(a) and (b) shows that the cost function is maximized for the true values of $\omega_c$ and $L = 2$.

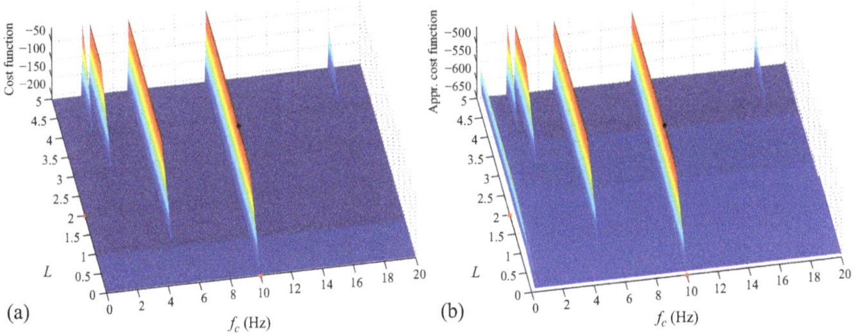


*Figure 1.16 Exact and approximate PSD ($\omega_s = 2\pi 50$ rad/s, $\omega_c = 2\pi 10$ rad/s, and $L = 2$). (a) Exact cost function for model order estimation in (1.44). (b) Approximate cost function for model order estimation in (1.45)*

The estimation of sideband number $L$ is of great interest since it allows determining the health-operating condition of the induction machine. In fact, two cases can be distinguished:

- $L = 0$: The sidebands responsible of the fault does not exist and the machine is operating correctly.
- $L \neq 0$: The machine is faulty and the sidebands amplitude must be computed in order to measure the fault severity.

Figure 1.17 shows simulation results for a signal without sidebands. It is obvious from this illustration that the exact method is more appropriate to discriminate the faulty and healthy cases by estimating model order $L$. In fact, the exact ML correctly estimates the model order $L = 0$. Unlike the exact approach, the approximate ML, which is based on the DFT, overestimates the number of sidebands ($L = 1$). This is due to side lobes in the Fourier transform that are due to windowing. These side lobes can be interpreted as fault characteristic frequency and lead to false interpretations. It must

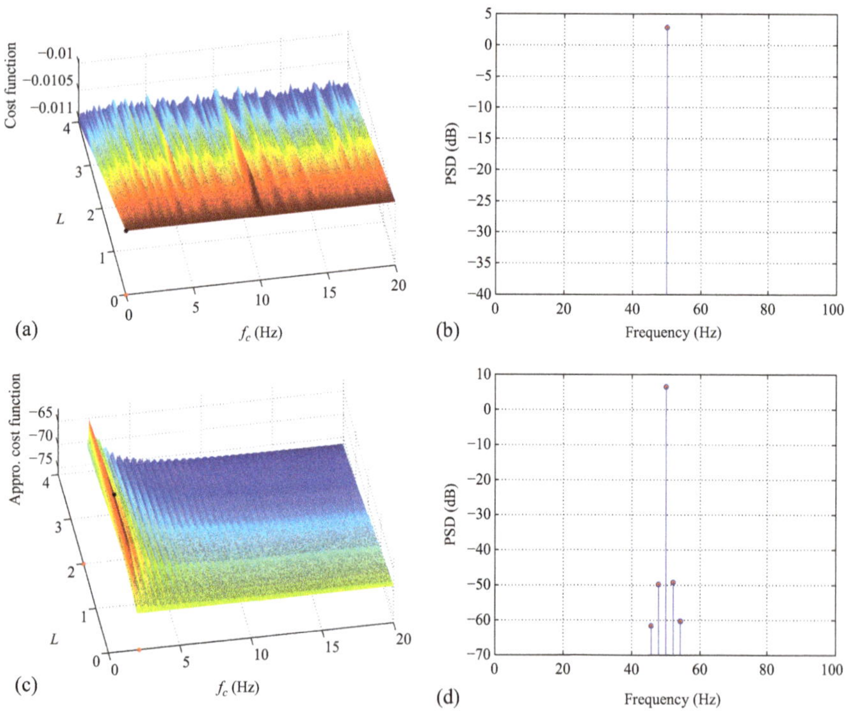


*Figure 1.17 Exact and approximate MLE cost function and related PSD estimation without sidebands. (a) Exact cost function for model order estimation in (1.44). (b) PSD estimation for exact method. (c) Approximate cost function for model-order estimation in (1.45). (d) PSD estimation for approximate method*

be stressed herein that the use of particular window function (Triangular, Hamming, Hanning, Blackman, etc.) can reduce the side-lobe amplitude but increase the width of the mainlobe, which may lead to false alarm. Consequently, it is preferable in case of approximate approach to compute the fault detection criteria and set a threshold beyond which the fault exists and the operator must be informed. From Figure 1.17(b) and (d), it can be seen that even if the model order is erroneously estimated, the PSD is correctly estimated and the sideband amplitudes correspond to side-lobe amplitudes in the DFT.

## 1.3 Fault detection and diagnosis

This section proposes to deal with automatic decision-making. It briefly describes artificial intelligence techniques that are used in the literature for faults diagnosis and classification. Then, it proposes the use of a constant false alarm rate (CFAR) detector, inspired by radar community, to track the frequency signature of the faults on the stator currents. Induction machine fault detection is formulated as a binary hypothesis test. The objective is to decide between two hypothesis; the motor is healthy or faulty. Then, GLRT is used to tackle the binary hypothesis testing problem.

### *1.3.1 Artificial intelligence techniques briefly*

MCSA-based fault detection consists of fault characteristics extraction based on signal processing methods followed by a fault detection stage. In the literature, the detection stage is often manually performed, based on visual inspection of stator current PSD or analysis of time-frequency representation. Several authors have proposed algorithms based on threshold detectors to help automatically detecting faults and measuring the severity [73,74]. Unfortunately, the implementation of these techniques requires the operator to manually set a threshold, based on the knowledge of the electrical machine that need to be monitored. To overcome this limitation, artificial intelligence (AI) and pattern recognition techniques have been investigated as useful tools to improve diagnosis, mainly during the decision process [75–78]. AI techniques include several sophisticated approaches such as ANN [79,80], SVM [81,82], Fuzzy logic [75,83], and combined approaches.

ANNs are computational models whose design is schematically inspired by the operation of biological neurons of human brain and is composed of simple arithmetical units connected in a complex architecture [78,84]. A neural network (NN) is generally composed of a succession of layers, each of which takes its inputs from the outputs of the previous one. Each layer ($i$) is composed of $N_i$ neurons, taking their inputs from the $N_{i-1}$ neurons of the previous layer.

Each synapse is associated with a synaptic weight, so that the $N_{i-1}$ are multiplied by this weight, then added by the neurons of level $i$. This operation is equivalent to multiplying the input vector by a transformation matrix. Putting the different layers of an NN one behind the other would be equivalent to cascading several transformation matrices and could be reduced to a single matrix, product of the others, if there were not at each layer the output function, which introduces a non-linearity at each stage.

This shows the importance of the judicious choice of a good output function. The ANNs need actual case examples (labelled data) used for learning, which is called learning database [85]. Learning database must be sufficiently large depending on the structure and complexity of the problem to be dealt with. However, a large learning database may lead to overfitting problem and thus degrades the NN performance. Indeed, overfitting causes the NN to lose its ability to generalize. Consequently, there is a trade-off between generalization and overtraining while training an ANN.

Several research papers have dealt with condition monitoring and fault diagnosis of electrical machines based on ANNs. The ANNs have been applied for several tasks such as pattern recognition, parameter estimation, operating condition clustering, faults classification, and incipient stage fault prediction [79,80].

SVMs are classifiers that are based on two key ideas, to deal with non-linear discrimination problems, and to reformulate the classification problem as a quadratic optimization problem. The first key idea is the concept of maximum margin. The margin corresponds to the distance between the separation boundary and the nearest samples, which are called support vectors. In SVM, the separation boundary is chosen as the one that maximizes the margin. The issue is to find this optimal separating boundary from a learning set, which is justified by the statistical learning theory. This is done by formulating the problem as a quadratic optimization problem, for which there are known algorithms. Moreover, to deal with cases where the data are not linearly separable, the second key idea of SVM is to transform the representation space of the input data into a larger (possibly infinite) dimensional space, in which a linear separation is likely to exist. This is achieved by means of a kernel function, which must meet the conditions of Mercer's theorem, and which has the advantage of not requiring explicit knowledge of the transformation to be applied for the space transfomation. Kernel functions allow transforming a scalar product in a large space, which is costly, into a simple point evaluation of a function, which is known as kernel trick.

SVMs are a set of supervised learning techniques used to solve problems such as data clustering, pattern recognition, and regression analysis [85]. The theoretic principle of SVM consists of two stages:

- Non-linear transform ($\phi$) of input data to high-dimensional space.
- Determination of optimal hyperplane or set of hyperplanes in a high- or infinite-dimensional space allowing to linearly separate the input data.

Similarly to ANNs, SVM requires a learning stage. In supervised learning, the learning database consists of examples, which are pairs consisting of an input objects and a desired output values. A supervised learning algorithm analyses the training data and produces an inferred function, which can be used to correctly determine the class labels for new unseen examples. Various applications in academia have proved that such techniques are well suited to deal with electrical machines diagnosis [81,82]. In [86], several statistical features have been extracted from vibration signals and used as input for SVM in order to perform a classification allowing to distinguish faulty (rolling elements faults) from healthy case for different fault severity.

Fuzzy logic is a form of probabilistic logic, which formalizes modes of reasoning that are approximate rather than exact. In contrary to traditional binary logic, fuzzy logic uses variables that may have true values that ranges in degree from 0 to 1. Fuzzy logic is much more general than traditional combinatory logic. Fuzzy logic systems for diagnosis purposes are able to process linguistic variables via fuzzy if-then rules. Fuzzy logic and the learning capabilities of ANNs or genetic algorithms can be combined to develop an adaptive fuzzy logic systems, which can adjust model parameters and afterwards enhance the global system performance [87]. In [88], NNs and fuzzy logic are combined together for the detection of stator inter-turn insulation and bearing wear faults in single-phase induction motor. Fuzzy logic for diagnosis can be applied to perform system modelling, predicting abnormal operating conditions, and faults classification [75,83,89].

These artificial intelligence-based techniques are black-box methods. Indeed, these technique parameters are difficult to tune in for industrial applications. Furthermore, fault detection performance critically depends on the chosen learning database. In fact, the training phase is critical for optimal operation as it may be misleading or produce results limited to a particular set of faults. Moreover, the learning database must be sufficiently large depending on the studied faults and the electrical machine operating conditions. However, a large learning database may lead to overfitting problems thus limiting the detector generalization capability [90].

## *1.3.2 Detection theory-based approach*

This section aims at demonstrating the usefulness of the combination of the ML and the detection theory for induction motor fault detection based on stator current processing. This problem is mainly referred to as hypothesis testing in the signal processing community. There are two possible hypotheses: $\mathscr{H}_0$ the machine is healthy (i.e. null hypothesis) and $\mathscr{H}_1$ the machine is faulty (i.e. alternative hypothesis). The objective is then to determine which of these two hypotheses best describes experimental measurements. To address this issue, a decision rule has been proposed. This decision rule is based on the Neyman–Pearson (NP) detector. The NP detector is based on the GLRT approach for which the unknown parameters are replaced by their estimates. The ML is used for frequency bins and amplitudes estimation. Then, the GLRT is used to perform the binary hypothesis testing. To measure the fault severity, a fault criterion is proposed, which is based on the amplitude of the frequency bins that are signature of the fault. Monte Carlo simulations have been performed in order to evaluate the statistical performance of the proposed approach.

### 1.3.2.1 Background on binary hypothesis testing

Let us assume a basic binary hypothesis testing problem. There are two hypotheses: $\mathscr{H}_0$ and $\mathscr{H}_1$. The PDF under each hypothesis is assumed to be completely known. The objective then is to design a good decision rule:

$$\Phi(x) = \begin{cases} 0 & \text{decide } \mathscr{H}_0 \\ 1 & \text{decide } \mathscr{H}_1 \end{cases} \tag{1.46}$$

For this hypothesis testing, there are four possible cases that can occur:

- **Detection:** Decide that $\mathscr{H}_1$ is true when $\mathscr{H}_1$ is true, characterized by the probability of detection $\mathscr{P}_D$ (see Figure 1.18).
- **False alarm:** Decide that $\mathscr{H}_1$ is true when $\mathscr{H}_0$ is true, measured through the probability of false alarm $\mathscr{P}_{FA}$ (see Figure 1.18).
- **Miss detection:** Decide that $\mathscr{H}_0$ is true when $\mathscr{H}_1$ is true, corresponding the probability of miss detection $\mathscr{P}_M = 1 - \mathscr{P}_D$.
- **Correct rejection:** Decide that $\mathscr{H}_0$ is true when $\mathscr{H}_0$ is true, corresponding to the probability of correct rejection $\mathscr{P}_R = 1 - \mathscr{P}_{FA}$.

In binary hypothesis testing, there are three main decision rules, which are Bayes, min-max, and NP criteria. NP decision rule maximizes the detection probability $\mathscr{P}_D$ for a given constraint on the false alarm $\mathscr{P}_{FA} = \alpha$. This rule can be formulated as the following objective function:

$$J = \mathscr{P}_D + \lambda(\mathscr{P}_{FA} - \alpha) \tag{1.47}$$

$$= \int_{R_1} p(\mathbf{x}; \mathscr{H}_1)d\mathbf{x} + \lambda\left(\int_{R_1} p(\mathbf{x}; \mathscr{H}_0)d\mathbf{x} - \alpha\right) \tag{1.48}$$

$$= \int_{R_1} (p(\mathbf{x}; \mathscr{H}_1) + \lambda p(\mathbf{x}; \mathscr{H}_0))d\mathbf{x} - \lambda\alpha \tag{1.49}$$

where $\lambda < 0$ is the Lagrange multiplier and $R_i = \{\mathbf{x}: \text{ decide } \mathscr{H}_i\}$ is the critical region that verifies:

$$\begin{cases} R_0 \cup R_1 = \mathbb{R} \\ R_0 \cap R_1 = \emptyset \end{cases} \tag{1.50}$$

This cost function is maximized if

$$p(\mathbf{x}; \mathscr{H}_1) + \lambda p(\mathbf{x}; \mathscr{H}_0) > 0 \tag{1.51}$$

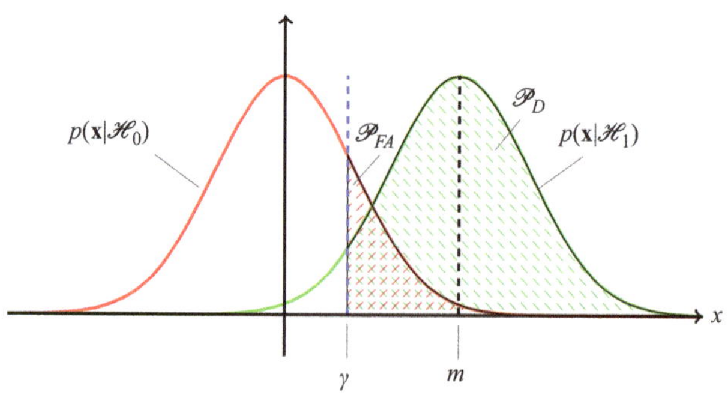


*Figure 1.18 Probability of detection and false alarm for Gaussian PDF*

Finally, NP decision rule, termed as likelihood ratio test (LRT), decides $\mathscr{H}_1$ if

$$\Gamma(\mathbf{x}) = \frac{p(\mathbf{x}; \mathscr{H}_1)}{p(\mathbf{x}; \mathscr{H}_0)} > -\lambda = \gamma \tag{1.52}$$

where $\gamma$ is a threshold that is computed based on the knowledge of the false alarm probability:

$$\mathscr{P}_{FA} = \int_{\{\mathbf{x}:\Gamma(\mathbf{x})>\gamma\}} p(\mathbf{x}; \mathscr{H}_0) d\mathbf{x} \tag{1.53}$$

LRT is a decision rule for a binary hypothesis testing that parameters characterizing each hypothesis are known. In the case where these parameters are unknown, the likelihood functions associated with the two considered hypotheses depend on one or more unknown parameters. Hence, the performance of the detector depends on the true value of PDFs parameters. This problem is called composite hypothesis testing. In general tests, there are two approaches to composite hypothesis testing: Bayesian approach and the GLRT approach. For the Bayesian formulation, the unknown parameters are assumed to be random quantities. In contrary, for the GLRT approach, the unknown parameters are first estimated and then used in the LRT. When applying the GLRT to solve an hypothesis test, two cases can be distinguished:

- **Clairvoyant detector** assumes all model parameters are known.
- **Blind detector** requires model parameters to be estimated before performing the test.

#### 1.3.2.2 GLRT for fault detection

The goal is to choose between two hypotheses: the induction motor is healthy ($\mathscr{H}_0$) or a fault is present ($\mathscr{H}_1$). The objective is to decide if the amplitudes $a_k$ at $\omega_k = \omega_s + k \times \omega_c$ with $k \neq 0$ are null ($\mathscr{H}_0$) or not ($\mathscr{H}_1$). Using the signal model in (1.5), this hypothesis test can be reformulated as follows:

$$\begin{aligned} &\mathscr{H}_0 : \mathbf{A}\theta = \mathbf{0},\ \sigma^2 > 0 \\ &\mathscr{H}_1 : \mathbf{A}\theta \neq \mathbf{0},\ \sigma^2 > 0 \end{aligned} \tag{1.54}$$

where $\mathbf{A}$ is an $r \times p$ matrix of rank $r$, with $r = 4 \times L$ and $p = 4 \times L + 2$. This matrix is an $p \times p$ identity matrix $\mathbf{I}_p$ for which the $L + 1$ and $3 \times L + 2$ rows have been removed. For instance, for $L = 1$ the matrix $\mathbf{A}$ is given by

$$\mathbf{A} = \begin{bmatrix} 1 & 0 & 0 & 0 & 0 & 0 \\ 0 & 0 & 1 & 0 & 0 & 0 \\ 0 & 0 & 0 & 1 & 0 & 0 \\ 0 & 0 & 0 & 0 & 0 & 1 \end{bmatrix} \tag{1.55}$$

Let us assume that stator current samples vector $\mathbf{x}$ has the PDF $p(\mathbf{x}; \theta_0, \mathscr{H}_0)$ under $\mathscr{H}_0$ and $p(\mathbf{x}; \theta_1, \mathscr{H}_1)$ under $\mathscr{H}_1$. GLRT decides $\mathscr{H}_1$ if

$$\mathbf{L}_G(x) = \frac{p(\mathbf{x}; \theta_1, \mathscr{H}_1)}{p(\mathbf{x}; \theta_0, \mathscr{H}_0)} > \gamma \tag{1.56}$$

Using the linear model in (1.9), the GLRT for this hypothesis testing problem [91], is to decide $\mathscr{H}_1$ if:

$$\mathscr{T}(x) = \frac{N-p}{r}(\mathbf{L}_G(x)^{\frac{2}{N}} - 1) > \gamma' \tag{1.57}$$

$$= \frac{N-p}{r}\frac{(\mathbf{A}\widehat{\theta}_1)^T[\mathbf{A}(\mathbf{H}^T\mathbf{H})^{-1}\mathbf{A}^T]^{-1}(\mathbf{A}\widehat{\theta}_1)}{\mathbf{x}^T(\mathbf{I} - \mathbf{H}(\mathbf{H}^T\mathbf{H})^{-1}\mathbf{H}^T)\mathbf{x}} > \gamma' \tag{1.58}$$

where $\widehat{\theta}_1 = (\mathbf{H}^T\mathbf{H})^{-1}\mathbf{H}^T\mathbf{x}$ is the MLE of $\widehat{\theta}$ under $\mathscr{H}_1$. The exact detection performance is given by

$$\mathscr{P}_{FA}: \ \mathscr{Q}_{F_{r,N-p}}(\gamma') \tag{1.59}$$

$$\mathscr{P}_D: \ \mathscr{Q}_{F'_{r,N-p}(\lambda)}(\gamma') \tag{1.60}$$

where $\mathscr{Q}_f(x)$ is the complementary cumulative distribution function for a $f$ random variable $x$. The symbols $\mathscr{P}_{FA}$ and $\mathscr{P}_D$ correspond to the probability of false alarm and the probability of detection, respectively. $F_{r,p}$ denotes an $F$ distribution, with $r$ numerator degrees of freedom and $p$ denominator degrees of freedom, and $F'_{r,N-p}(\lambda)$ denotes a non-central $F$ distribution with $r$ numerator degrees of freedom, $p$ denominator degrees of freedom, and non-centrality parameter $\lambda$. The non-centrality parameter is given by

$$\lambda = \frac{(\mathbf{A}\widehat{\theta}_1)^T[\mathbf{A}(\mathbf{H}^T\mathbf{H})^{-1}\mathbf{A}^T]^{-1}(\mathbf{A}\widehat{\theta}_1)}{\sigma^2} \tag{1.61}$$

The GLRT leads to a CFAR detector since $\mathscr{P}_{FA}$ does not depend on $\sigma^2$.

### 1.3.3 Simulation results

The proposed GLRT-based induction motor fault detection scheme is presented in Figure 1.19. The performance of the proposed fault detection scheme is assessed through synthesized signals. The Nelder–Mead simplex algorithm is initialized at $\omega_s = 2\pi 50$ rad/s and $\omega_c = 10$ rad/s, and the termination tolerance on $\omega_s$ and $\omega_c$ is set to $10^{-6}$. The performances are assessed with regards to SNR and signal acquisition length, $N$.

#### 1.3.3.1 Estimation performance

The frequency estimator is first tested under different signal acquisition length, $N$. The fundamental frequency is equal to $\omega_s = 2\pi 50.1$ rad/s, the fault-related frequency is equal to $\omega_c = 10.2$ rad/s and $L = 2$. Simulation parameters are given by Table 1.4. The sampling frequency is equal to $F_s = 1$ kHz, the SNR is equal to $SNR = 30$ dB, and $\mathscr{P}_{FA} = 10^{-3}$. The performances are evaluated in terms of the mean squared error (MSE). The MSE is estimated using $K = 1{,}000$ Monte Carlo trials by

$$MSE_{\omega_{sc}} = \frac{1}{K}\sum_{k=0}^{K-1}(\omega_{sc} - \widehat{\omega}_{sc})^2 \tag{1.62}$$

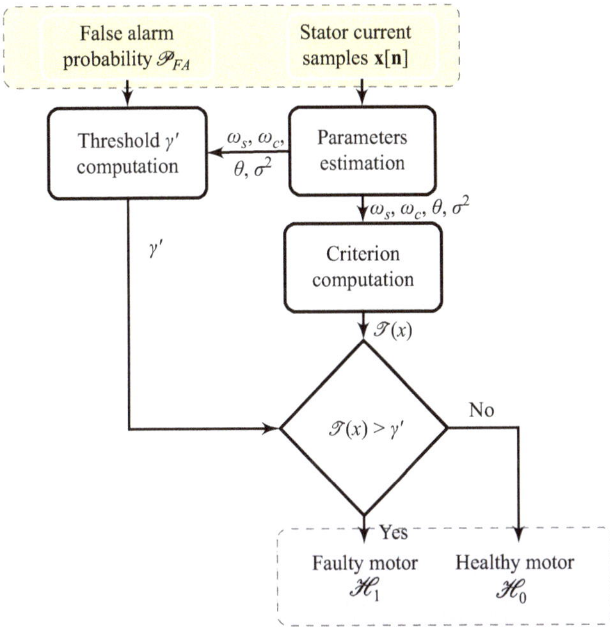


*Figure 1.19 Flow chart of the CFAR fault detector*

*Table 1.4 Simulation parameters (L = 2)*

| Parameter | $a_{-2}$ | $a_{-1}$ | $a_0$ | $a_1$ | $a_2$ |
|---|---|---|---|---|---|
| Value | 0.0004 | 0.018 | $\sqrt{2}$ | 0.0175 | 0.0003 |

Figure 1.20 displays the MSE versus data length $N$. It allows concluding that the fundamental frequency estimate $\widehat{\omega}_s$ is more accurate than the fault-related frequency estimate $\widehat{\omega}_c$. Moreover, it worth to note that the MLE performance is better as $N$ increases.

#### 1.3.3.2 Fault detection performance

In this subsection, the GLRT performance is investigated. The detection performance is studied through the receiver operating characteristic (ROC) curves. The ROC curves display the probability of detection $\mathscr{P}_D$ with respect to the probability of false alarm $\mathscr{P}_{FA}$ [92]. Figure 1.21(a) gives the obtained ROC curves for different fault degrees. The fault severity is controlled by increasing the amplitude of the frequency components around the fundamental frequency. It must be emphasized that an increase in fault severity leads to a higher $\mathscr{P}_D$ for the same value of $\mathscr{P}_{FA}$. For high fault severity, the ROC curve approaches the ideal case where the $\mathscr{P}_D$ is always equal to 1, except for $\mathscr{P}_{FA} = 0$.

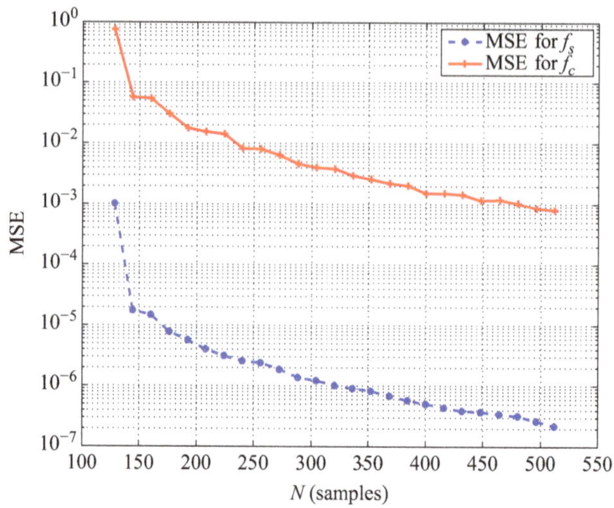


*Figure 1.20 Monte Carlo simulations: MSE for frequency estimates ($f_s = \frac{\omega_s}{2\pi}$ and $f_c = \frac{\omega_c}{2\pi}$) with respect to N (L = 2, SNR = 30 dB)*

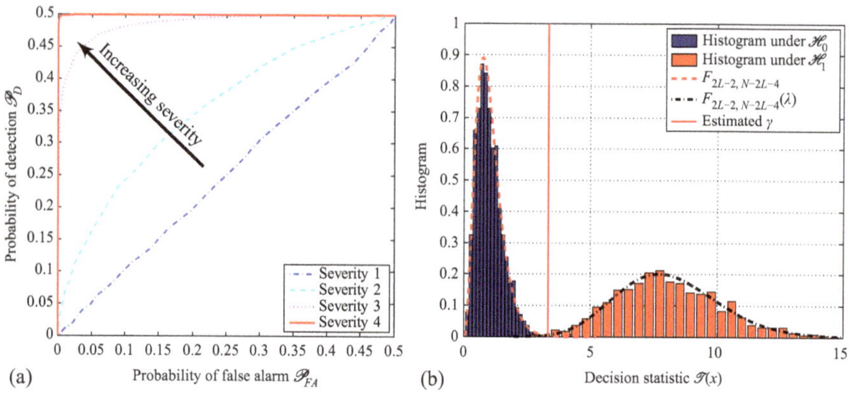


*Figure 1.21 ROC curves and histogram of the estimated $\mathscr{T}(x)$. (a) ROC curves for several fault degrees. (b) Histogram of $\mathscr{T}(x)$ for a set of healthy and faulty data*

The histogram of the $\mathscr{T}(x)$ for a set of healthy and faulty data is shown in Figure 1.21(b). For this histogram, $\omega_s$ and $\omega_c$ are assumed to be known. From this figure, it appears that the PDFs are distinct. A decision between the two hypotheses can simply be made by considering an adequate threshold. Figure 1.22(a) and (b) shows the GLRT performance for varying $N$ and SNR. It seems from this figure that the GLRT allows to reveal the existence of the fault even for low acquisition duration. Regarding the SNR, it appears that the GLRT is correctly performing for an

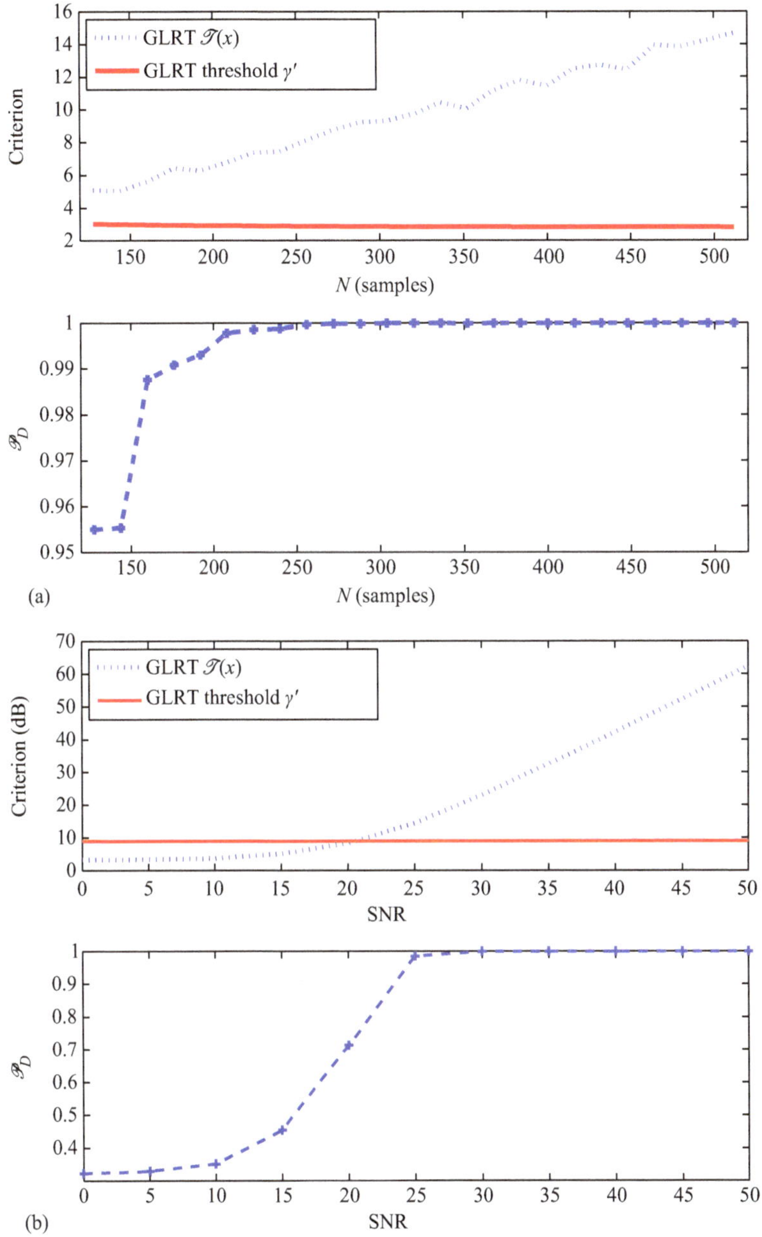


*Figure 1.22 GLRT $\mathscr{T}(x)$. (a) $\mathscr{T}(x)$ with respect to N for SNR = 30 dB. (b) $\mathscr{T}(x)$ with respect to SNR for N = 500 samples*

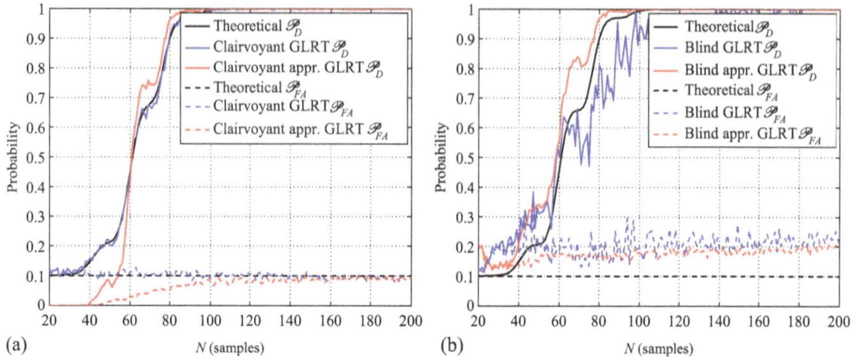


*Figure 1.23 Decision statistic performance. (a) Clairvoyant estimator: $\mathscr{P}_D$ and $\mathscr{P}_{FA}$ versus N (samples). (b) Bind estimator: $\mathscr{P}_D$ and $\mathscr{P}_{FA}$ versus N (samples)*

SNR higher than 25 dB. Fortunately, this is the case for signals issued from induction motor where the SNR is higher.

The detection performance is assessed from the estimation of the probability of detection $\mathscr{P}_D$ and probability of false alarm $\mathscr{P}_{FA}$ with respect to $N$ for blind and clairvoyant estimators (see Figure 1.23). The results depicted in Figure 1.21(a) show that $\mathscr{P}_D$ is always equal to 1 for $N \geq 100$ samples. The $\mathscr{P}_D$ for approximate GLRT has the same shape as the theoretical one for $N \geq 50$. Regarding the $\mathscr{P}_{FA}$, it is constant for the exact case while it can be considered as constant for the approximate GLRT for $N \geq 120$. Consequently, both the exact and approximate GLRTs allow detecting fault for low acquisition duration with CFAR.

## 1.4 Some experimental results

This section illustrates the behaviour of the exact and approximate blind detectors. Two mechanical faults and one electrical fault are considered. These faults include eccentricity fault, rolling-element bearing faults, and broken rotor bars. The stator currents are measured using a data acquisition card and processed offline on a standard desktop PC using MATLAB. Fundamental and fault-related frequencies are estimated from the raw data using ML principle. Then, fault detection is performed using GLRT.

### *1.4.1 Experimental set-up description*

- **Experimental set-up for mechanical faults:** Healthy machine and faulty ones with bearing and eccentricity faults have been tested. Each machine is a 230/400 V, 0.75 kW, three-phase induction machine with $N_p = 1$ and 2,780 rpm rated speed. Induction machines are fed by a PWM inverter with a varying fundamental frequency ranging from 0 to 60 Hz. The experimental test bed is depicted in Figure 1.24(a). The induction machines have two 6204-2 ZR type bearings (single row and deep groove ball bearings) with the following parameters: outside

(a) (b)

*Figure 1.24 Experimental set-up. (a) Machinery fault simulator. (b) Measurement devices*

diameter is 47 mm, inside diameter is 20 mm, and pitch diameter $D$ is 31.85 mm. Bearings have eight balls with an approximate diameter $d$ of 12 mm and a contact angle of 0°. For induction machine with misalignment fault, non-uniform air gap is introduced by acting on jack bolts on the circumference of each end bell. Bearing faults are artificially created by drilling holes of several diameters in the inner raceway (faults ranging in diameter from 0.178 to 1.016 mm).

During steady-state conditions, stator currents have been measured using closed-loop (compensated) current transducers using Hall effect. Stator current acquisition is performed by a 24-bit LabJack UE9 acquisition card with 20 kHz sampling frequency as illustrated in Figure 1.24(b).

- **Experimental set-up for rotor electrical fault:** A three-phase 5 kW induction motor with $N_p = 2$ and a nominal toque $\Gamma = 32$ N.m is considered. The induction motor is supplied by a standard industrial inverter using a constant voltage to frequency ratio control strategy. Induction motor is loaded using a DC motor with separate excitation. Motor broken bars have been achieved by drilling the rotor bars. It is worth to stress that broken rotor bar significantly increases currents flowing in the adjacent bars. These excessive currents increase the mechanical stresses on the adjacent bars and may consequently cause the breakage of the corresponding bars and may lead to catastrophic failure. In this study, one and two broken rotor bars have been considered for fault severity tracking.

### *1.4.2 Eccentricity fault detection*

Under eccentricity fault, the effective air-gap function varies in a sinusoidal manner with respect to angular position $\theta_s$ and time in the stator reference frame. In the literature, two types of eccentricity are commonly considered, which are static and dynamic eccentricities. They can jointly occur leading to the so-called mixed eccentricity [19]. Air-gap variations have an effect over the air-gap permeance function and consequently on the motor inductances. Moreover, eccentricity fault leads to an increase in oscillating torque components at $\omega_r = \left(\frac{1-s}{N_p}\right)\omega_s$ [19]. It has been proved that under eccentricity faults, the stator currents contain the frequencies given by Table 1.2:

$$\omega_k = \omega_s \left| 1 \pm k \left( \frac{1-s}{N_p} \right) \right| \tag{1.63}$$

For illustration purpose, stator current PSD under eccentricity fault is shown in Figure 1.25. This figure shows an increase in the fault characteristic frequencies around the fundamental frequency for several load conditions. Hence, monitoring strategy based on these components can be efficiently implemented for detection purposes.

The original stator current signal has been low-pass filtered and down-sampled to 400 Hz. Then, it has been processed using the proposed approach. The detector threshold $\gamma$ is obtained by setting $\mathscr{P}_{FA} = 0.1$. The evolution of the fault detection criteria $\mathscr{T}(\mathbf{y})$ with respect to load conditions for two rotational speeds is given in Figure 1.26. The fault criteria $\mathscr{T}(\mathbf{y})$ is below the threshold $\gamma$ for healthy motor

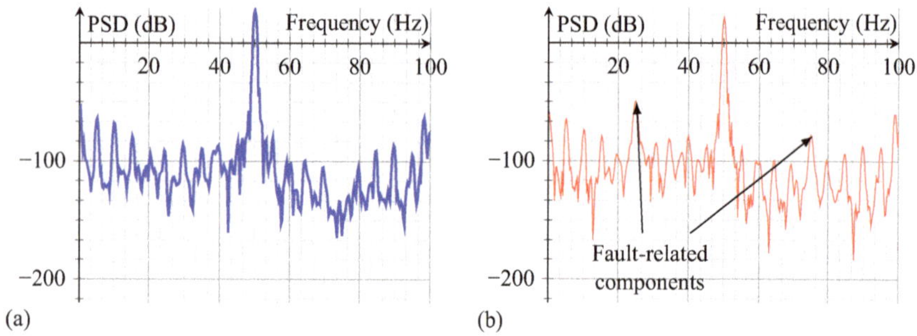


*Figure 1.25 Eccentricity: healthy and faulty stator current PSD. (a) PSD for healthy machine. (b) PSD for eccentricity failure*

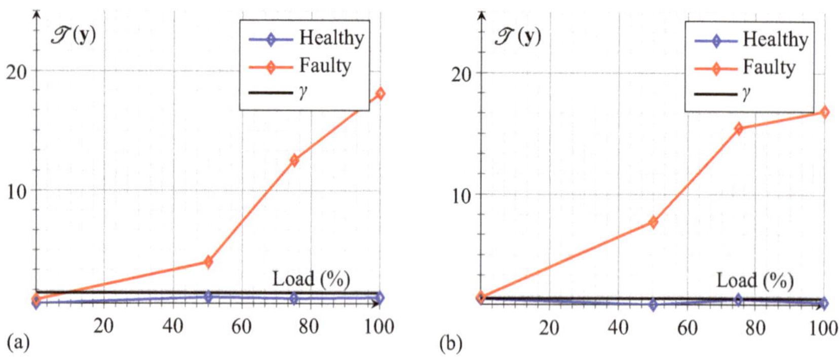


*Figure 1.26 Eccentricity fault: $\mathscr{T}(\mathbf{y})$ versus induction motor load for several rotational speeds. (a) Exact detector for 30 Hz. (b) Exact detector for 50 Hz*

whatever the load and speed conditions. This figure shows that the detector allows detecting eccentricity fault regardless of the rotational speed conditions. However, it seems that the detector is unable to detect the fault for unloaded motor. This is mainly due to the loading impact on the fault signature.

### *1.4.3 Bearing fault detection*

Since bearing supports the rotor of an induction machine, any bearing fault can induce two different effects [22,93]:

- Introduction of a particular radial rotor movement
- Load torque variations.

Bearing single-point defects can be supervised using frequency components introduced by fault. These frequencies depend on shaft rotational speed, fault location, and bearing dimensions as given by (1.1) and Table 1.2:

$$\omega_k = (\omega_s + k\omega_d) \tag{1.64}$$

The proposed fault detector sensitivity has been investigated for various bearing fault levels as summarized by Table 1.5. Stator current PSD for healthy and faulty induction machines is given in Figure 1.27. This figure shows that some frequency bins already exist on the stator current spectrum of healthy machines, which may be due to inherent eccentricity due the manufacturing stage, or caused by harsh operating

*Table 1.5 Bearing fault degree versus inner raceway hole diameter*

| Fault degree | 1 | 2 | 3 | 4 |
|---|---|---|---|---|
| Bearing hole diameter (inches) | 0.007 | 0.014 | 0.02 | 0.03 |

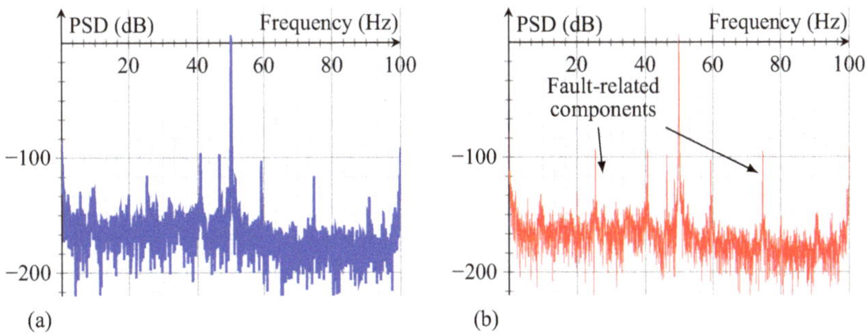


*Figure 1.27 Bearing defect: healthy and faulty stator current PSD. (a) PSD for healthy machine. (b) PSD for bearing failure*

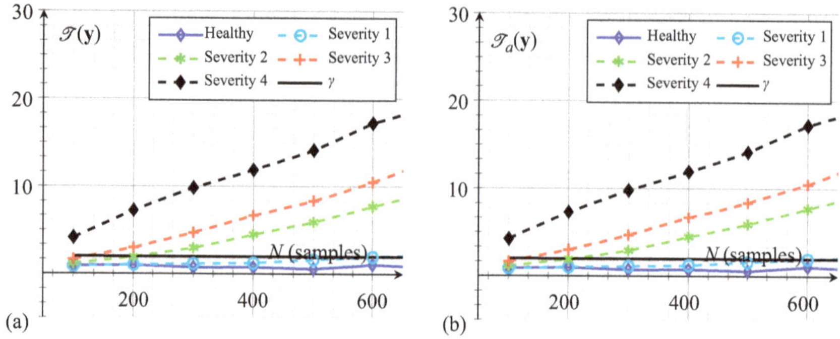


*Figure 1.28 Bearing faults: $\mathscr{T}(\mathbf{y})$ and $\mathscr{T}_a(\mathbf{y})$ versus N (L = 3). (a) Exact detector. (b) Approximate detector*

conditions. However, these components are extremely low for healthy machines and their amplitude increases for faulty conditions.

Similarly to eccentricity fault, raw data has been low-pass filtered and down-sampled to 400 Hz. Then, it has been processed using the proposed approach. The evolution of the exact and approximate fault detection criteria $\mathscr{T}(\mathbf{y})$ and $\mathscr{T}_a(\mathbf{y})$ with respect to $N$ for several fault degrees is given in Figure 1.28. The detector threshold $\gamma$ obtained by setting $\mathscr{P}_{FA} = 0.1$ is also shown. It can be observed that the fault criteria $\mathscr{T}(\mathbf{y})$ and $\mathscr{T}_a(\mathbf{y})$ are below the threshold $\gamma$ whatever the number of samples, $N$ for healthy cases. However, detector requires at least $N = 600$ samples to correctly detect the fault for faulty machine with bearing defect. For severe bearing fault situations, small stator current acquisition duration is required. Moreover, GLRT detector allows tracking fault severity.

### *1.4.4 Broken rotor bars fault detection*

Broken rotor bars fault leads to asymmetry of the rotating electromagnetic field in the air gap. Since stator currents are linked to the air-gap electromagnetic field, any broken rotor bar implies a significant effect on stator current waveforms [25]. Stator current PSD under broken rotor bars shows fault characteristic frequencies located at

$$\omega_k = (1 + 2ks)\omega_s \tag{1.65}$$

Induction machines under healthy and broken bars conditions have been supplied with an inverter with fundamental frequency equal to 50 Hz at 50% load. Under steady-state conditions, stator currents have been measured using a data acquisition board, with $F_s = 20$ kHz. The signal is further low-pass filtered and down-sampled to 400 Hz. The stator current PSD for healthy and faulty induction machines is given in Figure 1.29. This figure allows visualising fault-related frequency under broken rotor bars. These sidebands are too close to the fundamental frequency and not distinguishable using classical FFT for short data measurement.

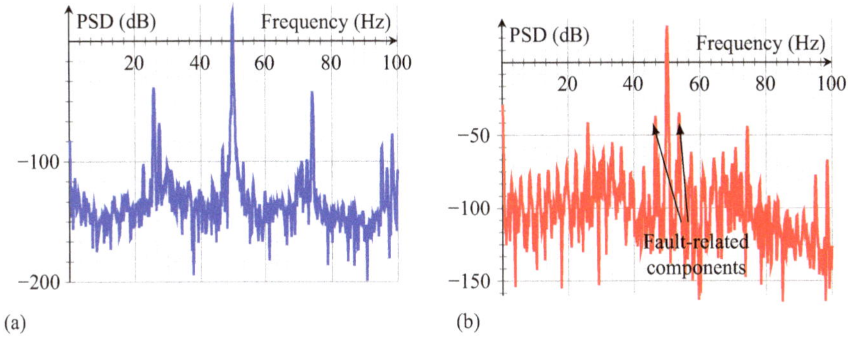


*Figure 1.29 Broken rotor bars: healthy and faulty stator current PSD. (a) PSD for healthy machine. (b) PSD for two broken rotor bars*

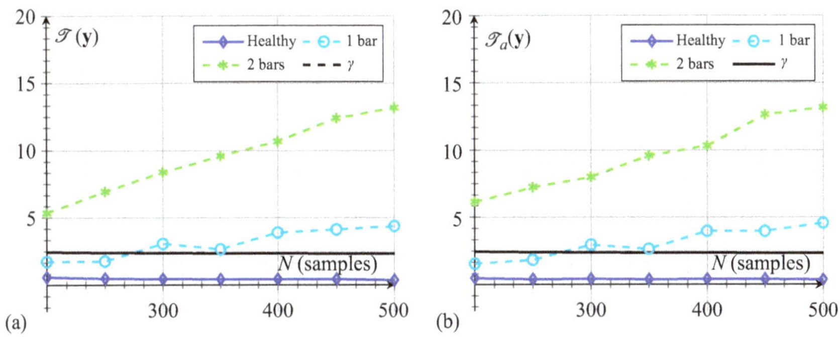


*Figure 1.30 Broken rotor bars: $\mathscr{T}(\mathbf{y})$ and $\mathscr{T}_a(\mathbf{y})$ versus N (L = 18). (a) Exact detector. (b) Approximate detector*

GLRT detector is used for broken rotor bars detection. Detector threshold $\gamma$ is computed by setting $\mathscr{P}_{FA} = 0.1$. The evolution of the exact and approximate fault detection criteria $\mathscr{T}(\mathbf{y})$ and $\mathscr{T}_a(\mathbf{y})$ with respect to $N$ for several fault degrees is given in Figure 1.30. This figure shows that the criteria $\mathscr{T}(\mathbf{y})$ and $\mathscr{T}_a(\mathbf{y})$ are below the threshold whatever the number of samples $N$ in case of healthy motor. However, for one broken bar, the detector requires at least $N = 400$ samples to correctly detect the fault. For two broken rotor bars, a smaller number of samples is required to detect the fault.

## 1.5 Conclusion

This chapter has described parametric spectral estimation methods that are further used for fault detection in induction machines through a stator current monitoring. Based on the knowledge of the fault characteristic frequencies, a stator current model has been proposed to enhance the PSD estimation performance. Then, parametric spectral estimation approaches based on MLE and approximate MLE have been developed for PSD estimation. Approximate MLE presents the advantages of being easy to implement as it can be implemented based on FFT on modern DSP boards and requiring lower computational cost. However, approximate MLE is less accurate as the side-lobe effects introduces some artefacts, which may lead to false alarms, especially for short acquisition duration.

For automatic fault detection purpose and decision-making, a GLRT detector has been proposed. The proposed theoretical and experimental results show that GLRT allows distinguishing faulty motor from healthy one and gives a measurement of the decision relevance. These results are promising and need to be implemented in real-world applications.

## References

[1] Benbouzid MEH. A review of induction motors signature analysis as a medium for faults detection. IEEE Transactions on Industrial Electronics. 2000;47(5):984–993.
[2] Bellini A, Filippetti F, Tassoni C, and Capolino GA. Advances in diagnostic techniques for induction machines. IEEE Transactions on Industrial Electronics. 2008;55(12):4109–4126.
[3] Elbouchikhi E, Choqueuse V, and Benbouzid M. Induction machine faults detection using stator current parametric spectral estimation. Mechanical Systems and Signal Processing. 2015;52:447–464.
[4] Elbouchikhi E, Amirat Y, Feld G, and Benbouzid M. Generalized likelihood ratio test-based approach for stator faults detection in a PWM inverter-fed induction motor drive. IEEE Transactions on Industrial Electronics. 2019;66(8):6343–6353.
[5] Matić D, Kulić F, Pineda-Sánchez M, and Kamenko I. Support vector machine classifier for diagnosis in electrical machines: Application to broken bar. Expert Systems with Applications. 2012;39(10):8681–8689.
[6] Ghate VN and Dudul SV. Cascade neural-network-based fault classifier for three-phase induction motor. IEEE Transactions on Industrial Electronics. 2011;58(5):1555–1563.
[7] D'Angelo MF, Palhares RM, Takahashi RH, Loschi RH, Baccarini LM, and Caminhas WM. Incipient fault detection in induction machine stator-winding using a fuzzy-Bayesian change point detection approach. Applied Soft Computing. 2011;11(1):179–192.
[8] Bonnett AH and Yung C. Increased efficiency versus increased reliability. IEEE Industry Applications Magazine. 2008;14(1):29–36.
[9] Joksimovic GM, Durovic MD, Penman J, and Arthur N. Dynamic simulation of dynamic eccentricity in induction machines-winding function approach. IEEE Transactions on Energy Conversion. 2000;15(2):143–148.
[10] Thomson WT, Rankin D, and Dorrell DG. On-line current monitoring to diagnose airgap eccentricity in large three-phase induction motors-Industrial case

histories verify the predictions. IEEE Transactions on Energy Conversion. 2006;14(4):1372–1378.

[11] Zhang P, Du Y, Habetler TG, and Lu B. A survey of condition monitoring and protection methods for medium-voltage induction motors. IEEE Transactions on Industry Applications. 2011;47(1):34–46.

[12] Qiao W and Lu D. A survey on wind turbine condition monitoring and fault diagnosis—Part I: Components and subsystems. IEEE Transactions on Industrial Electronics. 2015;62(10):6536–6545.

[13] Siddique A, Yadava G, and Singh B. A review of stator fault monitoring techniques of induction motors. IEEE Transactions on Energy Conversion. 2005;20(1):106–114.

[14] Nandi S, Toliyat HA, and Li X. Condition monitoring and fault diagnosis of electrical motors – a review. IEEE Transactions on Energy Conversion. 2005;20(4):719–729.

[15] Garcia-Perez A, de Jesus Romero-Troncoso R, Cabal-Yepez E, and Osornio-Rios RA. The application of high-resolution spectral analysis for identifying multiple combined faults in induction motors. IEEE Transactions on Industrial Electronics. 2011;58(5):2002–2010.

[16] Isermann R. Model-based fault-detection and diagnosis–status and applications. Annual Reviews in Control. 2005;29(1):71–85.

[17] Venkatasubramanian V, Rengaswamy R, Yin K, and Kavuri SN. A review of process fault detection and diagnosis: Part I: Quantitative model-based methods. Computers & Chemical Engineering. 2003;27(3):293–311.

[18] Blolt M. Condition monitoring of mechanical faults in variable speed induction motor drives, applications of stator current time-frequency analysis and parameter estimation. PhD Thesis. INPT, Toulouse; 2006.

[19] Blodt M, Regnier J, and Faucher J. Distinguishing load torque oscillations and eccentricity faults in induction motors using stator current Wigner distributions. IEEE Transactions on Industry Applications. 2009;45(6): 1991–2000.

[20] Bouzid MBK and Champenois G. New expressions of symmetrical components of the induction motor under stator faults. IEEE Transactions on Industrial Electronics. 2013;60(9):4093–4102.

[21] Knight AM and Bertani SP. Mechanical fault detection in a medium-sized induction motor using stator current monitoring. IEEE Transactions on Energy Conversion. 2005;29(4):753–760.

[22] Blodt M, Granjon P, Raison B, and Rostaing G. Models for bearing damage detection in induction motors using stator current monitoring. IEEE Transactions on Industrial Electronics. 2008;55(4):1813–1822.

[23] Faiz J, Ebrahimi BM, Akin B, and Toliyat HA. Finite-element transient analysis of induction motors under mixed eccentricity fault. IEEE Transactions on Magn. 2008;44(1):66–74.

[24] Joksimovic GM. Dynamic simulation of cage induction machine with air gap eccentricity. In: Proc. Inst. Elect. Eng. Elect. Power Appl., vol. 152. Cargèse, France; July 2005. pp. 803–811.

[25] Shi P, Chen Z, Vagapov Y, and Zouaoui Z. A new diagnosis of broken rotor bar fault extent in three phase squirrel cage induction motor. Mechanical Systems and Signal Processing. 2014;42(1):388–403.

[26] Gu F, Wang T, Alwodai A, Tian X, Shao Y, and Ball AD. A new method of accurate broken rotor bar diagnosis based on modulation signal bispectrum analysis of motor current signals. Mechanical Systems and Signal Processing. 2015;50:400–413.

[27] Saidi L, Fnaiech F, Henao H, Capolino GA, and Cirrincione G. Diagnosis of broken-bars fault in induction machines using higher-order spectral analysis. ISA Transactions. 2013;52(1):140–148.

[28] Blodt M, Bonacci D, Regnier J, Chabert M, and Faucher J. On-line monitoring of mechanical faults in variable-speed induction motor drives using the Wigner distribution. IEEE Transactions on Industrial Electronics. 2008;55(2): 522–533.

[29] Heller B and Hamata V. Harmonic Field Effects in Induction Machine. Elsevier Scientific Publishing Company, North-Holland, USA; 1977.

[30] Schoen RR, Lin BK, Habetler TG, Schlag JH, and Farag S. An unsupervised, on-line system for induction motor fault detection using stator current monitoring. IEEE Transactions on Industry Applications. 1995;31(6): 1280–1286.

[31] Stoica P and Moses R. Introduction to Spectral Analysis. Prentice Hall, Upper Saddle River, USA; 1997.

[32] Bellini A, Yazidi A, Filippetti F, Rossi C, and Capolino GA. High frequency resolution techniques for rotor fault detection of induction machines. IEEE Transactions on Industrial Electronics. 2008;55(12):4200–4209.

[33] Benbouzid MEH, Vieira M, and Theys C. Induction motors' faults detection and localization using stator current advanced signal processing techniques. IEEE Transactions on Power Electronics. 1999;14(1):14–22.

[34] Didier G, Ternisien E, Caspary O, and Razik H. Fault detection of broken rotor bars in induction motor using a global fault index. IEEE Transactions on Industry Applications. 2006;42(1):79–88.

[35] Yazidi A, Henao H, Capolino GA, Artioli M, and Filippeti F. Improvement of frequency resolution for three-phase induction machine fault diagnosis. In: Proc. 40th IAS Annual Meeting Conference Record of the 2005. Hong Kong, China; 2005. pp. 20–25.

[36] Cupertino F, de Vanna E, Salvatore L, and Stasi S. Analysis techniques for detection of IM broken rotor bars after supply disconnection. IEEE Transactions on Industry Applications. 2004;40(2):526–533.

[37] Kia SH, Henao H, and Capolino GA. A high-resolution frequency estimation method for three-phase induction machine fault detection. IEEE Transactions on Industrial Electronics. 2007;54(4):2305–2314.

[38] Xu B, Sun L, Xu L, and Xu G. Improvement of the Hilbert method via ESPRIT for detecting rotor fault in induction motors at low slip. IEEE Transactions on Energy Conversion. 2013;28(1):225–233.

[39] Kim Y-H, Youn Y-W, Hwang D-H, Sun J-H, and Kang D-S. High-resolution parameter estimation method to identify broken rotor bar faults in induction motors. IEEE Transactions on Industrial Electronics. 2013;60(9):4103–4117.

[40] Elbouchikhi E. On parametric spectral estimation for induction machine faults detection in stationary and non-stationary environments. Ph.D. Dissertation. Université de Bretagne Occidentale, Brest; November 2013.

[41] Cusido J, Romeral L, Ortega JA, Rosero JA, and Espinosa AG. Fault detection in induction machines using power spectral density in wavelet decomposition. IEEE Transactions on Industrial Electronics. 2008;55(2):633–643.

[42] Yazici B and Kliman GB. An adaptive statistical time-frequency method for detection of broken bars and bearing faults in motors using stator current. IEEE Transactions on Industry Applications. 1999;35(2):442–452.

[43] Rajagopalan S, Aller JM, Restrepo JA, Habetler TG, and Harley RG. Analytical-wavelet-ridge-based detection of dynamic eccentricity in brushless direct current (BLDC) motors functioning under dynamic operating conditions. IEEE Transactions on Industrial Electronics. 2007;54(3):1410–1419.

[44] Yang W, Tavner PJ, Crabtree CJ, and Wilkinson M. Cost-effective condition monitoring for wind turbines. IEEE Transactions on Industrial Electronics. 2010;57(1):263–271.

[45] Rajagopalan S, Aller JM, Restrepo JA, Habetler TG, and Harley RG. Detection of rotor faults in brushless DC motors operating under nonstationary conditions. IEEE Transactions on Industry Applications. 2006;42(6):1464–1477.

[46] Blodt M, Chabert M, Regnier J, and Faucher J. Current-based mechanical fault detection in induction motors through maximum likelihood estimation. In: Proceedings of the 2006 Conference IEEE IECON. Paris, France; 2006. pp. 4999–5004.

[47] Rajagopalan S, Restrepo JA, Aller JM, Habetler TG, and Harley RG. Nonstationary motor fault detection using recent quadratic time-frequency representations. IEEE Transactions on Industry Applications. 2008;44(3):735–744.

[48] Mohanty AR and Kar C. Fault detection in a multistage gearbox by demodulation of motor current waveform. IEEE Transactions on Industrial Electronics. 2006;53(4):1285–1297.

[49] Pons-Llinares J, Roger-Folch J, and Pineda-Sanchez M. Diagnosis of eccentricity based on the Hilbert transform of the startup transient current. In: Proceedings of SDEMPED'09. Cargese, France; August/September 2009. pp. 1–6.

[50] Pineda-Sanchez M, Riera-Guasp M, Roger-Folch J, Antonino-Daviu JA, and Perez-Cruz, J. Diagnosis of rotor bar breakages based on the Hilbert Transform of the current during the startup transient. In: Proceedings of IEMDC '09. Miami, FL, USA; May 2009. pp. 1434–1440.

[51] Li H, Fu L, and Zhang Y. Bearing fault diagnosis based on Teager energy operator demodulation technique. In: International Conference on Measuring Technology and Mechatronics Automation. Zhangjiajie, China; 2009. pp. 594–597.

[52] Pineda-Sanchez M, Puche-Panadero R, Riera-Guasp M, *et al.* Application of the Teager–Kaiser energy operator to the fault diagnosis of induction motors. IEEE Transactions on Signal Processing. 2013;28(4):1036–1044.

[53] Trajin B, Chabert M, Regnier J, and Faucher J. Hilbert versus Concordia transform for three phase machine stator current time-frequency monitoring. Mechanical Systems and Signal Processing. 2009;23(8):2648–2657.

[54] Choqueuse V, Benbouzid MEH, Amirat Y, and Turri S. Diagnosis of three-phase electrical machines using multidimensional demodulation techniques. IEEE Transactions on Industrial Electronics. 2012;59(4):2014–2023.

[55] Pires VF, Martin JF, and Pires AJ. Eigenvector/eigenvalue analysis of a 3D current referential fault detection and diagnosis of an induction motor. Energy Conversion and Management. 2010;51(5):901–907.

[56] Choqueuse V, Belouchrani A, Elbouchikhi E, and Benbouzid MEH. Estimation of amplitude, phase and unbalance parameters in three-phase systems: analytical solutions, efficient implementation and performance analysis. IEEE Transactions on Signal Processing. 2014;62(16):4064–4076.

[57] Rosero JA, Romeral L, Ortega JA, and Rosero E. Short-circuit detection by means of empirical mode decomposition and Wigner-Ville distribution for PMSM running under dynamic condition. IEEE Transactions on Industrial Electronics. 2009;56(11):4534–4547.

[58] Amirat Y, Choqueuse V, and Benbouzid M. EEMD-based wind turbine bearing failure detection using the generator stator current homopolar component. Mechanical Systems and Signal Processing. 2013;41(1):667–678.

[59] Amirat Y, Elbouchikhi E, Zhou Z, Benbouzid M, and Feld G. Variational mode decomposition-based notch filter for bearing fault detection. In: Proceedings of the 2019 IEEE IECON. Lisbon, Portugal; October 2019. pp. 1–6.

[60] Papoulis A and Pillai SU. Probability, Random Variables, and Stochastic Processes. Tata McGraw-Hill Education, New York, USA; 2002.

[61] Rosenblatt M. A central limit theorem and a strong mixing condition. Proceedings of the National Academy of Sciences of the United States of America. 1956;42(1):43.

[62] Stoica P and Besson O. Training sequence design for frequency offset and frequency-selective channel estimation. IEEE Transactions on Communications. 2003;51(11):1910–1917.

[63] Stoica P and Babu P. The Gaussian data assumption leads to the largest Cramér-Rao bound [lecture notes]. IEEE Signal Processing Magazine. 2011;28(3):132–133.

[64] Kay SM. Fundamentals of Statistical Signal Processing: Estimation Theory. Prentice-Hall Signal Processing Series, Upper Saddle River, USA; 1993. 17th Printing.

[65] Stoica P and Selen Y. Model-order selection: a review of information criterion rules. IEEE Signal Processing Magazine. 2004;21(4):36–47.

[66] Kay SM and Marple SL. Spectrum analysis – a modern perspective. Proceedings of the IEEE. 1981;69(11):1380–1419.

[67] Stoica P and Moses RL. Introduction to Spectral Analysis. Prentice-Hall, New Jersey; 1997.

[68] Stoica P and Nehorai A. MUSIC, maximum likelihood, and Cramer-Rao bound. IEEE Transactions on Acoustics, Speech, and Signal Processing. 1989;37(5):720–741.

[69] Kay SM. Modern Spectral Estimation: Theory and Application. Prentice Hall, Englewood Cliffs, New Jersey; 1998.

[70] Petersen KB and Pedersen MS. The Matrix Cookbook. Technical University of Denmark; November 2008.

[71] Wax M and Kailath T. Detection of signals by information theoretic criteria. IEEE Transactions on Acoustics, Speech, and Signal Processing. 1985;ASSP-33:387–392.

[72] Bouckaert RR. Probabilistic network construction using the minimum description length principle. In: Clarke M, Kruse R, Moral S (eds). Symbolic and Quantitative Approaches to Reasoning and Uncertainty. Springer, Berlin, Heidelberg; 1993.

[73] Trachi Y, Elbouchikhi E, Choqueuse V, and Benbouzid M. Induction machines fault detection based on subspace spectral estimation. IEEE Transactions on Industrial Electronics. 2016;63(9):5641–5651.

[74] Bouleux G. Oblique projection pre-processing and TLS application for diagnosing rotor bar defects by improving power spectrum estimation. Mechanical Systems and Signal Processing. 2013;41(1):301–312.

[75] Zidani F, Benbouzid MEH, Diallo D, and Nait-Said MS. Induction motor stator faults diagnosis by a current Concordia pattern-based fuzzy decision system. IEEE Transactions on Energy Conversion. 2003;18:469–475.

[76] Awadallah MA and Morcos MM. Application of AI tools in fault diagnosis of electrical machines and drives – an overview. IEEE Transactions on Energy Conversion. 2003;18(2):245–251.

[77] Filippetti F, Franceschini G, Tassoni C, and Vas P. AI techniques in induction machines diagnosis including the speed ripple effect. IEEE Transactions on Industry Applications. 1998;34(1):98–108.

[78] Bishop CM. Neural Networks for Pattern Recognition. Oxford University Press; 1995.

[79] Yang S, Li W, and Wang C. The intelligent fault diagnosis of wind turbine gearbox on artificial neural network. In: Proceedings International Conference on Condition Monitoring and Diagnosis. Beijing, China; April 2008.

[80] Salles G, Filippetti F, Tassoni C, Crellet G, and Franceschini G. Monitoring of induction motor load by neural network. IEEE Transactions on Power Electronics. 2000;15(4):762–768.

[81] Delgado M, Garcia A, Ortega JA, Cardenas JJ, and Romeral L. Multidimensional intelligent diagnosis system based on support vector machine classifier. In: Proceedings of the 2011 IEEE ISIE. Gdansk, Poland; June 2011. pp. 2127–2131.

[82] Samanta B. Gear fault detection using artificial neural networks and support vector machines with genetic algorithms. Mechanical Systems and Signal Processing. 2004;18(3):625–644.

[83] Zidani F, Diallo D, Benbouzid MEH, and Nait-Said R. A fuzzy-based approach for the diagnosis of fault modes in a voltage-fed PWM inverter induction

motor drive. IEEE Transactions on Industrial Electronics. 2008;55(2): 586–593.

[84] Bose NK and Liang P. Neural Network Fundamentals with Graphs, Algorithms and Applications. McGraw-Hill, New York, USA; 1996.

[85] Theodoridis S and Koutroumbas K. Pattern Recognition. Elsevier Academic Press; 2003.

[86] Barakata M, Lefebvre D, Khalil M, Mustapha O, and Druaux F. BSP-BDT classification technique: application to rolling elements bearing. In: Proceedings of the 2010 IEEE SYSTOL. Nice, France; October 2010. pp. 654–659.

[87] Altug S, Chen MY, and Trussell HJ. Fuzzy inference systems implemented on neural architectures for motor fault detection and diagnosis. IEEE Transactions on Industrial Electronics. 1999;46(6):1069–1079.

[88] Ballal MS, Khan ZJ, Suryawanshi HM, and Sonolikar RL. Adaptive neural fuzzy inference system for the detection of inter-turn insulation and bearing wear faults in induction motor. IEEE Transactions on Industrial Electronics. 2007;54(1):250–258.

[89] Wang H and Chen P. Fuzzy diagnosis method for rotating machinery in variable rotating speed. IEEE Sensors Journal. 2011;11(1):23–34.

[90] Bishop CM. Pattern recognition. Machine Learning. 2006;128.

[91] Kay SM. Fundamentals of Statistical Signal Processing: Detection Theory, vol. 2. Prentice Hall, Upper Saddle River, NJ, USA; 1998.

[92] Van Trees HL. Detection, Estimation, and Modulation Theory. John Wiley & Sons; 2004.

[93] Immovilli F, Bianchini C, Cocconcelli M, Bellini A, and Rubini R. Bearing fault model for induction motor with externally induced vibration. IEEE Transactions on Industrial Electronics. 2013;60(8):3408–3418.

*Chapter 2*

# The signal demodulation techniques

*Yassine Amirat[1] and Mohamed Benbouzid[2]*

Condition monitoring of electrical machines is a broad scientific area, the ultimate purpose of which is to ensure the safe, reliable and continuous operation of electrical machines. Hence, the task of fault detection is still an art, because induction machines are widely used in variable speed drives and in renewable energy conversion systems. A deep knowledge about all the phenomena involved during the occurrence of a failure constitutes an essential background for the development of any failure detection and diagnosis system. For the failure detection problem, it is important to know if a failure exists or not in the electric machine via the processing of available measurements. This chapter provides then an approach based on a electric machine current data collection and attempts to highlight the use of demodulation techniques for failure detection for stationary and nonstationary cases.

## 2.1 Introduction

Electrical machines have become unavoidable device in industrial and domestic applications, for producing mechanical power in drive trains or transforming it into electrical power in generation systems. So, it is to be expected that electrical machines are related to huge financial variables as well as safety and reliability. Despite electrical machines are robust devices, they remain subject to faults and downtime, hence, affecting their reliability performances. According to the defected component and the type of the electrical machine, faults can be classified in three categories:

- Stator-related fault: It includes electrical failures affecting the stator winding such as short circuits, inter-turn short circuits and open circuits [1].
- Rotor-related fault: It includes electrical failures affecting the rotor winding, commutators/slip rings/brushes failures for all rotor-wounded machines, and broken rotor bars and end rings for squirrel-cage machines, and permanent magnet demagnetization or cracks for permanent magnet motors.
- Mechanical-related fault: It includes bearing failures, rotor eccentricity and shaft misalignment.

[1]ISEN Yncréa Ouest, LABISEN, Brest, France
[2]Institut de Recherche Dupuy de Lôme, CNRS, University of Brest, Brest, France

The safety and reliability of electrical machines are related directly to these faults, hence affecting the operation and maintenance cost. So, new challenges arise particularly with regard to maintenance. In this context, cost-effective, predictive and proactive maintenance assume more importance. Condition monitoring systems (CMS) provide then an early indication of component incipient failure, allowing the operator to plan system repair prior to complete failure. Hence, CMS will be an important tool for lifting uptime and maximizing productivity, when cost-effective availability targets must be reached.

For this purpose, many techniques and tools are developed for condition monitoring of electrical machines in order to prolong their life span as reviewed in [2]. Some of the technologies used for monitoring include existing and pre-installed sensors, such as speed sensor, torque sensor, vibrations, temperature and flux density sensor. These sensors are managed together in different architectures and coupled with algorithms to allow an efficient monitoring of the system condition. A plethora of electrical machines faults and diagnostic methods are presented in the literature. The most favorable is the motor current signature analysis (MCSA) which is the analysis of the stator current harmonics index [3,4]. Most define the MSCA as the monitoring and spectra analysis of the stator current at steady state. Despite the method's origins, the name is very generic and should include the analysis of the stator current spectra under transient operation also. Anyway, this method has become favorable due to its unique characteristics such as remote monitoring [5], low implementation costs and equipment, and continuous and online monitoring capability. The advantage of signature analysis of the motor electrical quantities is that it is a noninvasive technique as those quantities are easily accessible during operation [6]. Moreover, stator currents are generally available for other purposes such as control and protection, avoiding the use of extra sensors [7]. Hence, most of the recent researches on induction machine faults detection have been focused on electrical monitoring with emphasis on current analysis [8,9].

Industrial survey on condition monitoring of induction motors show important features of failure rate and index the major faults of electrical machines can broadly be classified by the following [2,10]:

- Static and/or dynamic air-gap irregularities
- Broken rotor bar or cracked rotor end-rings
- Stator faults (opening or shorting of one coil or more of a stator phase winding)
- Abnormal connection of the stator windings
- Bent shaft (akin to dynamic eccentricity) which can result in a rub between the rotor and stator, causing serious damage to stator core and windings
- Bearing and gearbox failures.

The most common faults are bearing faults, stator faults, rotor faults and eccentricity or any combination of these faults. When analyzed statistically, about 40% of the faults correspond to bearing faults, 30–40% to stator faults, 10% to rotors faults, while remaining 10% belong to a variety of other faults. Frequencies induced by each fault depend on the particular characteristic data of the motor (like synchronous speed, slip frequency and pole-pass frequency) as well as operating conditions.

Moreover, in many industries context, bearing failures have been a persistent problem which accounts for a significant proportion of all failures in electrical machines; for example, bearing failure of electric drive or rotating electric generation system is the most common failure mode associated with a long downtime. Bearing failure is typically caused by improper lubrication, and occasionally manufacturing faults in the bearing components, and also some misalignment in the drive train, which gives rise to abnormal loading and accelerates bearing wear. A plethora of research works [11,12] states that due to the construction of rolling-element bearings, a defect generates precisely identifiable signature on vibration, and the generated frequencies present an effective route for monitoring progressive bearing degradation. On the other hand, experience and industrial feedback have demonstrated that vibration monitoring has made out its efficiency; and it is highly suitable for rolling-element bearings—however it represents an issue when requiring a good vibration baseline [13]. If no baseline is available, no history has been built up, making the detection of the specific frequencies impossible when the background noise has risen [12].

To overcome this issue, many alternatives have emerged in electric machines by analyzing the stator-side electrical quantities. These alternatives are known as MCSA, including the use of electrical current [13,14], or the instantaneous power factor [15]. For steady-state operations, current spectral estimation based on fast Fourier transform (FFT) and its extension, the short-time Fourier transform (STFT), have been widely employed, such as FFT-based bispectrum/bicoherence [9]. Due to frequency limitation of these techniques [16], high resolution technique: MUSIC (MUltiple SIgnal Classification) [17] and ESPRIT (Estimation of Signal Parameters via Rotational Invariance Techniques) [18,19] were afterwards investigated. However, these techniques have several drawbacks since they are difficult to interpret and it is difficult to extract variation features in time domain for nonstationary signals. To overcome this problem and under nonstationary behavior, procedures based on time-frequency representations (spectrogram, quadratic Wigner-Ville, etc.) [20–22] or time-scale analysis (wavelet) have been proposed in the literature of the electric machines community [23–25]. There are also parametric methods based on parameter estimation of a known model [16]. Nevertheless, these methods are formulated through integral transforms and analytic signal representations [26], so their accuracy depends on data length, stationarity and model accuracy.

Most of electric machine faults lead to current modulation (amplitude and/or phase) [27]. This is the particular case of bearing faults [28]. Indeed, a bearing fault is assumed to produce an air-gap eccentricity [21], and consequently, an unbalanced magnetic pull. Hence, this gives rise to torque oscillations, which lead to amplitude and/or phase modulation of the stator current [13,21,29].

So, for failure detection, a possible approach relies on the use of amplitude demodulation techniques; in other words, the fault detection relies on the extraction of the instantaneous amplitude (IA) and/or the instantaneous frequency (IF). Therefore, it is sufficient to demodulate the current for bearing faults detection. However, the demodulation techniques depend on the type and the dimension of the signal. In this chapter, we try to highlight the use of demodulation techniques for mono-dimensional and multidimensional signals and for mono-component and multicomponent signals.

## 2.2 Brief status on demodulation techniques as a fault detector

As mentioned, the investigation of demodulation techniques as a failure detection relies on the extraction of the IA and/or IF of the electrical quantities, and in most cases, the machine current is taken as a transducer of the fault. For demodulation, let us consider the complex (analytic signal) representation of such signals is given by

$$x(t) = a(t)e^{j\phi(t)} \tag{2.1}$$

where $a(t)$ and $\phi(t)$ are the IA and instantaneous phase, respectively. Signals with more complicated structure can be represented by a combination of signals of this type. A survey allowed to establish a road map for different demodulation techniques [30] and the choice of the demodulation technique depends on the type of the signal.

### 2.2.1 Mono-component and multicomponent signals

A mono-component signal is described in the time-frequency domain by one single "crest or ridge," corresponding to an elongated region of energy concentration [31,32]. Furthermore, interpreting the crest as a graph of IF versus time, the IF of a mono-component signal is a single-valued function of time. Consequently, such a mono-component signal can be expressed approximately as

$$z(t) = a(t)\cos(\phi(t)) \tag{2.2}$$

where

- $a(t)$, known as the IA, is real and positive;
- $\phi(t)$ is known as the instantaneous phase.

It will be noted that in the electrical community $z(t)$ has an analytic associate of the form given by

$$z(t) = a(t)e^{j\phi(t)}. \tag{2.3}$$

A multicomponent signal may be described as the sum of two or more mono-component signals such that

$$z(t) = \sum_{n=1}^{\infty} a_i(t)\cos(\phi(t)) \tag{2.4}$$

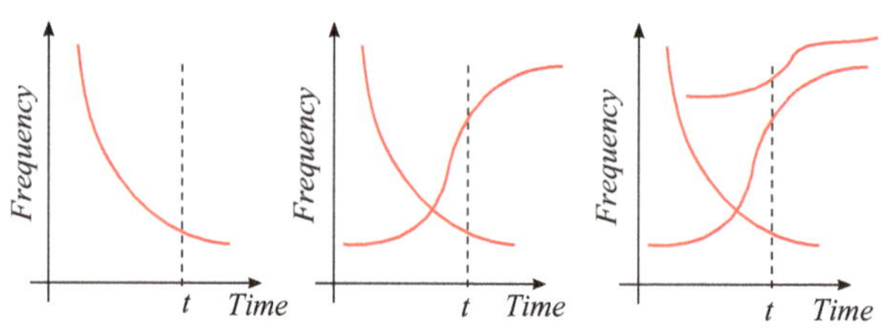


*Figure 2.1 Evolution of the IF for both mono-component and multicomponent signals*

The model described by (2.4) allows the extraction and separation of components from a given multicomponent signal using $(t,f)$ filtering methods [33]. Figure 2.1 shows the evolution of the IF of a mono-component signal and multicomponent signal with two and three components.

## *2.2.2 Demodulation techniques*

Most of electric machine faults lead to current modulation (amplitude and/or phase) [27]. This is the particular case of bearing faults [28]. So, for failure detection, a possible approach relies on the use of amplitude demodulation techniques; in other words, the fault detection relies on the extraction of the IA and IF.

### 2.2.2.1 Mono-dimensional techniques

As depicted in Figure 2.2, mono-dimensional techniques include synchronous demodulation, Hilbert transform (HT) and Teager–Kaiser energy operator (TKEO).

A mono-dimensional signal can be modeled in discrete form by

$$x(n) = a(n) \cdot \cos(\Phi(n)) \tag{2.5}$$

where $n = 0, \ldots, N-1$ is the sample index, with $N$ being the number of samples. In (2.5), frequency $\omega$ is equal to $2\pi f/F_e$ (where $f$ and $F_e$ are the supply and sampling frequency, respectively) and amplitude $a(n)$ is related to the fault. In this context, the

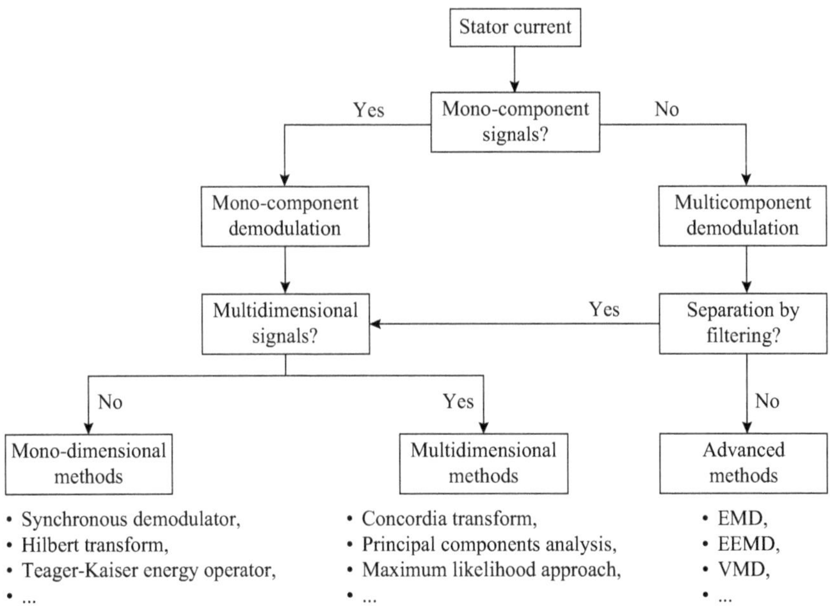


*Figure 2.2 Road map to choose the demodulation technique*

best path to extract feature extraction is the use of amplitude demodulation techniques. The signal $x(n)$ can be expressed in term of its IA and instantaneous phase as follows:

$$x(n) = a(n) \cdot \cos(\Phi(n)) \tag{2.6}$$

The signal $x(n)$ can be expressed in terms of two components: real component $y_1$ and imaginary component $y_2$ such as

$$y_1(n) = a(n) \cdot \cos(\Phi(n)) \qquad y_2(n) = a(n) \cdot \sin(\Phi(n)) \tag{2.7}$$

and $x(n)$ can be expressed by it is analytic signal representation as

$$x(n) = y_1(n) + jy_2(n) \tag{2.8}$$

#### 2.2.2.2 Multidimensional techniques

In electrical systems, a multidimensional signal refers to a multiphase systems; particularly in triphase systems, signals can be modeled in discrete form by

$$x_{0,1,2}(n) = a_{0,1,2}(n) \cdot \cos(\Phi_{0,1,2}(n)) \tag{2.9}$$

For instance, we assume a three-phase system that does not contain any harmonics, but in a noisy environment. The three-phase quantities can therefore be expressed by system (2.10):

$$\begin{cases} x_0(t) = a_0 \cos(\omega t + \alpha_0) \\ x_1(t) = a_1 \cos(\omega t + \alpha_1) \\ x_2(t) = a_2 \cos(\omega t + \alpha_2) \end{cases} \tag{2.10}$$

where $a_0$, $a_1$ and $a_2$ are the three magnitudes, and $\omega$ is the angular frequencies, and $\alpha_0, \alpha_1$ and $\alpha_2$ are the three initial phase angles of the corresponding phase.

The three-phase system can be expressed in a compact form as follows [34]:

$$x_m[k] = a_m \cos(k\omega_0 + \alpha_m) \tag{2.11}$$

where $\omega_0 = 2\pi \frac{f_0}{F_s}$ corresponds to the fundamental angular frequency, $m = 0, 1$ or 2 corresponds to the phase index for the three-phase electrical system, $f_0$ is the fundamental frequency, $F_s$ is the sampling frequency, $x_0[k]$, $x_1[k]$ and $x_2[k]$ are the electric signal of each phase, and $a_a$, $a_b$, $a_c$, $\alpha_a$, $\alpha_b$ and $\alpha_c$ are, respectively, the amplitudes and initial phases of each fundamental component of the three-phase system. Hence, the most common path to demodulate a multidimensional signal is the use the transformation of the three-phase quantities modeled by (2.11) to the corresponding complex phasor. The complex phasor for three-phase system can be expressed as follows:

$$x_m = x_\alpha + jx_\beta \tag{2.12}$$

where $x_\alpha$ and $x_\beta$ are the direct and quadrature components obtained by the use of $(abc)$ to $(\alpha\beta)$ transform. For multidimensional signal, the case of three-phase system, the three-phase transformations such as Concordia transform (CT) [35,36] and Park vector approach [37–39] have been indexed as a demodulation techniques.

## 2.3 Synchronous demodulation

Synchronous demodulation is an amplitude and phase demodulation technique. Figure 2.3 illustrates the principle of this demodulation technique, and it shows that the analyzed signal is multiplied with two reference signals $F_1$ and $F_2$.

Let a signal:

$$i(t) = a(t)\cos(2\pi f_p t + \varphi) \tag{2.13}$$

By multiplying the signal $i(t)$ by a carrier with pulsation $\omega_p$:

$$F_1(t) = i(t)\cos(2\pi f_p t) \tag{2.14}$$

$$F_2(t) = i(t)\sin(2\pi f_p t) \tag{2.15}$$

Using the trigonometric properties, we obtain:

$$F_1(t) = (a(t)/2)(\cos(4\pi f_p t + \varphi) + \cos(\varphi)) \tag{2.16}$$

$$F_2(t) = (a(t)/2)(\sin(4\pi f_p t + \varphi) + \cos(\varphi)) \tag{2.17}$$

To simplify the mathematical analysis, we use the frequency-domain representation of $F_2$ and $F_2$; this yields to

$$F_1(f) = (a(f)/2)\left(\frac{1}{2}(\delta(f - 2f_p) + \delta(f + 2f_p)) \cdot \mathrm{e}^{jf\varphi} + \cos(\varphi)\delta(f)\right) \tag{2.18}$$

$$F_1(f) = \frac{\cos(\varphi)}{2}a(f) + \frac{\mathrm{e}^{jf\varphi}}{4}(a(f - 2f_p) + a(f + 2f_p)) \tag{2.19}$$

In the same way, it can be shown that

$$F_2(f) = (a(f)/2)\left(\frac{1}{2}(\delta(f - 2f_p) + \delta(f + 2f_p)) \cdot \mathrm{e}^{jf\varphi} + \cos(\varphi)\delta(f)\right) \tag{2.20}$$

then

$$F_2(f) = \frac{\sin(\varphi)}{2}a(f) + \frac{j \cdot \mathrm{e}^{jf\varphi}}{4}(a(f + 2f_p) - a(f - 2f_p)) \tag{2.21}$$

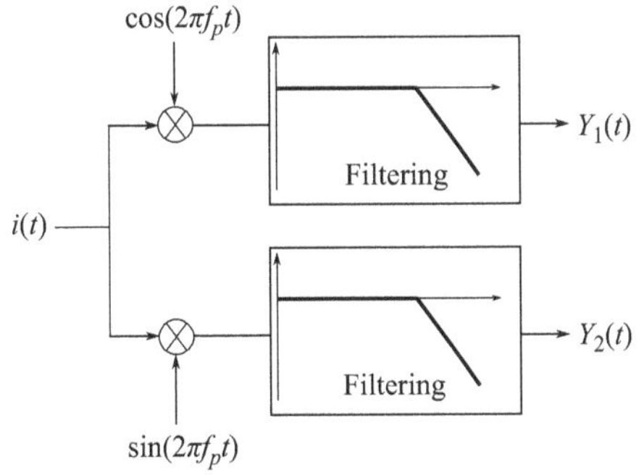


*Figure 2.3 Synchronous demodulation principle*

Under the assumption that the spectrum of $a(f)$ is frequency-bounded $[-f_{\max}, f_{\max}]$ with $f_{\max} < f_p$, it is possible to extract $a(f)$ with a low-pass filter of cutoff frequency $f_p$. Assuming that the low-pass filter is ideal (brickwall filter), the post-filter signals denoted by $F_1^{(pf)}(f)$ and $F_2^{(pf)}(f)$ can then be expressed as follows:

$$F_1^{(pf)}(f) = \frac{\cos(\varphi)}{2} a(f) \tag{2.22}$$

$$F_2^{(pf)}(f) = \frac{\sin(\varphi)}{2} a(f) \tag{2.23}$$

$$F_2^{(pf)}(f) = \frac{\sin(\varphi)}{2} a(f) \tag{2.24}$$

$$z(t) = \left(y_1^{(pf)}(t)\right)^2 + \left(y_2^{(pf)}(t)\right)^2 \tag{2.25}$$

$$z(t) = (a(t))^2 \cdot \left(\left(\frac{\cos(\varphi)}{2}\right)^2 + \left(\frac{\sin(\varphi)}{2}\right)^2\right) \tag{2.26}$$

$$z(t) = \frac{(a(t))^2}{4} \tag{2.27}$$

By this method, we can extract the IA of the signal. Except that, this approach has several drawbacks. First of all, its application requires to know exactly the frequency $f_p$. In particular, a poor knowledge of $f_p$ deteriorates considerably the estimation of the IA. Second, this technique requires the selection and calibration of a low-pass filter as well as the choice of a filter structure and a perfectly adapted cutoff frequency.

Synchronous demodulation has been applied for fault detection in electrical machines running at constant speeds. However, for machines rotating at variable speeds, synchronous demodulation requires a good knowledge of the law of evolution of the IF.

## 2.4 Hilbert transform

In order to estimate the IF and IA of a signal, a standard approach is to use the HT. The HT is a linear operator for which analytic signals can be derived if the Bedrosian theorem is verified from the signal $x(n)$. It is defined as the convolution (*) of the signal with the function $1/t$ [40]. If $\hat{x(t)}$ is the HT of a signal $x(t)$, the analytic signal introduced by [41] is given by the following equation:

$$z(t) = x(t) + j\hat{x(t)} \tag{2.28}$$

and $\hat{x(t)}$ is expressed by

$$\hat{x(t)} = x(t) * \frac{1}{\pi t} \tag{2.29}$$

For its discrete formulation, let us consider a discrete signal $x(n)$. The discrete HT (DHT) of $x(n)$ is given by the following [42]:

$$\mathscr{H}[x(n)] = \mathscr{F}^{-1}\{\mathscr{F}\{x(n)\} \cdot u(n)\} \tag{2.30}$$

where $\mathscr{F}\{\cdot\}$ and $\mathscr{F}\{\cdot\}$ correspond to the FFT and inverse FFT (IFFT), respectively, and where $u(n)$ is defined as

$$u(n) = \begin{cases} 1, & n = 0, \frac{N}{2} \\ 2, & n = 1, 2, \ldots, \frac{N}{2} - 1 \\ 0, & n = \frac{N}{2} - 1, \ldots, N - 1 \end{cases} \tag{2.31}$$

Let us define the analytic signal of $x(n)$, denoted $z(n)$, as

$$z_k(n) = x_k(n) + j\mathscr{H}[x_k(n)] \tag{2.32}$$

Using signal model (2.5), the amplitude envelope can be estimated by [42]:

$$|a(n)| \approx |z(n)| = \sqrt{x_k(n)^2 + \mathscr{H}[x_k(n)]^2} \tag{2.33}$$

and the instantaneous phase $\phi(n)$ can be estimated by

$$\phi(n) = Arg(z(n)) \tag{2.34}$$

## 2.5 Teager–Kaiser energy operator

The TKEO is an IA and IF demodulation technique for mono-component signal, and it estimates IA and IF without using the analytical signal $z(n)$. The estimation of IA and IF with TEO technique is based on the continuous energy separation algorithm, given by the following [43]:

$$|a(t)| \approx \frac{\psi[x(t)]}{\sqrt{\psi[\dot{x}(t)]}} \tag{2.35}$$

$$f(t) \approx \frac{1}{2\pi}\sqrt{\frac{\psi[\dot{x}(t)]}{\psi[x(t)]}} \tag{2.36}$$

with $\psi$ is the so-called TKEO:

$$\psi = [\dot{x}(t)]^2 - x(t)\ddot{x}(t)$$

where $x(t)$ is the analyzed signal and $\dot{x}(t)$ and $\ddot{x}(t)$ are its first and second derivatives, respectively.

It will be noted that, for discrete signals, the TKEO offers excellent time resolution because only three samples are required for the energy computation at each time instant, hence the result is highly depending on the sampling frequency. So, for discrete signals, the TKEO technique is performed by using the discrete-time energy separation algorithm developed in [44] and well known as (DESA-2). In this algorithm, the estimated IA and IF are given using the following equations:

$$|a(n)| \approx \frac{2\psi(x(n))}{\sqrt{\psi(x(n+1)-x(n-1))}} \tag{2.37}$$

$$f(n) \approx \frac{1}{2\pi}\arcos\left(1-\frac{\psi(x(n)-x(n-1))}{2\psi(x(n))}\right) \tag{2.38}$$

where the TKEO can be approximated by time differences as follows:

$$\psi = [x(n)]^2 - x(n+1)x(n-1)$$

## 2.6 Concordia transform

The CT converts the three-phase current to Park's space vector components $i_\alpha(n)$ and $i_\beta(n)$, as depicted by Figure 2.4.

The Park components are given by

$$\begin{bmatrix} i_\alpha(n) \\ i_\beta(n) \end{bmatrix} = \begin{bmatrix} \frac{2}{3} & -\frac{1}{3} & -\frac{1}{3} \\ 0 & \frac{1}{\sqrt{3}} & -\frac{1}{\sqrt{3}} \end{bmatrix} \begin{bmatrix} i_0(n) \\ i_1(n) \\ i_2(n) \end{bmatrix} \tag{2.39}$$

Several fault detectors based on CT have been proposed in literature [35–37, 45–47]. Recently, it has been shown that CT can be viewed as a demodulation technique for balanced system [35]. Indeed, under the assumption that the system is balanced, the Park components can be expressed as

$$\begin{aligned} i_\alpha(n) &= a(n)\cos(\omega n) \\ i_\beta(n) &= a(n)\sin(\omega n) \end{aligned}$$

Then, the amplitude can be estimated by

$$|a(n)| = \sqrt{i_\alpha^2(n) + i_\beta^2(n)} \tag{2.40}$$

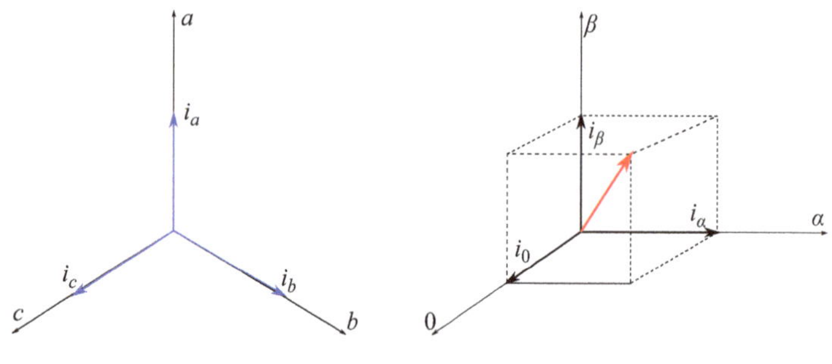


*Figure 2.4 CT principle*

It will be noted that for balanced system, the component $i_0$ is null. Therefore, CT can be considered as a low-complexity demodulating technique *if* the system is balanced. However if the system is unbalanced, and there is no assertion that during bearing fault the three-phase system remains balanced, (2.40) is no longer valid and depends on three modulating signals $i_a(n)$, $i_b(n)$ and $i_c(n)$, and the corresponding space phasor in its extended form is computed according to (2.41):

$$\vec{i}(n) = i_\alpha(n)\vec{u}_\alpha + i_\beta(n)\vec{u}_\beta + i_0(n)\vec{u}_0 \tag{2.41}$$

where $i_\alpha$, $i_\beta$ and $i_0$ are the components according to axis, respectively, and are the corresponding unit vectors, and the IA can be estimated by

$$|i(n)| = \sqrt{(i_\alpha(n))^2 + (i_\beta(n))^2 + (i_0(n))^2} \tag{2.42}$$

## 2.7 Fault detector

Several detectors based on the IA have been proposed in the literature [36,45,48–51]. However, most of these approaches use unnecessary and complicated classifiers, such as artificial neural networks, fuzzy logic and support vector machine, and most of them assume that a training database is available. This can be very difficult to obtain for many industrial applications. Indeed, it has been mentioned in a number of previously published papers that one of the main difficulties in real-word testing of developed condition monitoring technique is the lack of collaboration needed with industrial operators and manufacturers due to data confidentiality, particularly when failures are present [52], and can be difficult to obtain [53]. For this purpose, a statistical feature-based detector is proposed; it does not require any training sequence. The detector is based on the variance of $|a(n)|$ or $|a_k(n)|$, and the two basic parameters are the mean value $\mu$ and the standard deviation $\sigma$ [54].

### *2.7.1 Fault detector based on HT and TKEO demodulation*

After applying HT or TKEO independently on the three currents, we propose to exploit the information given by the three extracted envelopes. To avoid the edge effect problem of HT and TKEO, each envelope is truncated by removing $\alpha$ samples at the beginning and at the end of $|a_k(n)|$. The proposed criterion, $\sigma_H^2$ (i.e. $\sigma_{TKEO}^2$), is then equal to

$$\sigma_H^2 = \frac{1}{3(N-2\alpha)} \left( \sum_{k=0}^{2} \sum_{n=\alpha}^{N-\alpha-1} (|a_k(n)| - \mu_k)^2 \right) \tag{2.43}$$

where $\mu_k$ is the average of $|a_k(n)|$, i.e.

$$\mu_k = \frac{1}{(N-2\alpha)} \sum_{n=\alpha}^{N-\alpha-1} |a_k(n)| \tag{2.44}$$

In (2.43), the average is used to make the criteria $\sigma_H^2$, $\sigma_{TKEO}^2$ and $\sigma_C^2$ equivalent for balanced system. Indeed if $a_k(n) = a(n)$ for all $k = \{0, 1, 3\}$ and if the edge effects problems are neglected, then it can be shown that $\sigma_H^2 = \sigma_C^2 = \sigma_{TKEO}^2$ with $\alpha = 0$. This property no longer holds for unbalanced system. For healthy unbalanced system, envelopes $a_k(n)$ are different but they are all constant. It follows that $|a_0(n)| = \mu_0$, $|a_1(n)| = \mu_1$ and $|a_2(n)| = \mu_2$ and so $\sigma_H^2 = 0$. Therefore, we propose a simple hypothesis test to detect a fault under unbalanced condition:

- If $\sigma_H^2 < \gamma_H$, the machine is healthy.
- If $\sigma_H^2 > \gamma_H$, the machine is faulty.

Here $\gamma_H$ is a threshold which can be set subjectively. One should remark that this second hypothesis test is more powerful since it can be employed for balanced and unbalanced systems. In this section, the result of several simulations is presented to compare the performance of the proposed fault detectors. For each simulation, the amplitude envelope is estimated through CT, HT and TKEO. Then, depending of the demodulation technique, criteria $\sigma_C^2$, $\sigma_H^2$ or $\sigma_{TKEO}^2$ are computed to reveal the presence of a fault. The simulation has been performed for healthy and faulty machine.

### 2.7.2 Fault detector after CT demodulation

After applying CT, envelope $|a(n)|$ is extracted with (2.40). Then, we propose to compute the variance of $|a(n)|$ to detect a fault. This statistic criterion, denoted $\sigma_C^2$, is given by

$$\sigma_C^2 = \frac{1}{N}\sum_{n=0}^{N-1}(|a(n)| - \mu)^2 \tag{2.45}$$

where $\mu$ is the average of $|a(n)|$, i.e.

$$\mu = \frac{1}{N}\sum_{n=0}^{N-1}|a(n)| \tag{2.46}$$

The variance $\sigma_C^2$ measures the deviation of the amplitude around its mean $\mu$. This criterion can be used to detect amplitude modulation for balanced system. Indeed, if no fault is present, $|a(n)|$ is constant and so $|a(n)| = \mu$. Using (2.63), it follows that $\sigma_C^2 = 0$. On the contrary, for healthy machine $|a(n)| \neq \mu$, which also implies $\sigma_C^2 > 0$. Therefore, we can propose a simple hypothesis test to detect a fault under balanced assumption:

- If $\sigma_C^2 < \gamma_C$, the machine is healthy.
- If $\sigma_C^2 > \gamma_C$, the machine is faulty.

Here $\gamma_C$ is a threshold which can be set subjectively. For unbalanced system, one should note that this simple hypothesis test is no longer valid since $\sigma_C^2$ is not necessarily equal to 0 for healthy machine.

### 2.7.3 Synthetic signals

Several simulations are presented to compare the performance of the proposed fault detectors. For each simulation, the amplitude envelope is estimated through CT, HT and TKEO. Then, depending on the demodulation technique, criteria $\sigma_C^2$, $\sigma_H^2$ or $\sigma_{TKEO}^2$

are computed to reveal the presence of a fault. The simulation have been performed for healthy and faulty machine.

For this purpose, several simulations have been performed with amplitude modulated (AM) synthetic signals which are defined as follows [21]:

$$i_k(n) = \underbrace{(1 + \beta \sin(\omega_2 n + \psi_k))}_{a_k(n)} \cdot \cos(\omega n + \phi_k) \tag{2.47}$$

where $\beta$ is a fault index which is equal to 0 for healthy machines and greater than 0 for faulty ones. The parameters $\psi_k$ and $\gamma_k$ are calibrated depending on the balanced assumption. If the system is balanced, $\psi_k = \psi$ $(k = 0, 1, 2)$, where $\psi_k$ depends on $k$ for unbalanced system. Simulations have been run with a sampling frequency $F_e = 10\,\text{kHz}$ during 1 s with $\omega = 0.1534$ rad/s (supply frequency $f = 50\,\text{Hz}$) and $\omega_2 = 0.0307$ rad/s ($f_2 = 10\,\text{Hz}$). After HT demodulation, $\alpha = 10$ samples have been removed at the beginning and at the end of $|a_k(n)|$ to avoid edge effect problems. The fault index has been set to $\beta = 0.2$ to simulate faulty machine (see Figure 2.5 for time representation of $i_0(n)$).

#### 2.7.3.1 Balanced system ($\psi = 0$)

For balanced system, the amplitude envelopes are the same for the three currents. Figures 2.6 and 2.7 display $|a(n)|$ and $|a_0(n)|$ extracted with CT, HT and TKEO, respectively, for a healthy and faulty cases. One can notice that the three demodulation techniques lead to the same envelope. Table 2.1 shows the values of the fault detector criteria $\sigma_C^2$, $\sigma_H^2$ and $\sigma_{TKEO}^2$ for faulty and healthy machine. The three criteria lead to similar results, indeed $\sigma_C^2 = \sigma_H^2 = \sigma_{TKEO}^2 = 0$ for healthy machine (i.e. $\beta = 0$)

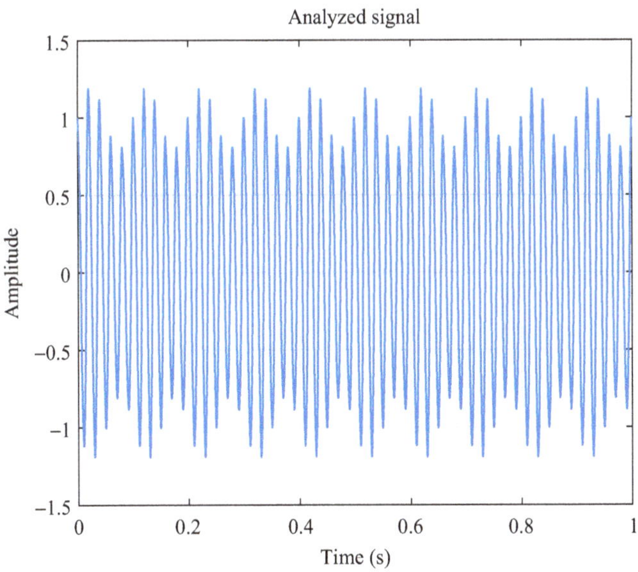


*Figure 2.5 Time representation of the current $i_0(n)$ for a faulty machine ($\beta = 0.2$)*

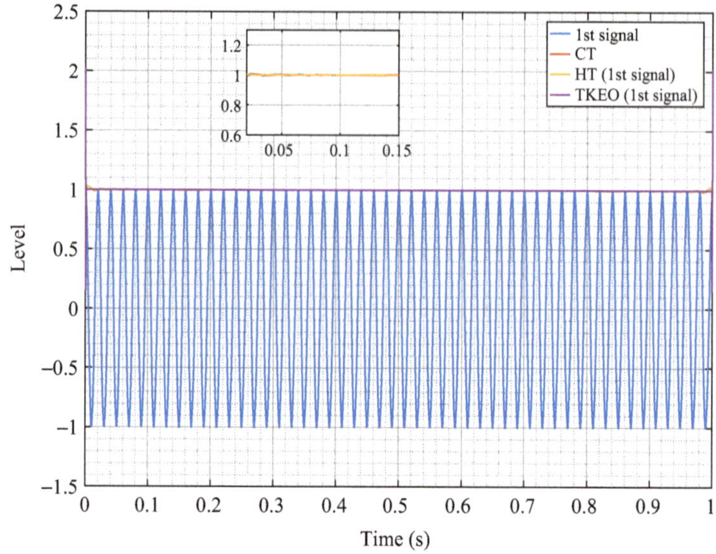


*Figure 2.6 Balanced system healthy machine: time representation of the envelopes after CT, HT and TKEO demodulation ($\beta = 0$)*

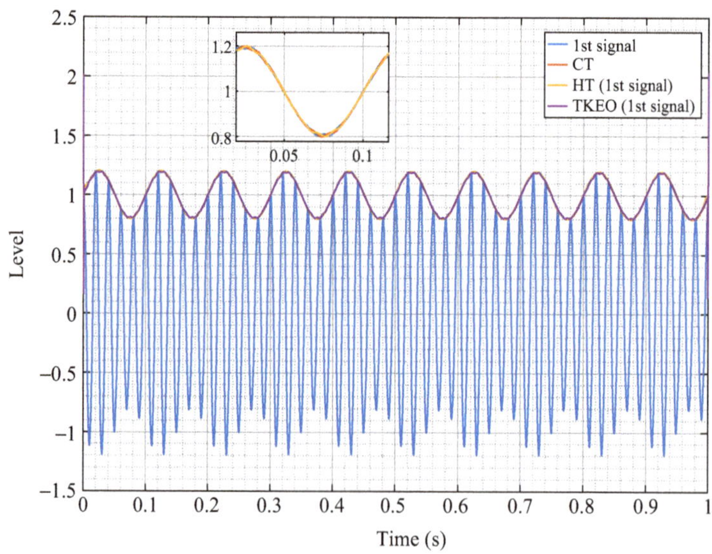


*Figure 2.7 Balanced system faulty machine: time representation of the envelopes after CT, HT and TKEO demodulation ($\beta = 0.2$)*

*Table 2.1 Fault detector for healthy and faulty machines*

| System | Demodulation | Fault detector | |
|---|---|---|---|
| | | **Healthy case** | **Faulty case** |
| Balanced and stationary | CT | $\sigma_C^2 = 0.000$ | $\sigma_C^2 = 0.020$ |
| | HT | $\sigma_H^2 = 0.000$ | $\sigma_H^2 = 0.020$ |
| | TKEO | $\sigma_{TKEO}^2 = 0.000$ | $\sigma_{TKEO}^2 = 0.018$ |
| Unbalanced and stationary | CT | $\sigma_C^2 = 0.000$ | $\sigma_C^2 = 0.005$ |
| | HT | $\sigma_H^2 = 0.000$ | $\sigma_H^2 = 0.020$ |
| | TKEO | $\sigma_{TKEO}^2 = 0.000$ | $\sigma_{TKEO}^2 = 0.018$ |
| Unbalanced and nonstationary | CT | $\sigma_C^2 = 0.000$ | $\sigma_C^2 = 0.005$ |
| | HT | $\sigma_H^2 = 0.001$ | $\sigma_H^2 = 0.021$ |
| | TKEO | $\sigma_{TKEO}^2 = 0.000$ | $\sigma_{TKEO}^2 = 0.017$ |

and $\sigma_C^2 = \sigma_H^2 = \sigma_{TKEO}^2 = 0.020$ for faulty ones (i.e. $\beta = 0.2$). Therefore, a fault can be easily detected in this context by setting the threshold of the fault detector to $\gamma_C = \gamma_H = \gamma_{TKEO} = 0.010$. From a practical point of view, one should note that CT demodulation must be preferred for balanced system since it has a lower complexity than HT and TKEO and does not suffer from edge-effect problems.

#### 2.7.3.2 Unbalanced system ($\psi_0 = 0, \psi_1 = 2\pi/3, \psi_2 = -2\pi/3$)

Let us simulate a system which is unbalanced under faulty condition. Figure 2.8 displays amplitude $a(n)$ and the envelope $|a_0(n)|$ extracted with CT, HT and TKEO respectively, for a faulty machine. As expected, CT is not able to demodulate the signals. Table 2.1 presents the values of the fault detector criterion $\sigma_C^2$, $\sigma_H^2$ and $\sigma_{TKEO}^2$ under healthy and faulty conditions. In our simulations, criterion $\sigma_H^2$ leads to the same values for balanced and unbalanced system whereas the value of $\sigma_H^2$ decreases under unbalanced condition. One can notice that the difference between healthy and faulty case is larger for $\sigma_H^2$. For fault detection, an hypothesis-test threshold equal to $\gamma_C = 0.0025$ for $\sigma_C^2$ and $\gamma_H = \gamma_{TKEO} = 0.010$ for $\sigma_H^2$ and $\sigma_{TKEO}^2$ lead to correct results in this context.

#### 2.7.3.3 Unbalanced system ($\psi_0 = 0, \psi_1 = 2\pi/3, \psi_2 = -2\pi/3$) under nonstationary supply frequency

To simulate nonstationary environment, supply frequency $f$ is assumed to vary linearly between 10 and 50 Hz, i.e.

$$\omega(n) = \frac{2\pi}{F_e}\left(\frac{40}{2N}n + 10\right) \tag{2.48}$$

Figure 2.9 displays amplitude $|a(n)|$ and the envelope $|a_0(n)|$ extracted with CT, HT and TKEO, respectively, for a faulty machine under nonstationary supply

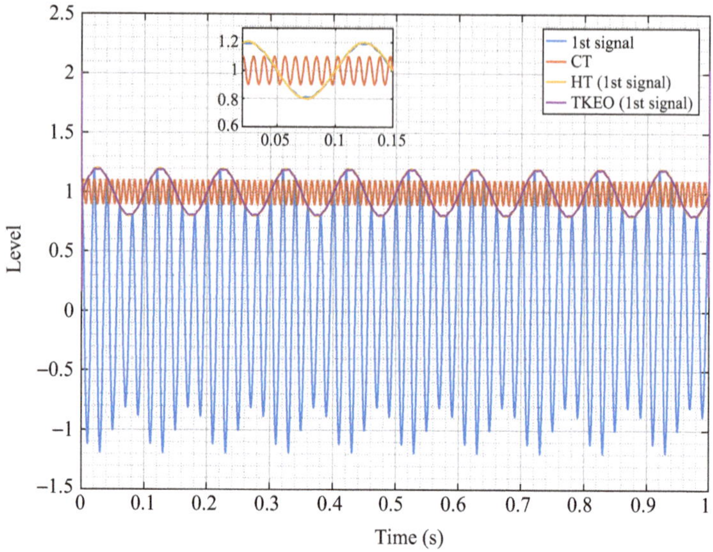


*Figure 2.8 Unbalanced system faulty machine: time representation of the envelopes after CT, HT and TKEO demodulation ($\beta = 0.2$)*

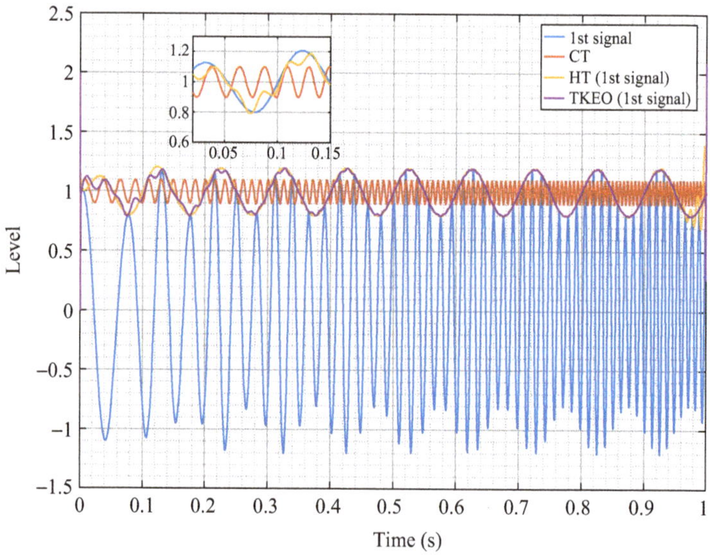


*Figure 2.9 Non-balanced system under nonstationary condition—faulty machine: time representation of the envelopes after CT, HT and TKEO demodulation ($\beta = 0.2$)*

frequency. As edge-effect problem occurs for HT (see Figure 2.9), $\alpha$ samples have been removed at the beginning and at the end of $|a_k(n)|$. Table 2.1 presents the values of the fault detector criterion $\sigma_C^2$, $\sigma_H^2$ and $\sigma_{TKEO}^2$. One should note that the values $\sigma_C^2$, $\sigma_H^2$ and $\sigma_{TKEO}^2$ do not depend on the stationary assumption in our context. Therefore, fault detectors based on amplitude demodulation seem to be well-suited for nonstationary scenario. In particular, these detectors do not need to employ complicated time-frequency representations (like spectrogram and Wigner–Ville) that suffer from artifact or poor resolution.

## 2.8 EMD method

Besides, in typical electric machines, stator current components are the supply fundamental, harmonics, additional components due to slot harmonics, saturation harmonics, other components from unknown sources such as environmental noise and design imperfection, and eventually effect introduced by bearing faults. In typical electric machines, the stator current is a multicomponets signal and can be expressed by a temporal model as

$$x(t) = \sum_{k=1}^{M} a_k(t) \sin(\phi_k(t)) \tag{2.49}$$

with $a_k(t) = a_k(1 + m_{ka} \sin(2\pi f_{ka} t + \varphi_{ka}))$ and $\phi_k(t) = 2\pi f_k t + m_{kp} \sin(2\pi f_{kp} t + \varphi_{kp})$. Here $m_{ka}$ and $m_{kp}$ are the AM index, and the PM index, respectively, that can be introduced by a fault as an AM/PM effect. This work considers only the AM effect. Therefore, $m_{kp} = 0$ and $\phi_k(t) = 2\pi f_k t$, where $f_k = k f_0$ with $f_0$ is the fundamental frequency and $k$ is the harmonic order. Hence, for fault detection, a possible approach relies on the use of amplitude demodulation techniques to extract fault-related features. In this multicomponent signal context, the empirical mode decomposition (EMD) is considered. The EMD is an emerging signal processing algorithm for signal demodulation. It has been first introduced in [55], and has since become an established tool for the analysis of nonstationary and nonlinear data [56]. This approach has focused considerable attention and has been widely used for rotating machinery fault diagnosis [22,54,57,58]. It is an adaptive time-frequency data analysis method for nonlinear and nonstationary signals [55], and behaves like an adaptive filter bank [59]. Compared to FFT or wavelets that decompose a signal into a series of *sine* functions or scaled mother wavelet, the EMD decomposes the multicomponent signal into a series of mono-components signal, known as *intrinsic mode function* denoted *IMFs*, and based on the local characteristic time-scale of the signal. This decomposition can be described as follows:

- Identification of all extrema of the logged current;
- Interpolation between minima (respectively maxima) ending up with some envelope $e_{\min}(n)$ (respectively $e_{\max}(n)$);

- Computation of the mean:

$$R(n) = \frac{e_{\min}(n) + e_{\max}(n)}{2} \tag{2.50}$$

- Extraction of the detail:

$$d_m(n) = i(n) - R(n) \tag{2.51}$$

- Iteration on the residue $R(n)$

In practice, this algorithm has to be refined by a sifting process until the detail $d_m$ can be considered as *IMF* [55]. To illustrate the EMD concept, let us assume the synthesized signal $x_{\text{syn}}(t)$ given by

$$x_{\text{syn}}(t) = a_1 \sin(\omega_1 t) + a_2 \sin(\omega_2 t), \tag{2.52}$$

where $a_1$ and $a_2$ are the amplitudes of the first and the second component respectively, while $\omega_1$ and $\omega_2$ are pulsations of those components. By decomposing $x_{\text{syn}}(t)$ through the EMD algorithm, the result is depicted in Figure 2.10. It appears clearly that the two components are presented by the first and second *IMFs*. Unfortunately, real signals

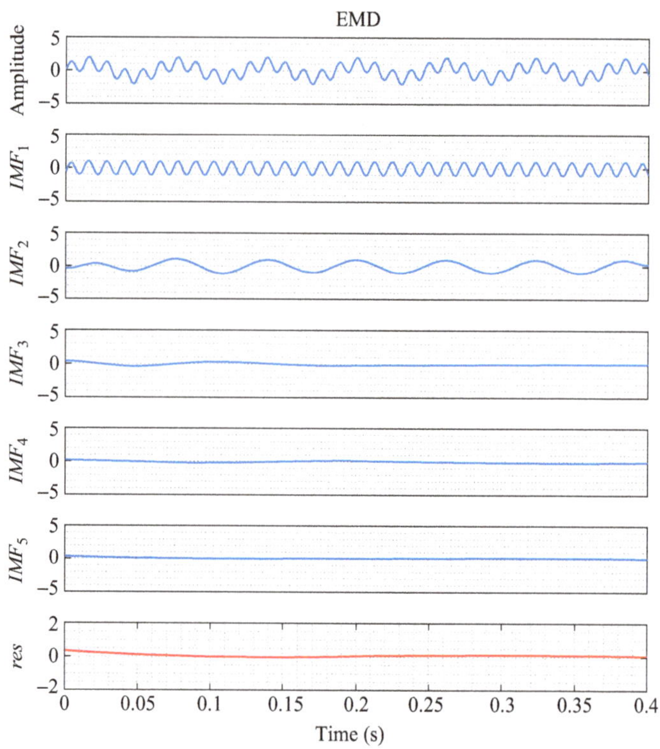


*Figure 2.10 EMD for uncorrupted synthetic signal*

are not immunized from noises, and in order to have a look on the behavior of the EMD on the added noise signal, let us consider that signal is corrupted by an added white Gaussian noise (AWGN), then $x_{\text{syn}}(t)$ can be expressed by

$$x_{\text{syn}}(t) = a_1 \sin(\omega_1 t) + a_2 \sin(\omega_2 t) + \text{AWGN}, \tag{2.53}$$

The corresponding local time oscillations or *IMFs* and residue are depicted in Figure 2.11.

The first observation is that the corresponding *IMFs* are shifted from the fourth to fifth *IMFs*; this is due to the AWGN to the original signal, hence high-frequency oscillations are introduced at the first, second and third *IMFs*. The second observation is the occurrence of the second component into at least two consecutive *IMFs*. This phenomenon is the mode mixing, as mentioned before. Consequently, it is difficult to really understand what the EMD provides as *IMFs*, and are devoid of a physical meaning [55]. Other drawbacks are indexed in literature, such as the *ad hoc* process on which it is based [59], sensitivity to noise, and the fact that it suffers from mode mixing. To overcome the mode-mixing problem, the Ensemble EMD (EEMD) was introduced.

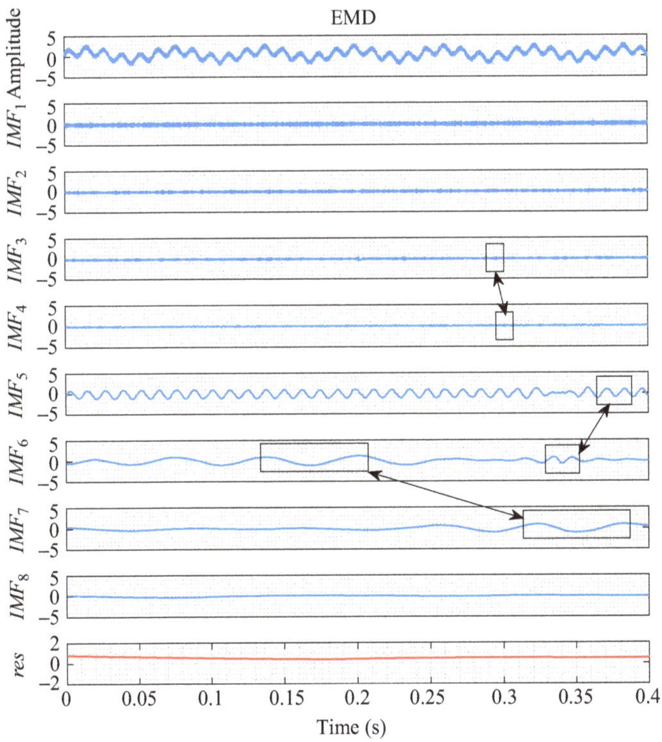


*Figure 2.11 EMD for corrupted synthetic signal by an AWGN*

## 2.9 Ensemble EMD principle

As mentioned next, the main drawbacks of the EMD are that it is based on an *ad hoc* process [59], it is mathematically difficult to model, it is noise sensitive and it suffers from mode mixing. Consequently, it is difficult to understand what the EMD provides as *IMFs* that are devoid from a physical meaning [55]. To deal with this drawbacks, an EEMD was proposed in [60,61], and has become a tool for the analysis of nonstationary and nonlinear data [56] in a wide of range applications in signal processing [62] and fault detection [22,57,58]. It is an improved EMD and is described as a noise-assisted data analysis method. Indeed, it deals with several EMD decompositions of the original signal corrupted by different artificial noises. The final EEMD is then the average of each EMD and defines true *IMFs* as the mean of an ensemble of trials. The EEMD algorithm is depicted in Figure 2.12 and its implementation is described step by step in [54].

The EEMD reliability depends on the choice of the ensemble number denoted by $M$ and the add-noise amplitude $a$. These two parameters are linked by the following [60]:

$$e = \frac{a}{\sqrt{M}} \tag{2.54}$$

where $e$ is the standard deviation error, and it is defined as the discrepancy between the input signal and the corresponding *IMF*.

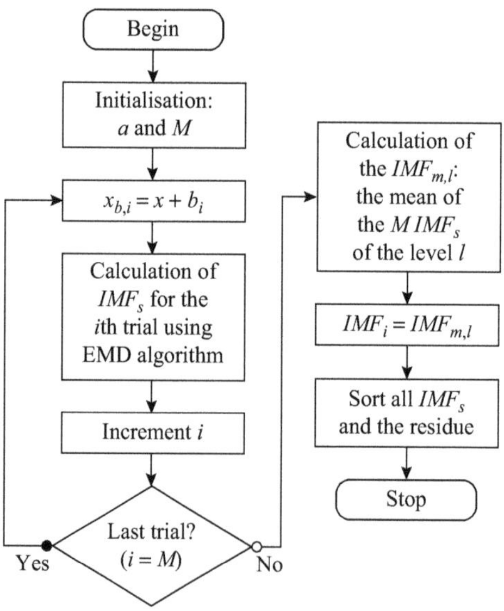


*Figure 2.12 EEMD process for signal decomposition*

So, through EEMD algorithm, a signal $x(t)$ can be expressed as a sum of $k$ modes or *IMFs* as follows:

$$x(t) = \sum_{i=1}^{k} IMF_i(t) + res(t) \tag{2.55}$$

Figures 2.13 and 2.14 illustrate the decomposition of free-noise signal and corrupted signal, respectively.

Let us consider the series $x(n)$ $(n = 1, \ldots, N)$ is the acquired stator current. Under the multicomponent assumption, the sampled current $x(n)$ can be decomposed as

$$x(n) = \sum_{i=1}^{j} IMF_i(n) + res(n) \tag{2.56}$$

where $IMF_i(n)$ is the $i$th intrinsic mode function, $res(n)$ is the residue and $j$ the total number of *IMFs*. In practice, *IMFs* are unknown and must be extracted from the

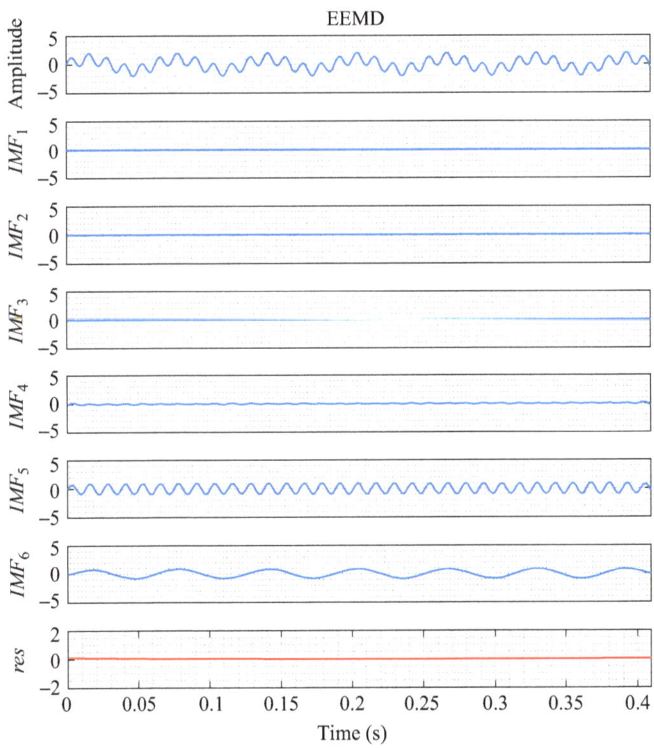


*Figure 2.13 EEMD for uncorrupted synthetic signal*

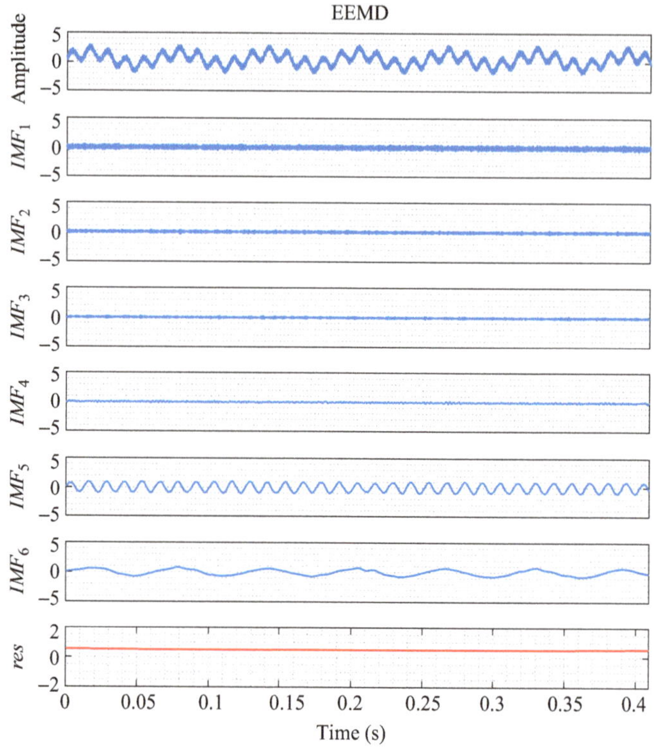


*Figure 2.14 EEMD for corrupted synthetic signal by an AWGN*

stator current $x(n)$. However, at least one *IMF* is related or representative of the main component. Consequently, $x(n)$ can be expressed by

$$x(n) = \sum_{i=1}^{c-1} IMF_i(n) + IMF_c(n) + \sum_{i=c+1}^{j} IMF_i(n) + res(n) \tag{2.57}$$

where $IMF_c(n)$ is the closest *IMF* to the original signal $x(n)$.

So, the main issue that rises is how to extract this *IMF*. To answer this question, in [63–65], a mode decomposition-based notch filter was developed.

## 2.10 EEMD-based notch filter

As mentioned in previous subsection, the decomposition of signal $x(t)$ through EEMD leads to a sum of modes as expressed in (2.57); among these modes, at least one mode is representative to the original signal, and this mode is the dominant mode denoted by $IMF_d(n)$. Assuming that the occurrence of a fault introduces a new component in the original signal, a specific mode denoted by $IMF_e$ is introduced in the mode decomposition of this original signal.

The aim of the notch filter is to cancel the dominant *IMF*, and the result denoted by $x(n)_{cEEMD}$ can therefore be used to detect bearing failure.

### 2.10.1 Statistical distance measurement

The statistical distance quantifies the distance between two statistical quantities, which can be two random variables, or two probability distributions or samples. Various approaches have been indexed in statistics literature and investigated in various fields, particularly for fault detection and diagnostic [54,66]. The statistical tool known as Pearson's correlation is used to measure the distance, and to give a weight to dependency between two temporal series $x(n)$ and $y(n)$ [67]. This dependency is weighted by a coefficient denoted by $r(x, y)$ and defined by (2.58); a value of this coefficient close to $-1$ or 1 indicates that $x(n)$ and $y(n)$ are highly correlated positively or negatively, respectively, while a value around 0 indicates that there is no dependency between $x(n)$ and $y(n)$ [65].

$$r(x,y) = \frac{\sum_n [(x(n) - m_x) * (y(n) - m_y)]}{\sqrt{\sum_n (x(n) - m_x)^2} \cdot \sqrt{\sum_n (y(n) - m_y)^2}} \tag{2.58}$$

where $m_x$ and $m_y$ are the means of $x$ and $y$, respectively.

### 2.10.2 Dominant-mode cancellation

The cancelation of the dominant *IMF* is illustrated in Figure 2.15.

The algorithm for the cancellation of the dominant *IMF* can be sketched as consisting of three steps [65]:

- Step 1: The analyzed signal is decomposed into a set of *IMFs* through EEMD,
- Step 2: Pearson's correlation coefficient is calculated using (2.58) as many times as there are *IMFs* then $r_d \approx 1$ indexes the $IMF_d$,
- Step 3: Then, the indexed $IMF_d$ is removed from the analyzed signal $x(n)$ and the result denoted by $x(n)_{dEEMD}$ can therefore be used to detect bearing failure.

To measure the strength of the association between two variables using the Pearson's correlation, let us consider $X(n)$ as line current and $Y(n)$ as the *IMF*, so

$$X(n) = x(n) \tag{2.59}$$

and

$$Y_i(n) = IMF_i(n) \tag{2.60}$$

or

$$Y_i(n) = Mode_i(n) \tag{2.61}$$

where $(i = 1, \ldots, j)$ corresponds to the *IMF* rank and $j$ is the total number of *IMFs*. Then, the Pearson's correlation coefficient $r_i$ is computed for each pair $(X(n), Y_i(n))$; as result, the score $r_d \approx 1$ indexes the dominant *IMF* denoted by $IMF_d$. After determining

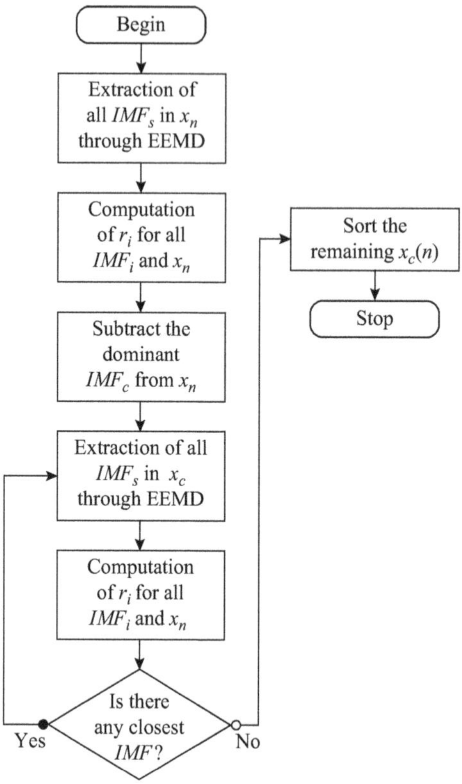


*Figure 2.15 Closest IMF subtraction principle [54]*

the $IMF_d$, it is canceled from the original signal $x(n)$, and the remaining signal $x_c(n)$ expressed by (2.62) can be investigated for bearing failure detection.

$$x_{c_{EEMD}}(n) = x(n) - IMF_d(n), \tag{2.62}$$

The cancellation process is repeated until there is no correlation between the main signal $x(n)$ and the *IMFs* contained in $x_c(n)$.

## 2.10.3 Fault detector based on EEMD demodulation

As mentioned previously for CT, HT and TKEO, the variance of $x_c(n)$ is investigated as a fault detector. This statistics criterion, denoted by $\sigma^2$, measures the deviation of the amplitude around its mean $\mu$. It is given by

$$\sigma^2 = \frac{1}{N}\sum_{n=0}^{N-1}(x_c(n) - \mu)^2 \tag{2.63}$$

where $\mu$ is the average of $x_c(n)$. To avoid the EEMD edge-effect problem, $x_c(n)$ is then truncated by removing $\alpha$ samples at the beginning and at the end of $x_c(n)$. Hence, the proposed criterion $\sigma^2$ is expressed by the following [68]:

$$\sigma^2 = \frac{1}{(N-2\alpha)} \left( \sum_{n=\alpha}^{N-\alpha-1} (x_c(n) - \mu)^2 \right) \tag{2.64}$$

and

$$\mu = \frac{1}{(N-2\alpha)} \sum_{n=\alpha}^{N-\alpha-1} x_c(n) \tag{2.65}$$

The hypothesis test to detect a fault can therefore be formulated as follows: If $\sigma^2 > \gamma$, the machine is faulty, where $\gamma$ is a threshold. For ideal acquisition conditions $\gamma = 0$, but in real-world applications there is always an add noise to measurements, then $\gamma$ can be set subjectively.

## *2.10.4 Synthetic signals*

In this validation step, simulations have been performed with AM synthetic signals. According to (2.49), and since additional components could be considered as noise in the context of bearing faults detection [69], the AM synthetic signal corrupted by an additive noise $\delta$ is defined as

$$x(n) = \underbrace{(1 + \beta \sin(\omega_2 n + \psi))}_{a(n)} \cdot \cos(\omega n + \phi) + \delta(n). \tag{2.66}$$

where $n = 0, \ldots, N-1$ is the sample index, $N$ being the number of samples, and $\phi$ is the phase parameter. In (2.66), frequency $\omega$ is equal to $2\pi f/F_e$ and $\omega_2$ is equal to $2\pi f_2/F_e$ (where $f$, $f_2$ and $F_e$ are the supply, fault and sampling frequency, respectively) and amplitude $a(n)$ is related to the fault. It should be noted that the additive noise $\delta(n)$ is supposed to be a zero mean and Gaussian noise process. This assumption is an approximation of the electrical noise picked up in the wiring and signal conditioning circuits [70], and it is widely considered in the measurement and electrical engineering communities [71]. The modulation index $\beta$ is the fault index, then $\beta = 0$ is for the healthy case and $\beta > 0$ is for the faulty one. Simulations have been carried out with a sampling frequency $F_e = 10$ kHz, a supply frequency $f = 50$ Hz and $f_2 = 100$ Hz. In order to simulate healthy and faulty cases, the modulation index has been set, respectively, to $\beta = 0.0$ for healthy case, and $\beta = 0.1$, $0.15$ and $0.2$ for different severity of the fault (Figure 2.16).

Figures 2.17 and 2.18 show the EEMD result of the synthetic signal $x(n)$ for both healthy and faulty cases, respectively.

It clearly shows that at least one *IMF* is close to the original signal. In order to quantify the strength of the association between $x(n)$ and each *IMF*, the Pearson's correlation coefficient $r_i$ is computed and results are depicted in Table 2.2. In this case, $IMF_5$ is the closest to the main signal. It is then subtracted, and the variance

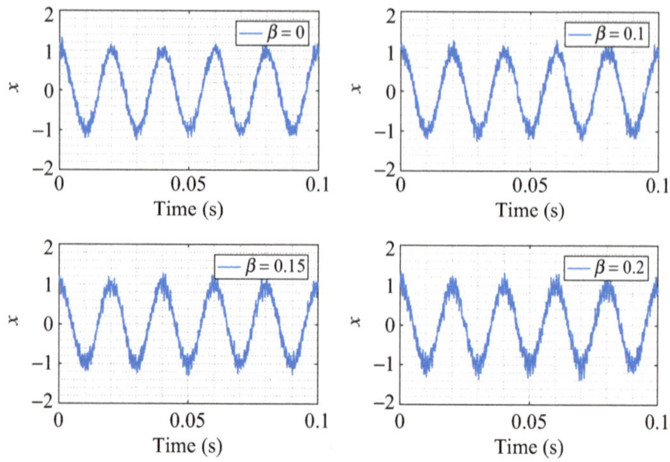


*Figure 2.16 Time representation of the synthetic signal for different modulation index*

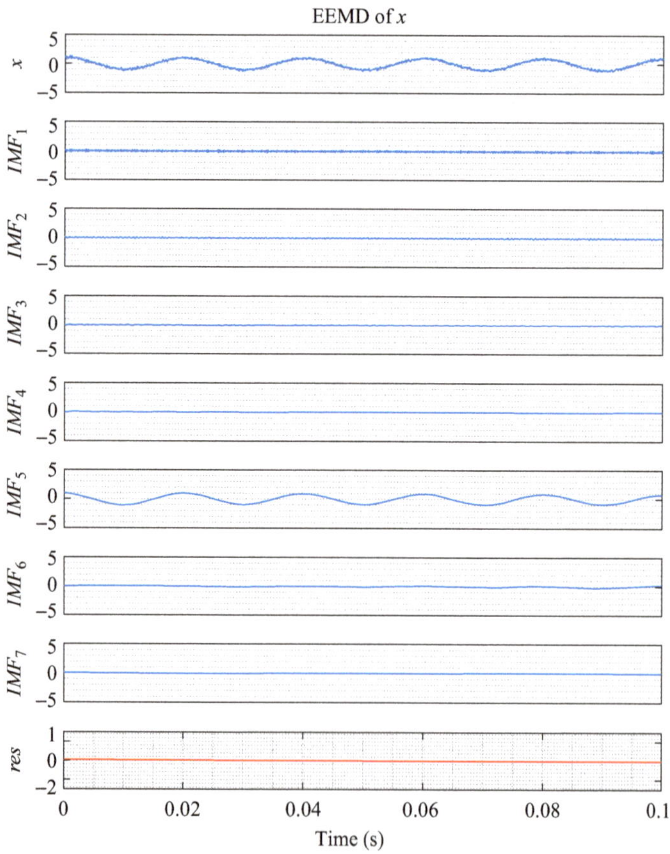


*Figure 2.17 EEMD for modulated synthetic signal: $\beta = 0$*

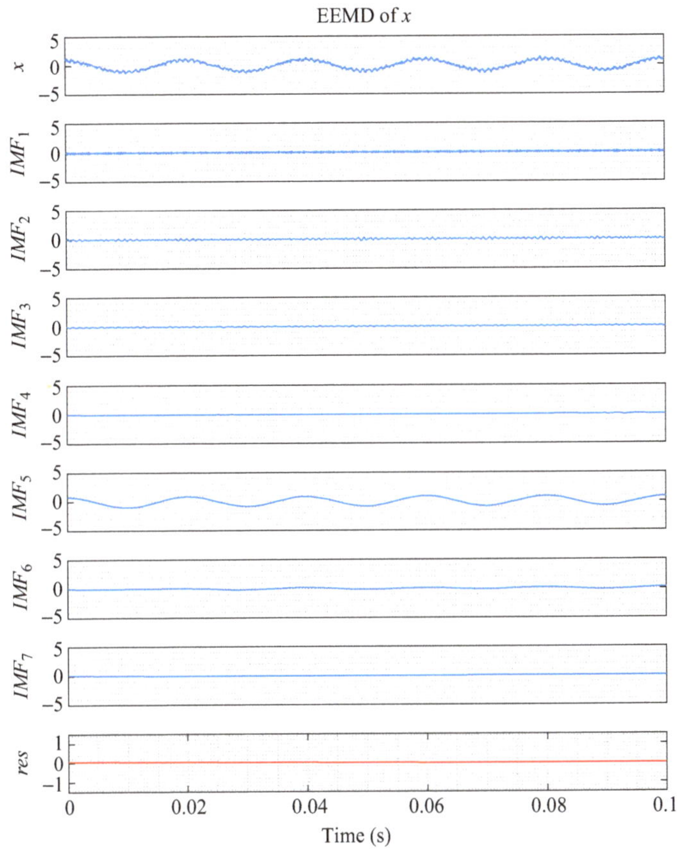


*Figure 2.18 EEMD for modulated synthetic signal: $\beta = 0.2$*

*Table 2.2 Coefficients of Pearson's correlation of synthetic signal for EEMD*

| IMF rank | $\beta = 0.0$ | $\beta = 0.1$ | $\beta = 0.15$ | $\beta = 0.2$ |
|---|---|---|---|---|
| $IMF_1$ | 0.1129 | 0.1182 | 0.1189 | 0.1193 |
| $IMF_2$ | 0.0873 | 0.1143 | 0.1324 | 0.1543 |
| $IMF_3$ | 0.0688 | 0.0811 | 0.1141 | 0.1387 |
| $IMF_4$ | 0.0598 | 0.0583 | 0.0284 | 0.0493 |
| $IMF_5$ | 0.9883 | 0.9820 | 0.9825 | 0.9771 |

of the remaining signal is computed and results for both algorithms are presented in Table 2.2. It is clearly shown that the fault criterion $\sigma^2$ rises with the modulation index $\beta$, as presented in Table 2.3. For healthy case ($\beta = 0$) and due to the added noise $\delta(n)$, $\sigma^2$ is not equal to 0.

Table 2.3 The variance ($\sigma^2$) of $x_c$ for EEMD

| $\beta = 0.0$ | $\beta = 0.1$ | $\beta = 0.15$ | $\beta = 0.2$ |
|---|---|---|---|
| $\sigma^2 = 0.0044$ | $\sigma^2 = 0.0057$ | $\sigma^2 = 0.0067$ | $\sigma^2 = 0.0076$ |

## 2.11 Summary and conclusion

In this chapter, we have proposed a review on fault detection based on demodulation techniques. First, the motor currents are demodulated using CT, HT and TKEO. Then, a hypothesis test based on the statistical variance of the demodulated envelope is performed to discriminate between healthy and faulty machines. The results of several simulations have shown that the mentioned methods perform well in stationary and nonstationary scenarios. Furthermore, results have shown that, even if CT is computationally attractive compared to HT and TKEO, this low-complexity demodulation technique can be inappropriate for the diagnosis of unbalanced system, and CT, HT and TKEO are inappropriate for multicomponent signals. Second, for multicomponent signals, the EMD-based notch filter is described; the core of this notch filter is a data-driven strategy combined to a statistical tool. The filtering operation was carried out following three steps: the first step concerns the decomposition of the phase machine current into *IMFs* using EEMD, then at the second step the dominant mode is subtracted from the original signal, and finally in the last step relays on the use of a statistical feature as a fault detector. The results of several simulations have shown that the proposed method performs well for amplitude-modulated signal regardless of the mode rank.

## References

[1] Cardoso AJM. Diagnosis and Fault Tolerance of Electrical Machines, Power Electronics and Drives. The Institution of Engineering and Technology, United Kingdom. IET Energy Engineering; 2018.

[2] Thorsen OV and Dalva M. Failure identification and analysis for high-voltage induction motors in the petrochemical industry. IEEE Transactions on Industry Applications. 1999;35(4):810–818.

[3] Thomson WT and Stewart ID. Online current monitoring for fault diagnosis in inverter-fed induction motors. In: Third International Conference on Power Electronics and Variable-Speed Drives; 1988. pp. 432–435.

[4] Kliman GB, Koegl RA, Stein J, Endicott RD, and Madden MW. Noninvasive detection of broken rotor bars in operating induction motors. IEEE Transactions on Energy Conversion. 1988;3(4):873–879.

[5] Antonino-Daviu J, Corral-Hernandez J, Climente-Alarcò V, *et al.* Case stories of advanced rotor assessment in field motors operated with soft-starters and frequency converters. In: IECON 2015—41st Annual Conference of the IEEE Industrial Electronics Society; 2015. pp. 001139–001144.

[6] Frosini L, Harlişca C and Szabó L. Induction machine bearing fault detection by means of statistical processing of the stray flux measurement. IEEE Transactions on Industrial Electronics. 2015;62(3):1846–1854.
[7] Seera M and Lim CP. Online motor fault detection and diagnosis using a hybrid FMM-CART model. IEEE Transactions on Neural Networks and Learning Systems. 2014;25(4):806–812.
[8] Leite VCMN, da Silva JGB, Veloso GFC, *et al.* Detection of localized bearing faults in induction machines by spectral kurtosis and envelope analysis of stator current. IEEE Transactions on Industrial Electronics. 2015;62(3):1855–1865.
[9] Benbouzid MEH. A review of induction motors signature analysis as a medium for faults detection. IEEE Transactions on Industrial Electronics. 2000;47(5):984–993.
[10] Zhang P, Du Y, Habetler TG, *et al.* A survey of condition monitoring and protection methods for medium-voltage induction motors. IEEE Transactions on Industry Applications. 2011;47(1):34–46.
[11] Wang W and Jianu OA. A smart sensing unit for vibration measurement and monitoring. IEEE/ASME Transactions on Mechatronics. 2010;15(1):70–78.
[12] Tavner P, Ran L, Penman J, and Sedding H. Condition Monitoring of Rotating Electrical Machines. IET Power and Energy series. IET, London; 2008.
[13] Schoen RR, Habetler TG, Kamran F, *et al.* Motor bearing damage detection using stator current monitoring. IEEE Transactions on Industry Applications. 1995;31(6):1274–1279.
[14] Frosini L, Harlişca C and Szabó L. Stator current and motor efficiency as indicators for different types of bearing fault in induction motor. IEEE Transactions on Industrial Electronics. 2010;57(1):244–251.
[15] Ibrahim A, Badaoui ME, Guillet F, *et al.* A new bearing fault detection method in induction machines based on instantaneous power factor. IEEE Transactions on Industrial Electronics. 2008;55(12):4252–4259.
[16] Elbouchikhi E, Choqueuse V and Benbouzid MEH. Induction machine faults detection using stator current parametric spectral estimation. Mechanical Systems and Signal Processing. 2015;52–53:447–464.
[17] Kia SH, Henao H and Capolino GA. A high-resolution frequency estimation method for three-phase induction machine fault detection. IEEE Transactions on Industrial Electronics. 2007;54(4):2305–2314.
[18] Stoica P and Moses R. Introduction to Spectral Analysis. Prentice Hall, Upper Saddle River, New Jersey; 1997.
[19] Elbouchikhi E, Choqueuse V and Benbouzid M. Induction machine bearing faults detection based on a multi-dimensional MUSIC algorithm and maximum likelihood estimation. ISA Transactions. 2016;63:413–424.
[20] Blodt M, Bonacci D, Regnier J, *et al.* On-line monitoring of mechanical faults in variable-speed induction motor drives using the Wigner distribution. IEEE Transactions on Industry Applications. 2008;55(2):522–533.
[21] Blodt M, Regnier J and Faucher J. Distinguishing load torque oscillations and eccentricity faults in induction motors using stator current Wigner distributions. IEEE Transactions on Industry Applications. 2009;45(6):1991–2000.

[22] Antonino-Daviu JA, Riera-Guasp M, Pineda-Sanchez M, *et al.* A critical comparison between DWT and Hilbert-Huang-based methods for the diagnosis of rotor bar failures in induction machines. IEEE Transactions on Industry Applications. 2009;45(5):1794–1803.

[23] Cusido JC, Romeral L, Ortega JA, *et al.* Fault detection in induction machines using power spectral density in wavelet decomposition. IEEE Transactions on Industrial Electronics. 2008;55(2):633–643.

[24] Riera-Guasp M, Antonio-Daviu JA, Roger-Folch J, and Molina Palomares MP. The use of the wavelet approximation signal as a tool for the diagnosis of rotor bar failure. IEEE Transactions on Industry Applications. 2008;44(3):716–726.

[25] Kia SH, Henao H and Capolino GA. Diagnosis of broken-bar fault in induction machines using discrete wavelet transform without slip estimation. IEEE Transactions on Industry Applications. 2009;45(4):1395–1404.

[26] Mandic DP, ur Rehman N, Wu Z, *et al.* Empirical mode decomposition-based time-frequency analysis of multivariate signals: the power of adaptive data analysis. IEEE Signal Processing Magazine. 2013;30(6):74–86.

[27] Tavner PJ. Review of condition monitoring of rotating electrical machines. IET Electric Power Applications. 2008;2(4):215–247.

[28] Stack JR, Harley RG and Habetler TG. An amplitude modulation detector for fault diagnosis in rolling element bearings. IEEE Transactions on Industrial Electronics. 2004;51(5):1097–1102.

[29] Riley CM, Lin BK, Habetler TG, *et al.* A method for sensorless on-line vibration monitoring of induction machines. IEEE Transactions on Industry Applications. 1998;34(6):1240–1245.

[30] Elbouchikhi E, Choqueuse V and Benbouzid MEH. Condition monitoring of induction motors based on stator currents demodulation. International Review on Electrical Engineering. 2015;10(6):1–6.

[31] Boualem B. Chapter 1. Time-frequency and instantaneous frequency concepts. In: Time-frequency Signal Analysis and Processing. 2nd ed. Academic Press, Oxford; 2016. pp. 31–63.

[32] Boashash B. Estimating and interpreting the instantaneous frequency of a signal. II. Algorithms and applications. Proceedings of the IEEE. 1992;80(4): 540–568.

[33] Delprat N, Escudie B, Guillemain P, *et al.* Asymptotic wavelet and Gabor analysis: extraction of instantaneous frequencies. IEEE Transactions on Information Theory. 1992;38(2):644–664.

[34] IEEE. IEEE Standard Definitions for the Measurement of Electric Power Quantities under Sinusoidal, Nonsinusoidal, Balanced, or Unbalanced Conditions. IEEE Press; 2010.

[35] Trajin B, Chabert M, Regnier J, *et al.* Hilbert versus Concordia transform for three-phase machine stator current time-frequency monitoring. Mechanical Systems & Signal Processing. 2009;23(8):2648–2657.

[36] Onel IY and Benbouzid MEH. Induction motor bearing failure detection and diagnosis: Park and Concordia transform approaches comparative study. IEEE/ASME Transactions on Mechatronics. 2008;13(2):257–262.

[37] Cruz SMA and Cardoso AJM. Stator winding fault diagnosis in three-phase synchronous and asynchronous motors, by the extended Park's vector approach. IEEE Transactions on Industry Applications. 2001;37(5): 1227–1233.
[38] Cardoso AJM, Cruz SMA and Fonseca DSB. Inter-turn stator winding fault diagnosis in three-phase induction motors, by Park's vector approach. IEEE Transactions on Energy Conversion. 1999;14(3):595–598.
[39] Cardoso AJM and Saraiva S. Computer-aided detection of airgap eccentricity in operating three-phase induction motors by Park's Vector Approach. IEEE Transactions on Industry Applications. 1993;28(5):897–901.
[40] Cizek V. Discrete Hilbert transform. IEEE Transactions on Audio and Electroacoustics. 1970;18(4):340–343.
[41] Gabor D. Theory of communication. Journal Institution of Electrical Engineers London. 1996;93(3):429–457.
[42] Oppenheim AV, Schafer RW and Padgett WT. Discrete-Time Signal Processing. 3rd ed. Prentice Hall, Upper Saddle River, New Jersey; 2009.
[43] Maragos P, Kaiser J and Quartieri T. On amplitude and frequency demodulation using energy operators. IEEE Transactions on Signal Processing. 1993;41(4):1532–1550.
[44] Maragos P, Kaiser J and Quatieri T. Energy separation in signal modulations with application to speech analysis. IEEE Transactions on Signal Processing. 1993;10(41):3024–3051.
[45] Nejjari H and Benbouzid MEH. Monitoring and diagnosis of induction motors electrical faults using a current Park's vector pattern learning approach. IEEE Transactions on Industry Applications. 2000;36(3):730–735.
[46] Jaksch I. Fault diagnosis of three-phase induction motors using envelope analysis. In: Proceedings of SDEMPED. Atlanta, USA; 2003. pp. 289–293.
[47] Diallo D, M E H Benbouzid DH and Pierre X. Fault detection and diagnosis in an induction machine drive: a pattern recognition approach based on Concordia stator mean current vector. IEEE Transactions on Energy Conversion. 2005;20(3):512–519.
[48] Ocak H and Loparo KA. A new bearing fault detection and diagnosis schema based on hidden Markov modeling of vibration signals. In: Proceedings of IEEE International Conference on Acoustics, Speech, and Signal Processing (ICASSP). Salt Lake City, USA; 2001. pp. 3141–3144.
[49] Miao Q and Makis V. Condition monitoring and classification of rotating machinery using wavelets and hidden Markov models. Mechanical Systems and Signal Processing. 2007;21(2):840–855.
[50] Guo L, Chen J and Li X. Rolling bearing fault classification based on envelope spectrum and support vector machine. Journal of Vibration and Control. 2009;15(9):1349–1363.
[51] Saidi L, Ali JB, Bechhoefer E, *et al.* Wind turbine high-speed shaft bearings health prognosis through a spectral Kurtosis-derived indices and SVR. Applied Acoustics. 2017;120:1–8.

[52] Yang W, Tavner PJ, Crabtree CJ, *et al.* Cost-effective condition monitoring for wind turbines. IEEE Transactions on Industrial Electronics. 2010;57(1): 263–271.

[53] Kusiak A. Renewables: share data on wind energy. Nature. 2016;522(1): 19–21.

[54] Amirat Y, Benbouzid MEH, Wang T, *et al.* EEMD-based notch filter for induction machine bearing faults detection. Applied Acoustics. 2018;133: 202–209. Available from: http://www.sciencedirect.com/science/article/pii/S0003682X17308125.

[55] Huang NE, Shen Z, Long SR, *et al.* The empirical mode decomposition and Hilbert spectrum for nonlinear and nonstationary time series analysis. Proceedings of Royal Society, London. 1998;454:903–995.

[56] Tanaka T and Mandic DP. Complex empirical mode decomposition. IEEE Letters on Signal Processing. 2007;14(2):101–104.

[57] Yu D, Cheng J and Yang Y. Application of EMD method and Hilbert spectrum to the fault diagnosis of roller bearings. Mechanical Systems and Signal Processing. 2005;19(2):259–270.

[58] Amirat Y, Choqueuse V and Benbouzid MEH. EEMD-based wind turbine bearing failure detection using the generator stator current homopolar component. Mechanical Systems and Signal Processing. 2013;41(1):667–678.

[59] Gilles J. Empirical wavelet transform. IEEE Transactions on Signal Processing. 2013;61(16):3999–4010.

[60] Wu ZH and Huang NE. Ensemble empirical mode decomposition: a noise-assisted data analysis method. Advances in Adaptive Data Analysis. 2009;1: 1–41.

[61] Torres ME, Colominas MA, Schlotthauer G, *et al.* A complete ensemble empirical mode decomposition with adaptive noise. In: Proceedings of 2011 IEEE International Conference on Acoustics, Speech and Signal Processing (ICASSP); 2011. pp. 4144–4147.

[62] Huang NE and Shen SSP. Hilbert-Huang Transform and Its Applications. 2nd ed. World Scientific, Interdisciplinary Mathematical Sciences, Singapore; 2014.

[63] Amirat Y, Benbouzid MEH, Wang T, *et al.* Bearing fault detection in wind turbines using dominant intrinsic mode function subtraction. In: Proceedings of the 2016 IEEE IECON. (Florence) Italy; 2016. pp. 6961–6965.

[64] Amirat Y, Elbouchikhi E, Zhou Z, *et al.* Variational mode decomposition-based notch filter for bearing faults detection. In: Proceedings of the 2019 IEEE IECON. (Lisbon) Portugal; 2019. pp. 1–6.

[65] Amirat Y, Elbouchikhi E, Delpha C, *et al.* Chapter 4. Modal decomposition for bearing fault detection. In: Electrical Systems 1: From Diagnosis to Prognosis. John Wiley & Sons, Ltd, London; 2020. pp. 121–168. Available from: https://onlinelibrary.wiley.com/doi/abs/10.1002/9781119720317.ch4.

[66] Harmouche J, Delpha C and Diallo D. Improved fault diagnosis of ball bearings based on the global spectrum of vibration signals. IEEE Transactions on Energy Conversion. 2015;30(1):376–383.

[67] Proakis JG and Manolakis DG. Digital Signal Processing. 3rd ed. Prentice Hall, Upper Saddle River, New Jersey; 1996.

[68] Amirat Y, Choqueuse V and Benbouzid M. Condition monitoring of wind turbines based on amplitude demodulation. In: 2010 IEEE Energy Conversion Congress and Exposition; 2010. pp. 2417–2421.

[69] Zhou W, Habetler TG and Harley RG. Bearing fault detection via stator current noise cancellation and statistical control. IEEE Transactions on Industrial Electronics. 2008;55(12):4260–4269.

[70] Phadke AG. Synchronized phasor measurements—a historical overview. In: IEEE/PES Transmission and Distribution Conference and Exhibition. vol. 1; 2002. pp. 476–479.

[71] Komaty A, Boudraa AO, Augier B, *et al.* EMD-based filtering using similarity measure between probability density functions of IMFs. IEEE Transactions on Instrumentation and Measurement. 2014;63(1):27–34.

# Chapter 3
# Kullback–Leibler divergence for incipient fault diagnosis

*Claude Delpha*[1] *and Demba Diallo*[2]

This chapter discusses the issue of incipient fault detection and diagnosis (FDD). After a general introduction, the requirements for FDD methods are defined under the three criteria of robustness, sensitivity, and simplicity. A methodology of FDD is also introduced in four main steps: modelling, preprocessing, features extraction, and features analysis. After the definition of incipient fault based on the levels of fault, signal, and environmental nuisances, a paradigm is drawn between information-hiding domain and FDD. We will show that dissimilarity measure of probability density function (PDF) used for data hiding is efficient for incipient fault detection. The methodology is illustrated through incipient crack detection in a conductive material using eddy currents and short intermittent open-circuit duration in three-level neutral-point-clamped inverter. The chapter also discusses fault detection threshold optimal setting and fault severity estimation.

## 3.1 Introduction

Sustainability of industrial activities requires preservation of both human and physical capital through the assessment of Reliability, Availability, Maintainability, and Safety of all equipment involved in each process. Worldwide, requirements are becoming increasingly stringent in terms of safety, and competition also requires continuous improvement in process efficiency to cut costs and increase efficiency and life cycle. Reliability, availability, and maintainability are usually treated first at the design stage with, for example, hardware redundancy, easy access to facilities, or/and appropriate material selection. Maintainability is addressed by the availability of replacement parts and maintenance policy. For now more than a decade, maintenance policies have evolved from event-based maintenance, scheduled maintenance, and on-demand maintenance to condition-based maintenance (known as CBM). This evolution is illustrated in Figure 3.1.

[1]Laboratoire des Signaux et Systèmes, Université Paris Saclay, CNRS, CentraleSupelec, Gif/Yvette, France
[2]Group of Electrical Engineering Paris, Université Paris-Saclay, CentraleSupelec, CNRS, Gif/Yvette, France

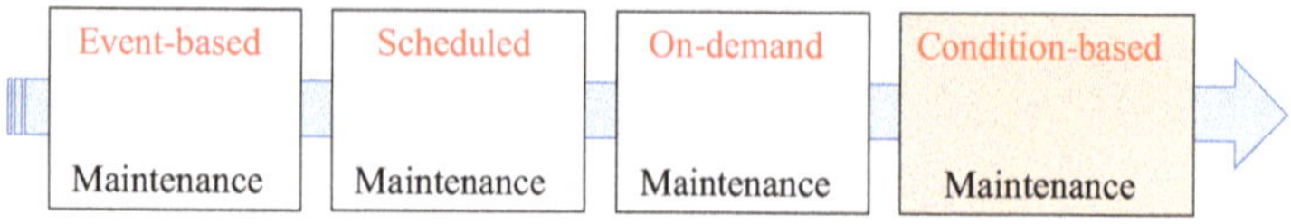


*Figure 3.1 Evolution of maintenance policies*

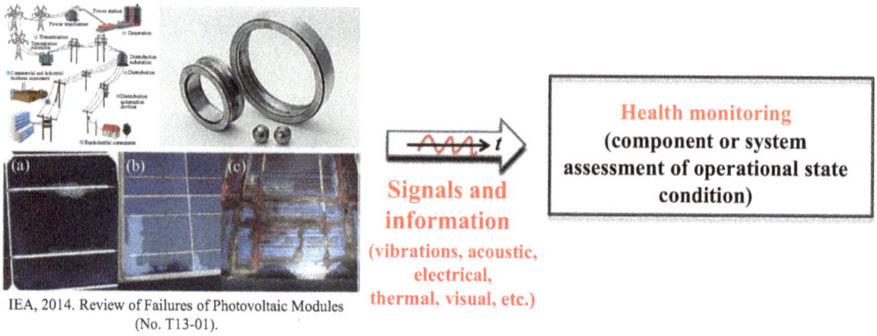


*Figure 3.2 Health monitoring*

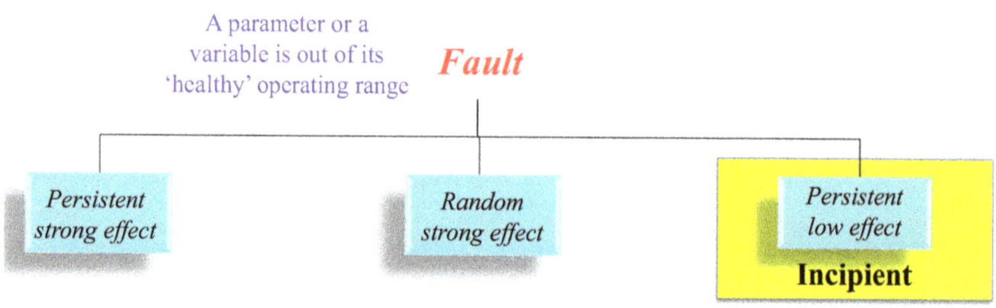


*Figure 3.3 Classification of fault types*

Condition-based maintenance requires knowledge of the current health status of equipment with the objectives to

- guarantee safety, security, and uninterrupted service whatever the environmental conditions;
- make the right decision in any situation (even for non-expert technician on site).

Through the continuous processing and analysis of measures or/and estimates (signals and information), the decision should be made whether a fault has occurred or not (see Figure 3.2).

Fault types can be classified in three groups as displayed in Figure 3.3 depending on their effect.

If a fault has occurred, further actions are required to identify the fault type, estimate its severity, and engage safe degraded operation prior to maintenance action. The FDD methodology is shown in Figure 3.4 with the respective challenges for each step [1].

Therefore, an efficient FDD method is a compromise between sensitivity, robustness, and simplicity as defined in Figure 3.5. Based on these three criteria displayed as a triptych, three compromises have to be considered:

- Robustness and sensitivity must allow to evaluate the accuracy of the method; it means evaluate the minimum of diagnosis confusion.
- Robustness and simplicity will allow to evaluate the efficiency of the method, i.e. the ability of the method to detect easily the fault.
- Simplicity and sensitivity will correspond to the reliability evaluation.

For each application, the compromise and consequently the method chosen will depend on the specifications.

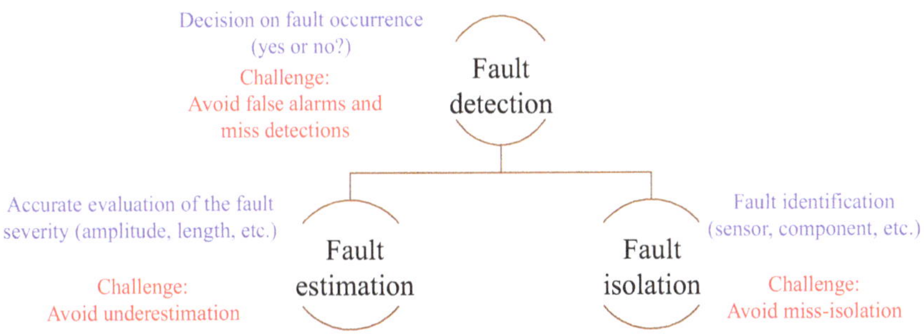


*Figure 3.4 The items of FDD*

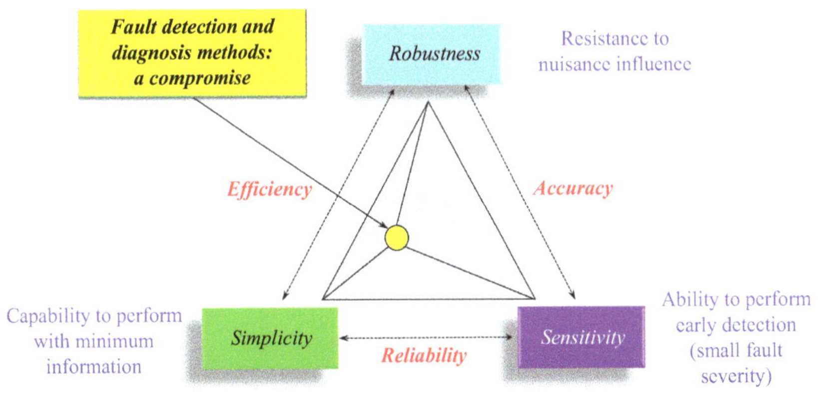


*Figure 3.5 Design requirements for FDD methods*

## 3.2 Fault detection and diagnosis

### 3.2.1 Methodology

FDD is a topic that has been studied for a while [2–7]. The different FDD's methodologies found in the literature can be decomposed into four steps: modelling, preprocessing, features extraction, and features analysis as displayed in Figure 3.6.

1. The first step corresponds to the knowledge building or modelling. The models can be built from laws of physics [8,9], natural language processing, or data history. Physics-based or analytical models are very convenient and powerful when they are accurate enough to represent all the interactions between inputs, outputs, internal states, and parameters. However usually they require making assumptions and are sensitive to uncertainties. Besides, the parameters may be dependent on operating conditions through non-linear relations and phenomenon like ageing may not even be taken into account. Therefore, the residuals computed by observers or analytical redundancy relations, for example, are sensible to all those discrepancies between the model and the real physical system [10]. The decision on fault occurrence may be flawed. Models derived from natural language processing are strongly dependant on the information collected from experts, technicians, or technical documents. Also the labelling and structuring of data is a complex operation. As a consequence, this approach may be suitable if the input data from human experts and technicians is 100% reliable. The third way to obtain a model is to take benefit of the increasing amount of data now available in most processes [11,12]. This rich information is a valuable input for continuous monitoring.
2. The second step is of particular importance. Preprocessing consists of transforming the input data to eliminate or reduce the environmental nuisances and the outliers, and project the data into the most suitable information domain where the fault signatures are the strongest. Several tools are available like denoising, normalization, Fourier transform, principal component analysis (PCA), wavelets, Concordia, Fourier, Wavelet, and Hilbert [13–18]. The chosen tools depend on each application: diversity and quantity of data, stationary or non-stationary signals, dimensionality, required fault detection performances, etc.
3. The third step corresponds to the features extraction. After preprocessing the raw input data, the transformed information is used to extract the fault signatures. As displayed in Figure 3.6, several tools are available depending on the domain in which the information lays, the user's expertise, the computational cost, and the desired performances (sensitivity, robustness, or simplicity).
4. In the last step, the features are analysed to make the decision whether a fault has occurred or not. Here also, there are several tools available to the users. The selection of the most relevant tool is done with the objective of amplifying the differences between healthy and faulty data for efficient separation and diagnosis. The selection also depends on the user's expertise, the features dimensionality and representation, the computational cost and the desired performances. A straightforward threshold-logic-based approach can be used to

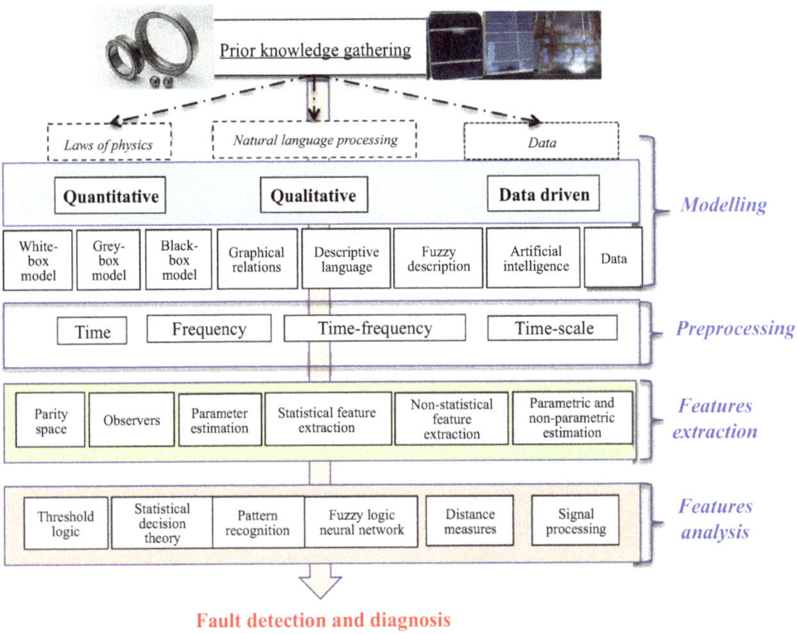


*Figure 3.6 Flow chart of FDD methodology*

distinguish healthy condition from a faulty one. If there are different faulty conditions, artificial neural network (ANN) [19], clustering or other classification techniques such as PCA, linear discriminant analysis (LDA) [20–26], support vector machines (SVMs) [27] can be used.

## 3.2.2 Application example of the methodology

Let us consider in the following the detection of cracks in a conductive material using eddy currents. The method consists of applying an AC voltage at the terminals of the primary coil. The measured induced voltage (or impedance) is the combination of the original magnetic field and the induced one that depends on the geometrical and magnetic properties of the material under inspection. The measured induced voltage or impedance is sensitive to the distortion of the magnetic field due to crack occurrence as displayed in Figure 3.7 [28].

Therefore the variation of the impedance $\Delta Z = Real + jImag$ can be used as fault signature. In the following, only the real part will be under consideration. The cracks are produced using an electric discharge machine. Figure 3.8 gives a description of the sample.

The test bed is displayed in Figure 3.9 with on the left side the impedance analyser and on the right side the three-axis robot used for scanning the material.

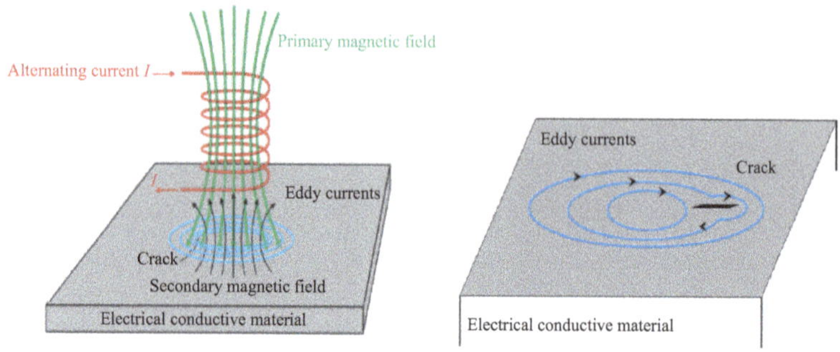


*Figure 3.7 Eddy currents*

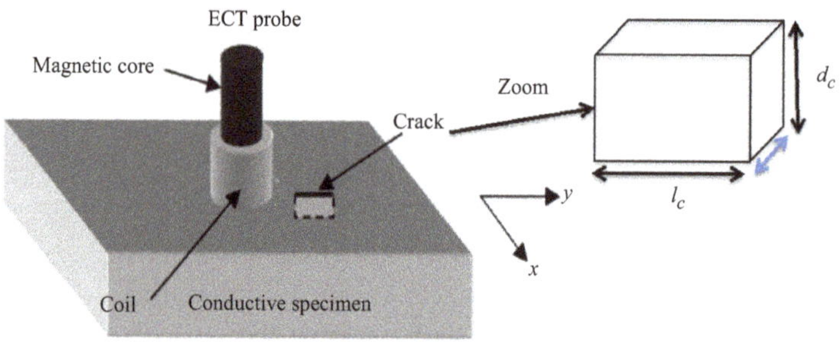


*Figure 3.8 Specimen of the material*

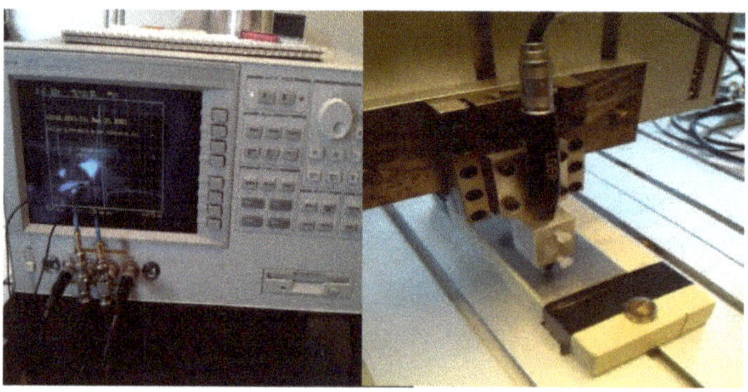

*Figure 3.9 Experimental test bed for non-destructive evaluation*

The impedance variation with a crack of dimensions ($l_c = 0.4$ mm and $d_c = 0.6$ mm) is displayed in Figure 3.10 where the fault effect is clearly visible despite the environmental nuisances.

However, when the crack has smaller dimensions, the fault effect is less visible as shown in Figure 3.11, and its detection becomes more difficult. The

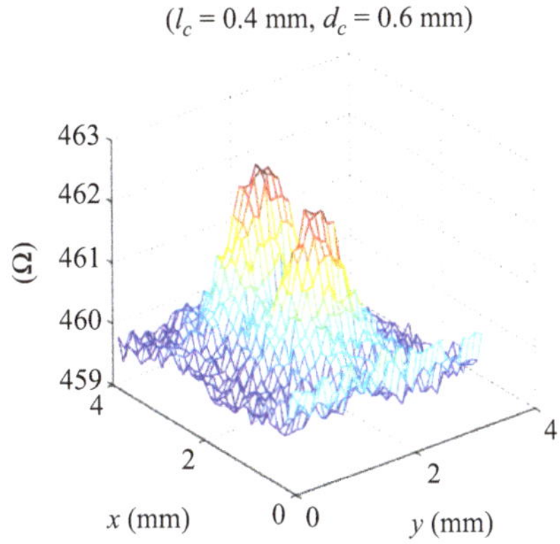


*Figure 3.10 Impedance variation due to crack*

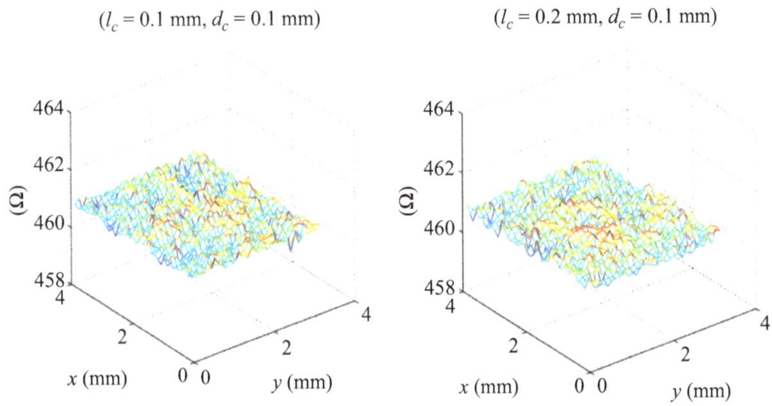


*Figure 3.11 Impedance variations in presence of small cracks*

impedance variation is clearly concealed in the environmental nuisances due to noise measurement, surface roughness, and variations of the lift-off.

The aforementioned FDD methodology in four steps is applied for the crack with the following dimensions ($l_c = 0.1$ mm and $d_c = 0.1$ mm) as described in the flow chart of Figure 3.12.

To introduce variability, Monte-Carlo simulations are done with 50 realizations of each condition (healthy and faulty). The three statistical moments (variance, skewness, and kurtosis) are used as fault signatures. The results are displayed in Figure 3.13.

With the variance and the skewness, the fault cannot be detected. With the kurtosis, the fault detection performances are very poor. We can conclude that with a 'small' fault, the detection capability with these features is very low. This is a real

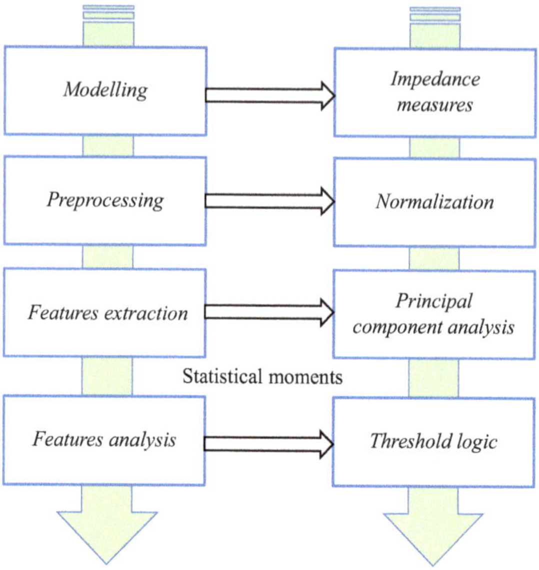


*Figure 3.12 Flow chart of the applied FDD methodology*

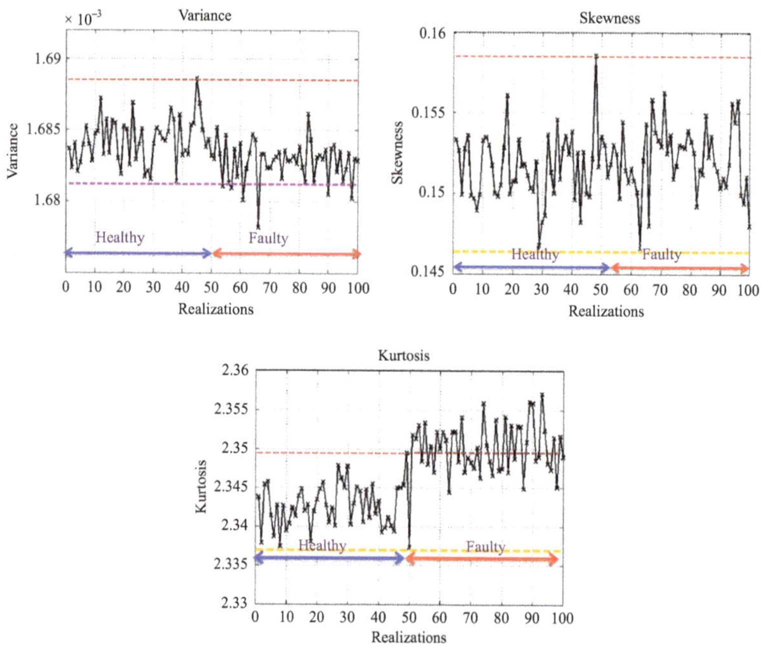


*Figure 3.13 FDD results with the three statistical moments*

challenge because if not detected, 'small' fault may keep on increasing gradually, and will finally lead to failure. So suitable incipient FDD methods should be designed while taking into account accurate setting of the threshold and coping with modelling errors and uncertainties.

## 3.3 Incipient fault

So incipient fault should be defined in relation with the levels of the useful signal and the nuisance level.

Let us define $\sigma_X^2$, $\sigma_V^2$, and $\sigma_F^2$ the powers of the signal, the noise, and the fault respectively.

Signal-to-noise ratio, $SNR = 10 \log \frac{\sigma_X^2}{\sigma_V^2}$,

Signal-to-fault ratio, $SFR = 10 \log \frac{\sigma_X^2}{\sigma_F^2}$, and

Fault-to-noise ratio, $FNR = 10 \log \frac{\sigma_F^2}{\sigma_V^2}$.

The incipient fault is defined as a fault which power level is in the same order of magnitude as the noise power level and at the same time much smaller than the signal power level. Figure 3.14 is the graphical representation of the linear relation between the three relative powers.

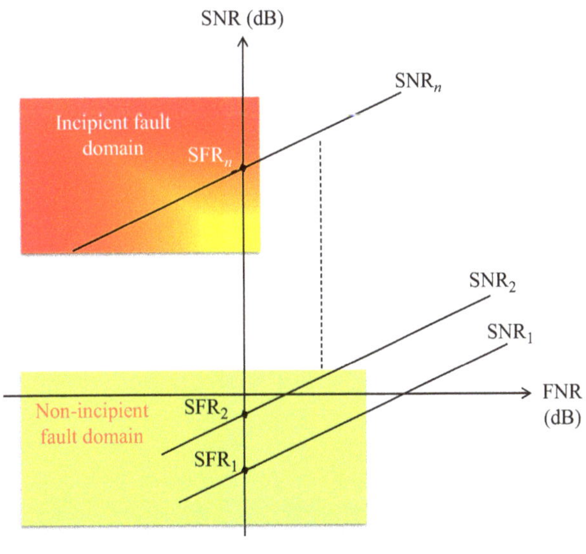


*Figure 3.14 Incipient fault domain definition*

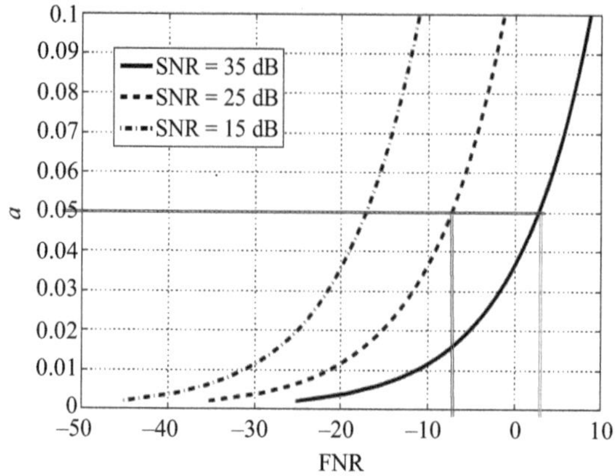


*Figure 3.15 Incipient fault amplitude*

Finally, a fault is incipient if the following conditions are fulfilled:

$$FNR \leq 0, \qquad SFR \gg 0, \qquad \text{and} \qquad SNR \gg 0$$

Figure 3.15 illustrates the incipient fault amplitude versus the FNR for several values of the SNR. As a conclusion, an incipient fault is strongly related to the environmental nuisances.

## 3.4 FDD as hidden information paradigm

### *3.4.1 Introduction*

The whole world has become more and more connected with a huge amount of digital information (music, photo, video, etc.) flowing from one side to another every second, but unfortunately along with digital piracy [29]. Despite the development of digital rights management, piracy has extended its network. To counteract digital piracy, several techniques have been proposed and are currently used and developed by important majors in the world. These techniques have been developed to preserve the integrity of the information, the intellectual property, and the privacy. They have mainly emerged from the signal processing, telecommunication, and sometimes computer science communities. Information hiding (namely known as data hiding) is a specific domain that is mainly related to multimedia security process. In this domain, the goal is to embed hidden information (namely a watermark $W$) in a host signal $X$ [30–32]. This embedded information can be a specific encoded message $m$ or identifier (a copyright information for example) to be inserted inside the host signal $X$ that can be an image, a video, or a song. The watermarked signal $S$ is transmitted into

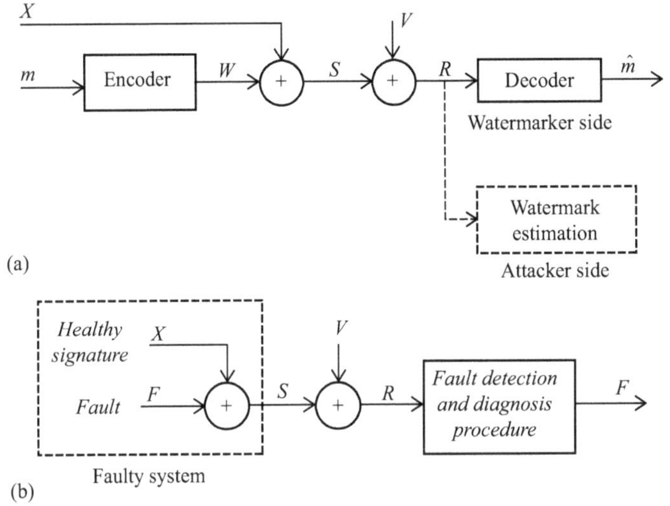


*Figure 3.16 Hidden information paradigm. (a) General data hiding scheme. (b) FDD scheme*

a communication channel and can be subjected to modifications often modelled by a noise $V$. From the receiver side, the signal $R$ corresponds to $S$ affected by $V$. With this signal $R$, the receiver proceeds to the extraction of the embedded information and then decodes the hidden message $\hat{m}$. This process is summarized in Figure 3.16(a) (note that the symbol '+' denotes mixing operations).

The main objective of the malicious user (attacker) is to estimate and steal the embedded information. By drawing a parallel between digital piracy and FDD, one can notice a reverse paradigm: detection of fault occurrence requires the extraction of fault information embedded (hidden) in the measured or estimated signals. The fault is considered as hidden information to be detected and characterized, whatever the distortions due to environmental nuisances. Figure 3.16(b) is a graphical representation of this paradigm.

In most of the cases, the channel is considered as an open network like the Internet. Therefore to prevent the illegal use of the signal by unauthorized malicious users (attackers), the embedded information must be designed and protected judiciously and efficiently from distortions, estimation, etc. For this purpose, a data-hiding scheme is characterized by three criteria: *robustness*, *capacity*, and *transparency* (Figure 3.17).

The *robustness* is the ability of the hidden information to withstand transformations and distortions in the channel. The *capacity* is the maximum of information that it is possible to embed and extract without errors for a given channel distortion level.

*Transparency* is the ability to perceptually and statistically detect the hidden information in the considered signal. These performances have to be tuned as the result of a trade-off depending on the target application [33] and also on the specifications. Thus

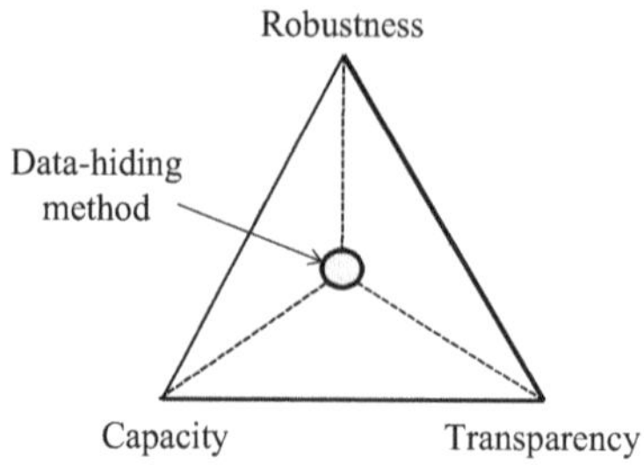


*Figure 3.17 Triptych for data hiding method*

in the point of view of the data hider (watermarker), transparency is crucial [34]: if the attacker is not able to differentiate the watermarked signal from the non-watermarked one, he will not be tempted to corrupt it. In this domain, perceptual aspects are treated with perceptual masking models. Considering the statistical transparency, it is mentioned that if the probability of false alarm (PFA) for the attacker is maximized, the statistical transparency will be minimized [35]. From the attacker's point of view, even perceptual masking is efficient enough to enable the perception of the hidden information; the statistical study of the watermarked signal can reveal significant details on the watermark information allowing extraction and characterization of the hidden information. For example, while using a basic quantization-based watermarking scheme, we can notice significant distortions on the watermarked signal PDF compared to the original one [34]. Two PDFs obtained from images are plotted in Figure 3.18(a). With such distortion, the attacker could be alerted to the presence of hidden information. He can be tempted to steal it. Thus, to avoid this situation, it is preferred to have the watermarked and the original PDF as close as possible (see Figure 3.18(b)). This statistical proximity is evaluated using distance measures.

By drawing the parallel with data hiding as described earlier, FDD can be considered as a hacking procedure. This methodology will be evaluated in case of incipient fault that produces slight modifications in the PDFs as shown in Figure 3.19. With no loss of generality, let us define for any process or component (electrical, mechanical, chemical, etc.):

- the healthy signal $X$ as the host signal in data hiding,
- and the fault $F$ as the embedded information.

The main difference is that this fault information is unknown: it is undesirable additional information. This fault is mixed with the host signal to produce the faulty signal $S$. It can be considered corrupted by additional noise $V$ so that $R = S + V$. The FDD methodology is designed to extract and characterize the modifications in the signal or its statistics. The decision can be made on fault occurrence and its severity estimated. Any FDD methodology must be evaluated against the following three performance criteria: *efficiency*, *accuracy*, and *reliability*.

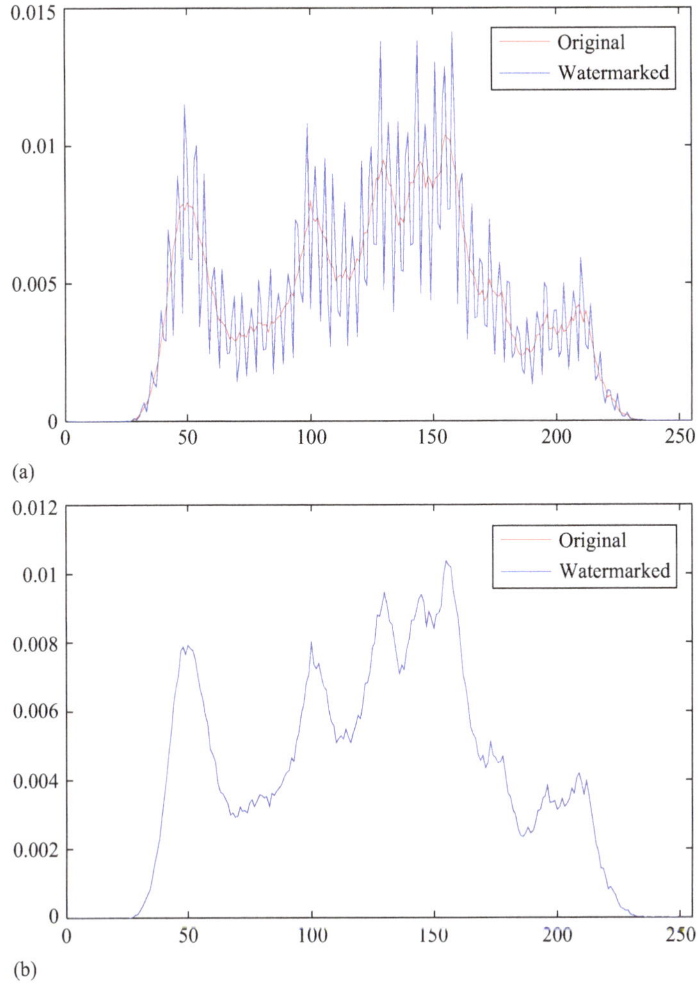


*Figure 3.18 PDFs: (a) quantization-based watermarking scheme; (b) improved watermarking*

Inspired from information hiding, the methodology is the trade-off between three properties: *robustness*, *sensitivity*, and *simplicity* as displayed in Figure 3.20.

1. Robustness: It corresponds to the ability of the method to properly detect and diagnose a fault with minimum detection misses and false alarms. It is evaluated through error probability for the detection of a fault taking into account the FNR and the SFR. In fine, lower will be the miss detection and false alarms, more robust will be the method.

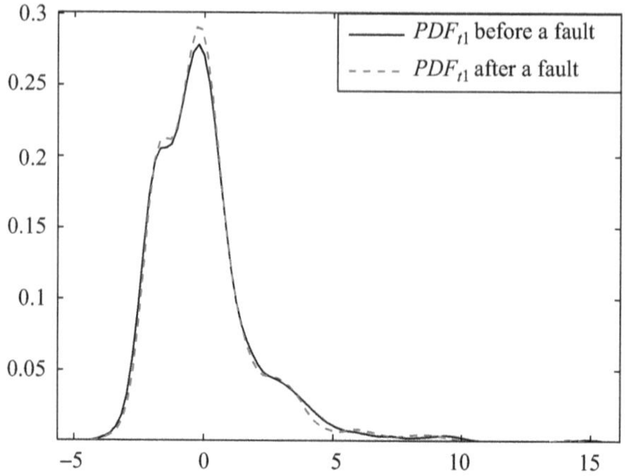


*Figure 3.19 PDFs: healthy and incipient fault*

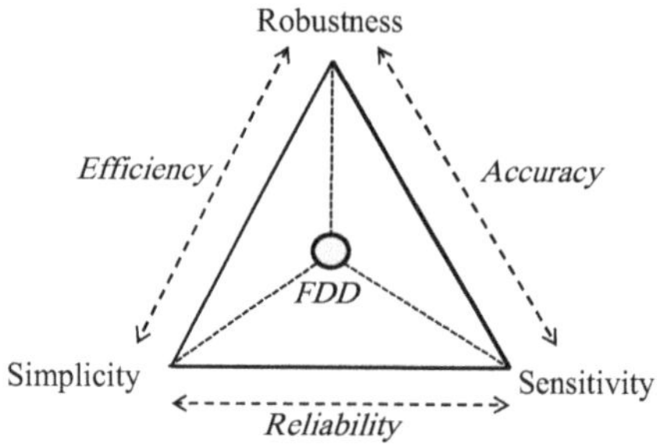


*Figure 3.20 Triptych for FDD methodology*

2. Sensitivity: It is the ability to detect faults at their earliest stage. It can be quantified as the incipiency level with the minimum detectable fault in the environment characterized by the signal, and the corruption noise levels: more incipient, i.e. smaller will be the fault, and more sensitive the method must be. In fact, this fault severity has to be evaluated by varying the FNR and SFR.

In case of a non-incipient fault, meaning that SFR is medium, the most interesting FNR conditions will be around 0 dB; that is, the noise and the fault levels are almost identical. In case of very incipient faults, more severe detection conditions, corresponding to FNR lower than 0 dB and high SFR values, will have to be considered. This means that the environmental noise level is higher than the fault's one, and the fault level is very small compared to the signal's one. A sensor gain drift of ~1–10% or a pitch of 180 m on a ball bearing with a diameter of 8 mm (corresponding to a 2% degradation) can be considered as incipient faults.

3. SIMPLICITY: It corresponds to the lowest amount of information needed for efficient FDD. It has a direct impact on the calculation cost for the implementation of the fault diagnosis procedure. This criterion will be directly linked to the number of descriptive variables necessary to create a pattern or signature describing the faulty signal $S$ or the noisy faulty one $R$ in non-parametric approaches. For this performance criterion, the trade-off between time-consuming, the number of samples, and the number of sensors will have to be found.

After this description, one can notice that the three criteria are somehow opposite. For example, maximizing simplicity will minimize robustness and sensitivity. That's why a trade-off is required. As for information hiding, this can lead to a non-cooperative optimization problem. In this case, to maximize the robustness, we need to minimize the false alarm probability ($P_{FA}$). Moreover to maximize the sensitivity, we need, for example, to minimize the error ($P_E$) and miss detection ($P_{MD}$) probabilities for the smaller fault size to be detected with the nuisance's parameters. Nevertheless, to maximize also the simplicity, the minimum number of used features has to be more relevant as possible. While the number of features decreases, the false alarm and miss detection probabilities increase. As an example, for incipient fault detection, the probability of missed detection ($P_{MD}$) is plotted against the PFA ($P_{FA}$) for different FNR values (see Figure 3.21). This highlights the difficulties to obtain minimum $P_{FA}$ with minimum $P_{MD}$ in severe FNR conditions.

Based on the aforementioned paradigm, the analysis of statistical proximity between PDFs is expected to be a powerful method to discriminate healthy from incipient fault conditions. In the particular case of incipient faults, small modifications are nested in the considered faulty signal $S$. Generally, these modifications

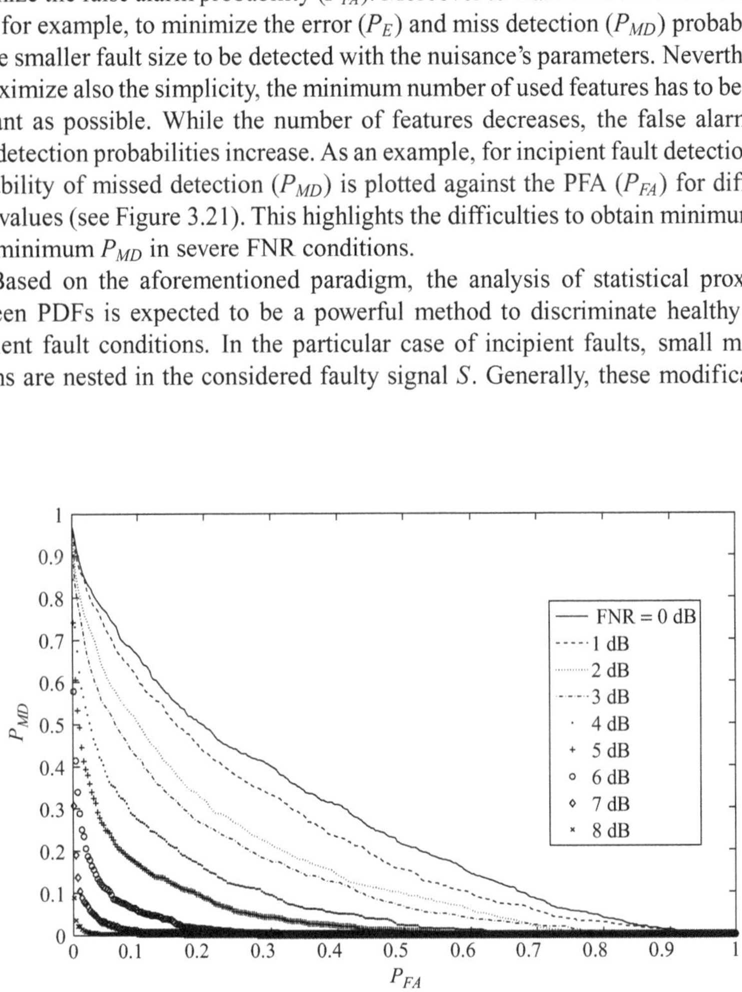


*Figure 3.21 PMD versus PFA*

are very difficult to detect mostly when noise is additionally mixed with the faulty signal. In this case, the fault is considered perceptually transparent but we can, like an attacker in data hiding domain, use the evaluation of the statistical transparency or apply statistical steganalysis to detect and then characterize the incipient fault.

In data hiding, the class of methods for detecting the presence of a message using the statistical transparency lack are called steganalysis methods [36]. In case of incipient FDD that is not perceptually detectable, we propose the use of steganalysis-based techniques.

### *3.4.2 Distance measures*

The FDD methodology based on data hiding paradigm is recalled in Figure 3.22. The data input is the dissimilarity measured between the PDFs.

There are several techniques for statistical dissimilarity measurement. In the information domain [37], one can cite

- entropy-type measurement that expresses the amount of information in a distribution,
- non-parametric techniques that measure the affinity between two PDFs like Kullback–Leibler divergence (KLD).

For distance measures [38], one can cite the following ones:

- the f-divergence among which Hellinger distance (HD), $\chi^2$-Divergence, Kolmogorov distance, total variation (TV), Bhattacharyya distance (BD), Matusita distance (MD), and KLD;
- the mean distance like Shannon or quadratic entropy.

A comparative review in [38] has shown that 'Kullback–Leibler divergence (KLD) and Hellinger distance (HD) take a key part for proving theoretical results as well as solving applied problems'. Moreover, for two PDFs $h$ and $q$, the following relations have been established [39]:

- $2TV^2(h,q) \leq KLD(h,q)$, Pinskers's inequality;
- $BD \leq HD^2(h,q) \leq KLD(h,q)$;
- $2MD^2(h,q) \leq KLD(h,q)$.

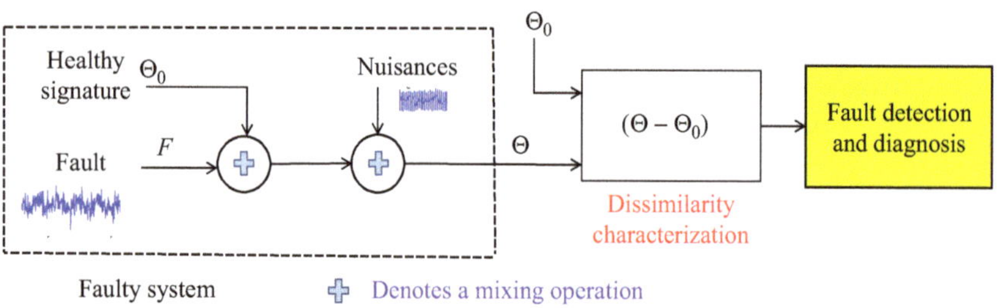


*Figure 3.22 Dissimilarity measurement for FDD*

Finally, one can conclude that KLD is the most sensitive dissimilarity measure to small variations.

### 3.4.3 *Kullback–Leibler divergence*

The KLD, or the relative entropy, is a well-known probabilistic tool that has proved its worth in machine learning, neuroscience, pattern recognition [40], and anomaly detection [41,42]. KLD has already proved its efficiency for detecting incipient faults in several applications [22,43,44]. Its main goal is the evaluation of the divergence of two signals based on their probability distribution functions (PDFs).

Theoretically, for two PDFs $f(x)$ and $g(x)$ of continuous random variable $x$, Kullback and Leibler have defined the Kullback–Leibler information from $f$ to $g$ [35] as

$$I(f||g) = \int f(x) \log \frac{f(x)}{g(x)} d(x) \tag{3.1}$$

The KLD is defined as the symmetric version of the Information [45] denoted as

$$KLD(f, g) = I(f||g) + I(g||f) \tag{3.2}$$

Following (3.2), the KLD is assumed non-negative and null if and only if the two distributions are strictly the same. One of the main constraints of this technique is that the two distributions have to share the same support set.

## 3.5 Case studies

In the last two decades, the use of embedded electronics and electrical systems in sensible applications, like transportation or renewable energy applications, has drastically increased [46]. For obvious safety reasons and economical ones, health monitoring and thus FDD are mandatory to ensure safety, reliability, and availability. It also contributes to the reduction of maintenance costs. In the electrical or mechanical engineering communities, many studies have been fruitfully done and some techniques were successful, for example, with the spectral analysis of vibration signals or currents flowing in electrical machines windings [47–49].

### 3.5.1 *Incipient crack detection*

We are in this section addressing the incipient crack fault detection using the KLD. The fault signature is displayed in Figure 3.11. The flow chart is presented in Figure 3.23 and the probability distributions for healthy and faulty cases are displayed in Figure 3.24. One can notice that they are very close. KLD will be used to measure the dissimilarity between the two signatures. It will be compared to the distribution's mean value.

Fifty Monte-Carlo realizations are done for both healthy and faulty cases. For each criterion denoted $Cr$, the detection threshold is set to $\mu_{Cr} + 3\sigma_{Cr}$ where $\mu_{Cr}$ and $\sigma_{Cr}$

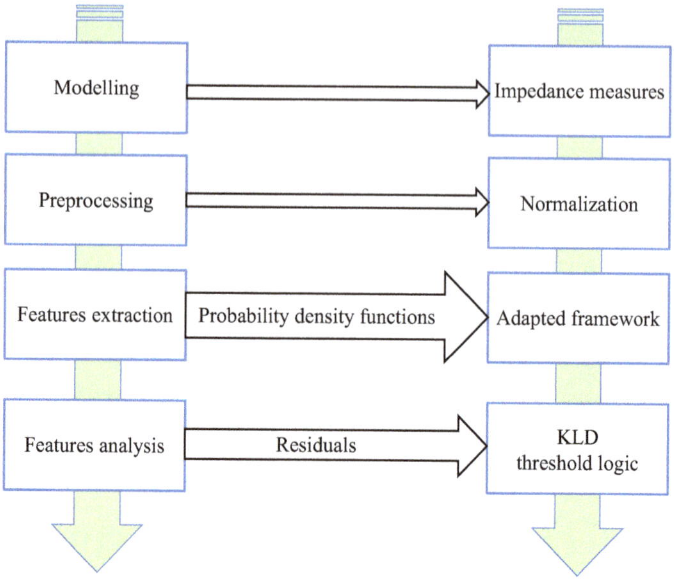


*Figure 3.23 Incipient fault detection flow chart*

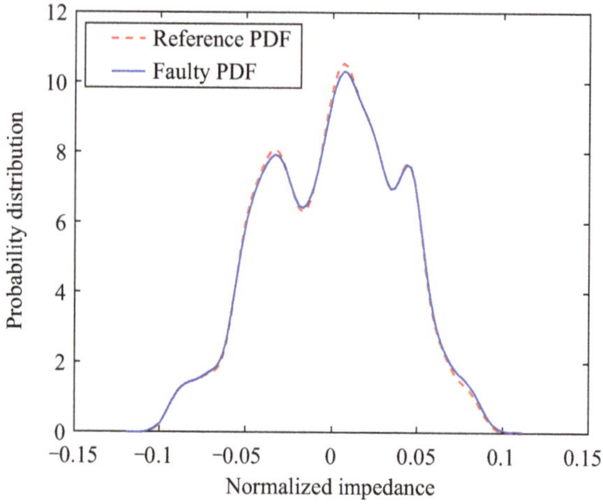


*Figure 3.24 Probability distributions*

are the mean value and standard deviation of the criterion's distribution, respectively. The KLD and the mean value for the smallest crack are plotted in Figure 3.25. For this incipient fault, KLD clearly exhibits the best performances with a significant step variation at fault occurrence.

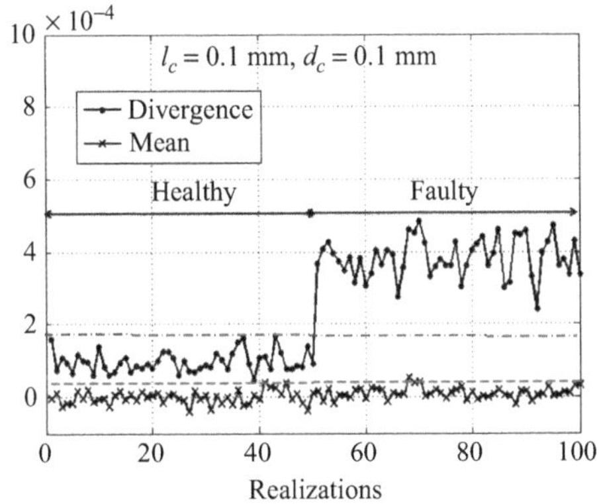


*Figure 3.25 Fault detection results*

*Table 3.1 Fault detection performances*

| $(l_c, d_c)$ (mm) | $Sens_{KLD}$ | $Sens_{Mean}$ |
|---|---|---|
| (0.1, 0.1) | 5.31 | 0.26 |
| (0.2, 0.1) | 5.73 | 0.47 |
| (0.1, 0.2) | 7.37 | 1.12 |
| (0.2, 0.2) | 23.8 | 3.64 |

To evaluate the fault detection performances for each criterion denoted $Cr$, we have defined a sensitivity coefficient $Sens$ as

$$Sens = \frac{<Cr>_{\text{for } R>50} - <Cr>_{\text{for } R<50}}{Max(Cr)_{\text{for } R<50}} \tag{3.3}$$

where $<Cr>$ is the mean value of the criteria.

Under the assumption that the threshold is set to the peak amplitude in healthy condition, the false alarm probability is null ($P_{FA} = 0$), and we can derive the following relations [50]:

$$\text{if } Sens = 1, \quad \text{then} \quad P_{MD} = 0.5 \tag{3.4}$$

$$\text{if } 1 < Sens < 2, \quad \text{then} \quad 0 < P_{MD} < 0.5 \tag{3.5}$$

The fault detection performances are summarized in Table 3.1. The results show that for incipient cracks, the KLD outperforms the mean value.

### *3.5.2 Incipient fault in power converter*

The neutral-point-clamped inverter shown in Figure 3.26 is one of the most efficient multilevel power converters [51]. In industrial applications using variable-speed AC drives, different studies have shown that about 38% of the faults are due to failures in the power device [52]. Therefore, as multilevel inverters have a higher number of power switches and capacitors, their reliability is an issue. Among the different fault types, intermittent ones can be considered as incipient because the immediate effect can be negligible, but its repetition may lead to failures. Intermittent fault is also the most difficult one to detect because it occurs randomly with different durations and its severity can also vary from incipient to severe. In the following, we'll present intermittent fault detection of power switches in a three-level neutral point clamped (NPC) feeding a speed controlled induction machine drive.

The flow chart of the methodology is described in Figure 3.27. The input data are the three-phase currents, already available for control purposes, flowing out of the inverter into the electrical machines windings. The switching frequency is set at 10 kHz.

It has been shown in [53] that the open-switch fault (OSF) detection was particularly tedious at low speed and high torque. Another issue is that in closed-loop, the current and speed controllers, as long as the fault dynamics lie in their bandwidth, will try to mitigate the fault effect. Therefore, the fault effect can be attenuated, and its detection becomes all the more difficult, as the fault is incipient. In Figure 3.28, the KLD results are displayed for three different durations of the OSF. As mentioned previously, Monte-Carlo simulations are done to introduce the variability that is naturally present in every process. The first 500 realizations represent the healthy conditions (no faults), and the last ones represent the faulty conditions.

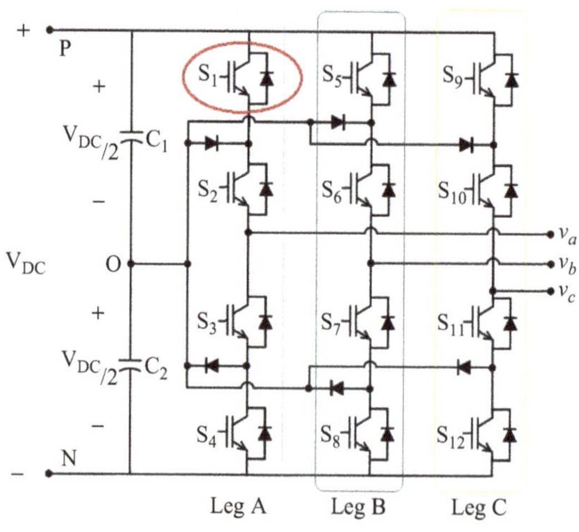


*Figure 3.26 Three-level NPC inverter*

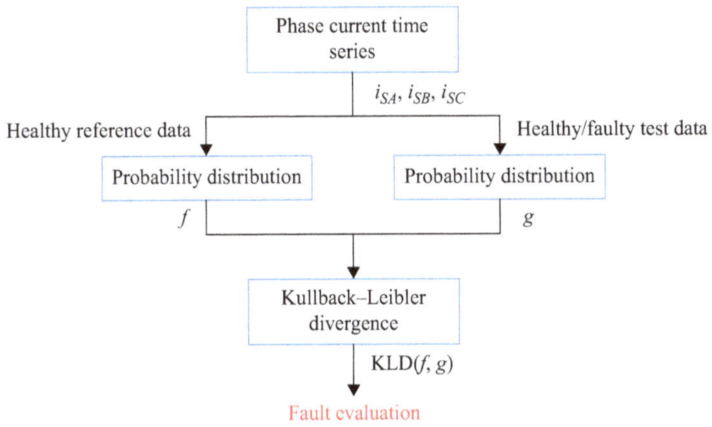


*Figure 3.27 Fault detection procedure*

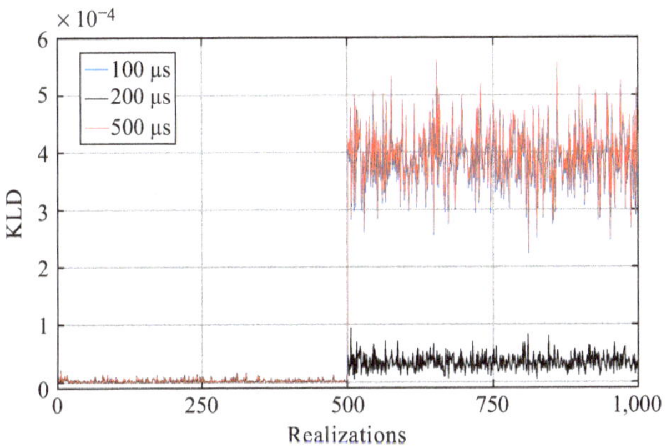


*Figure 3.28 KLD results for three different durations of the OSF at 20 rad/s, 50% of load and SNR = 20 dB*

Each fault detection method should be evaluated regarding the two following probabilities:

- The probability of detection ($P_D$) that represents the ability for correctly detecting a fault when it occurs.
- The PFA ($P_{FA}$) that measures the probability of considering a healthy situation as a fault.

These probabilities are calculated and plotted as the receiver operating characteristics (ROC) curve [54]. The performances are obtained considering all the possible detection threshold values. It has been shown in Section 3.3 that incipient fault definition is closely related to the environmental nuisances. This is confirmed in Figure 3.29

(corresponding to the most incipient fault with a duration of 100 μs), where the degradation of the performances can be noticed as the noise level increases (reduction of the SNR).

This result clearly shows the utmost importance of the threshold setting in relation with the noise level, as it determines the fault detection performances.

### 3.5.3 Threshold setting

It has been shown that fault detection is hypothesis test: healthy or faulty. The decision is linked to the crossing of a threshold. The setting of this threshold is one of the trickiest questions in fault detection. A high threshold will avoid false alarm (low PFA) but is unsuitable for incipient fault detection as the sensitivity is too high. On the contrary, a small threshold leads to a high detection capability but will unfortunately generate lots of false alarms. Therefore, threshold setting is necessarily a compromise that can be formulated as minimizing the Bayes risk defined as an optimization cost function for given environmental nuisances:

$$Threshold = argmin(P_{FA} + P_{MD})_{FNR} \tag{3.6}$$

In [55,56], under the assumption that the KLD has a Gaussian distribution, an analytical model of the cost function has been derived and solved using a deterministic optimization technique. The threshold denoted $h$ is defined as

$$h = \mu_{KLD} + \alpha \times \sigma_{KLD} \tag{3.7}$$

where $\alpha$ is the threshold factor. Figure 3.30 shows the evolution of $P_{FA}$ and $P_{MD}$ versus FNR and the threshold factor. It confirms the non-cooperative nature of the threshold setting.

Figure 3.31 displays the Bayes risk versus the threshold factor. These results show that the optimal setting depends on the noise level. The usual setting found in the literature ($\alpha = 2$) is no longer optimal when the noise level decreases (FNR increases).

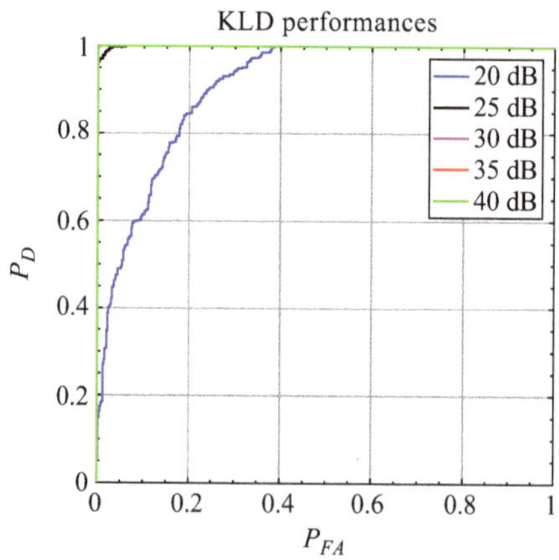


*Figure 3.29 ROC curves for several noise levels*

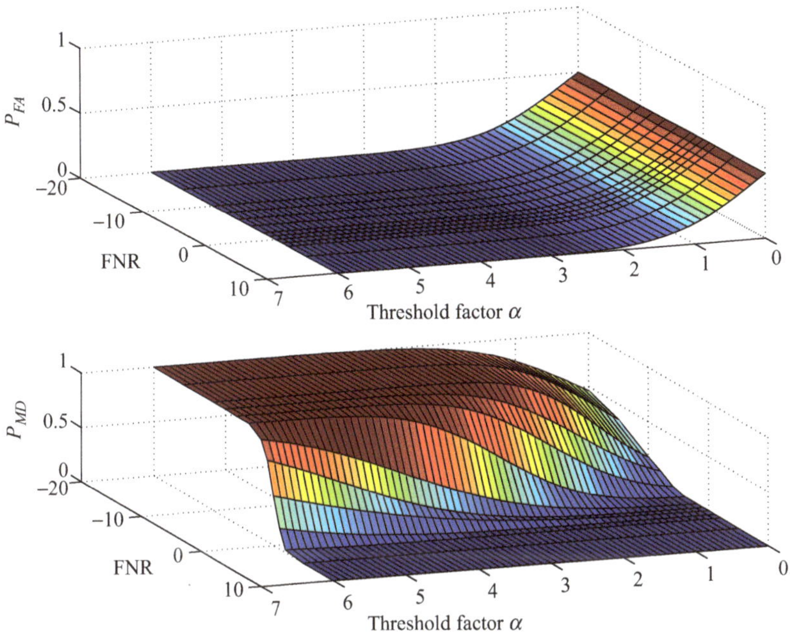


*Figure 3.30* *$P_{FA}$ and $P_{MD}$ versus FNR and threshold factor*

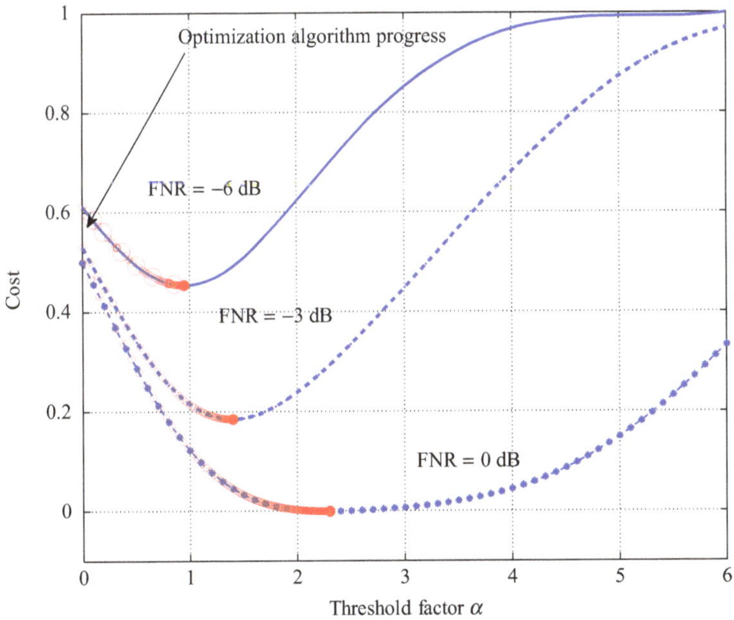


*Figure 3.31* *Cost optimization versus threshold factor*

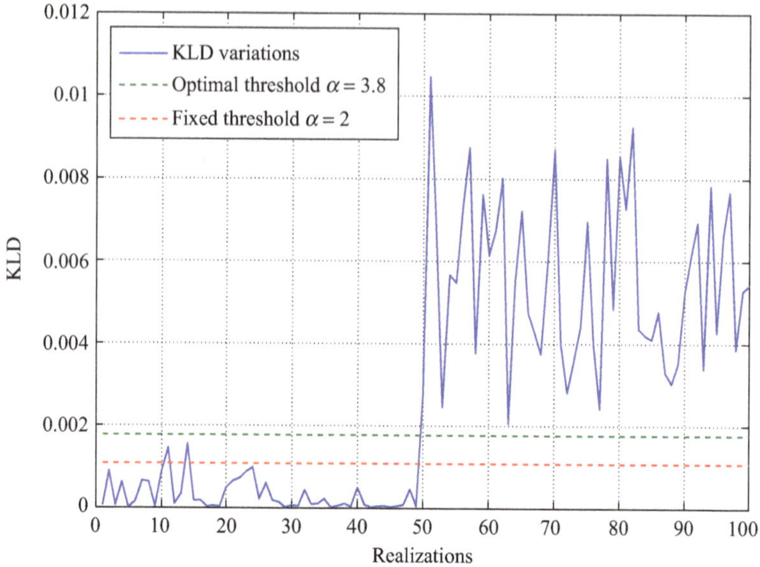


*Figure 3.32 Optimal threshold factor for incipient crack detection*

Figure 3.32 illustrates optimization results in case of incipient crack detection ($l_c = 0.1$ mm and $d_c = 0.1$ mm). One can observe that with the optimal factor both PFA and probability of missed detection are null.

If the KLD distribution is unknown, it can be approximated with kernel density estimator [57]. The cost (Bayes risk) function can be numerically computed and optimized with deterministic or stochastic techniques.

In conclusion, the results in this subsection has put into evidence the importance of taking into account the environmental nuisances when setting the threshold for incipient fault detection.

### 3.5.4 Fault-level estimation

Once the fault is detected, it could be interesting for maintenance or fault-tolerant control purposes to estimate its characteristics. One of the most important ones is the fault amplitude. The issue is to retrieve the fault amplitude from the feature used for fault detection. In the previous sections, we have shown that KLD is an efficient feature for incipient fault detection. Figure 3.33 displays the evolution of KLD for several incipient fault amplitudes [21]. It shows that the KLD value depends on the fault severity.

If an analytical model of the KLD is available, it can be used to retrieve the fault amplitude from the estimated KLD. The KLD is computed from the PDFs, retrieved from history process data through kernel density estimators. Once the KLD value is estimated, the fault amplitude can be deduced from the analytical model of the KLD [21]. An example of a relative estimation error is plotted in Figure 3.34. On

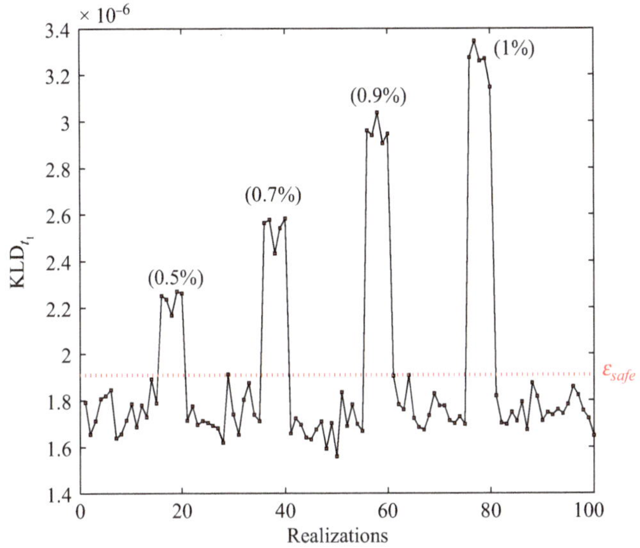


*Figure 3.33 KLD for different fault severities*

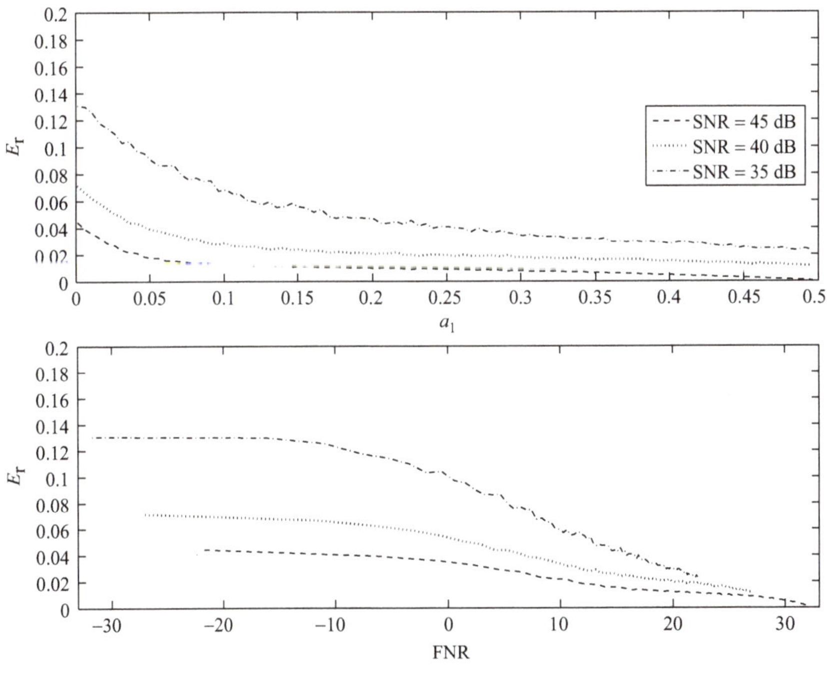


*Figure 3.34 Fault amplitude relative estimation error*

the top figure, the relative error is plotted against the actual fault amplitude. On the bottom figure, the relative error is plotted against the FNR. One can observe that the fault amplitude is always overestimated, which is better suited for safety issues. One can also notice that the fault estimation is efficient, even for incipient faults, meaning low FNR values and higher SNR.

## 3.6 Trends for KLD capability improvement

KLD detection and estimation capabilities in incipient fault detection has been proved. Nevertheless, some limitations reduce the detection capability: most of them are due to environmental environments influences or the considered model for data.

As an example, in [56] the authors highlight the detection capability changes with the influence of the noise. As displayed in Figure 3.35, lower will be the detection limits when the noise severity decreases (SNR increases): fault with more incipient severity are detectable.

For the estimation, the same type of behaviour can be observed (see Figure 3.33). Moreover, for including the effect of the considered model for the data, the authors in [58] have highlighted the effect of the noise severity in the KLD estimation using Gamma distributed data. These results are summarized in Figure 3.36.

To cope with these limitations, some works in the literature have been proposed and give some interesting results for future trends in the KLD evaluation for incipient fault diagnosis. As an example, in [59], a focus on the benefit of the multivariate KLD (MKLD) for incipient fault evaluation in a noisy environment is done. The MKLD

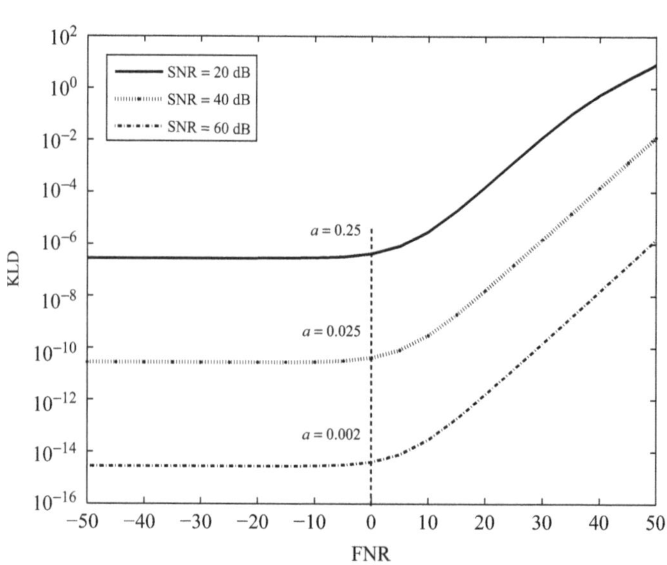


*Figure 3.35 KLD detection capabilities in presence of noise*

poses better detection performances in more severe noise environments for a given fault severity (Figure 3.37).

This leads to abilities to detect smallest fault severities in the same noise environment (Figure 3.38). The detection sensitivity offered by the multivariate approach is then higher than the univariate one. The complexity of this approach is somewhat increased but usual calculation tools could be sufficient in a practical implementation.

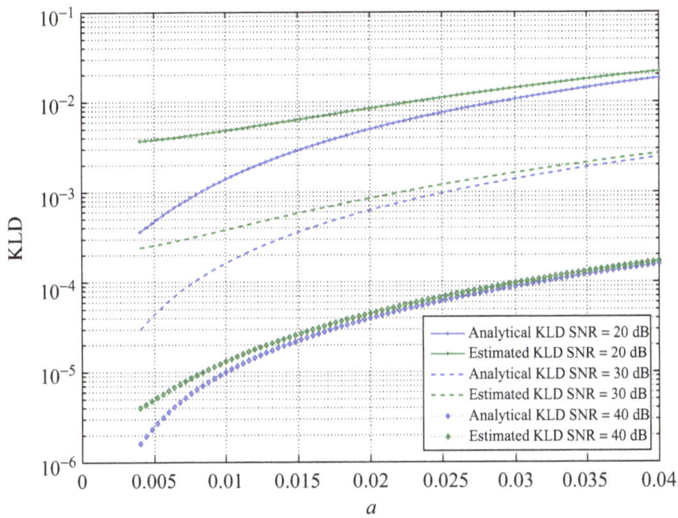


*Figure 3.36 KLD modelling effect*

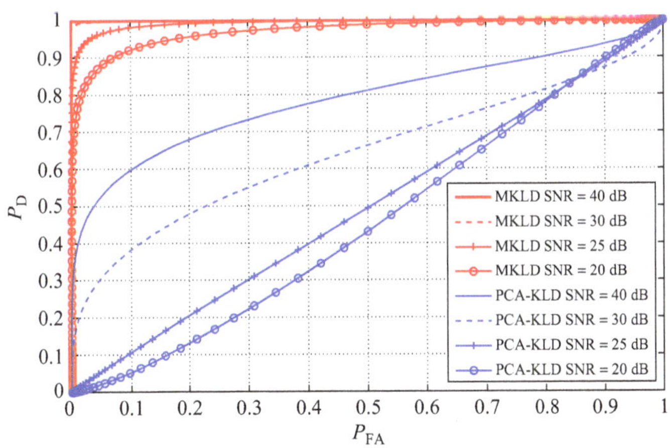


*Figure 3.37 MKLD vs. PCA-KLD detection performance comparison*

In terms of estimation, good performance is also obtained for the multivariate approach (Figure 3.39). The obtained results with the highest considered noise severity (SNR = 25 dB) allow to obtain performances at least equivalent to those obtained in the univariate case for the SNR = 35 dB.

Other trends in the improvement of the detection capability of the KLD concerns the use of a kernel-based dimension reduction tool like the KPCA (kernel PCA) to be able to consider non-linear data in real application studies. The benefit of this technique has been shown in different application cases [60,61]. One major complexity in that case is the choice of the best kernel function fitting the data and its optimal tuning. To the best of our knowledge, no general methodologies exist for this purpose. It has then to be done specifically for the considered data and application.

More recently, incipient fault diagnosis has been improved with the use of Jensen–Shannon divergence (JSD) [62,63]. This divergence is based on KLD evaluation and

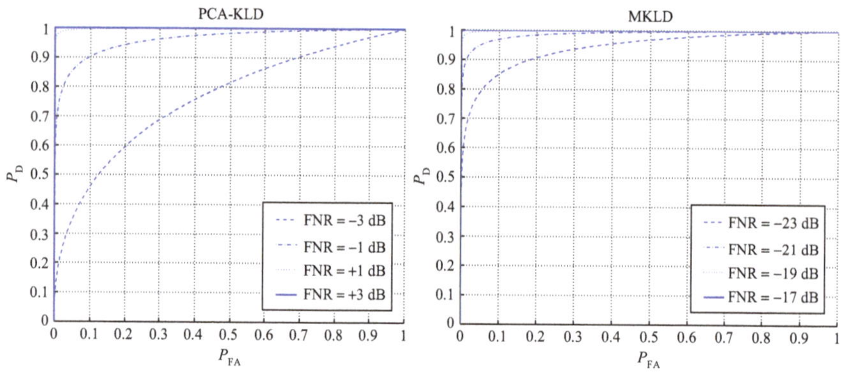


*Figure 3.38 KLD and MKLD detection performances versus FNR*

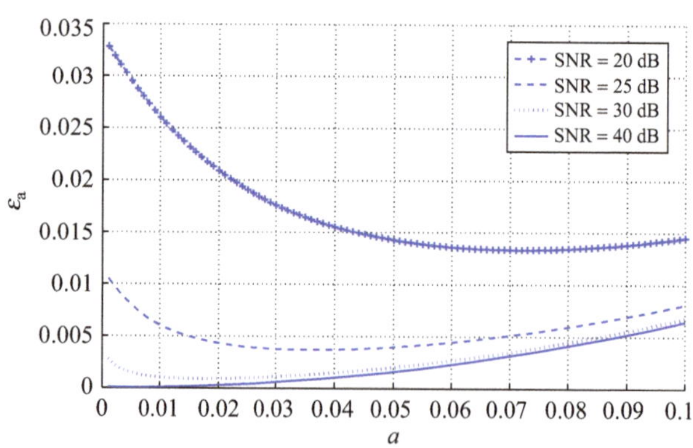


*Figure 3.39 MKLD estimation error*

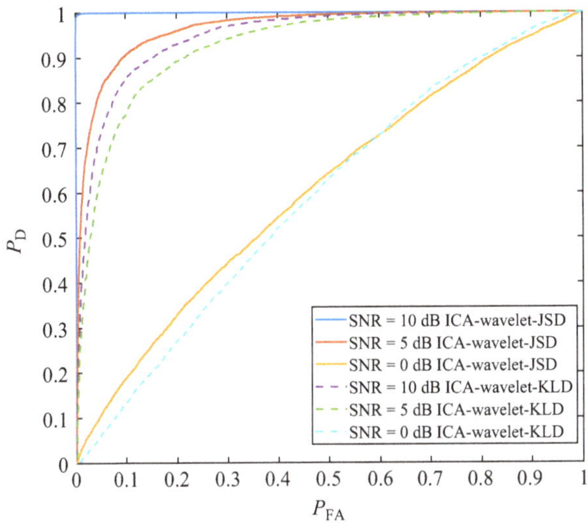


*Figure 3.40 KLD versus JSD for incipient crack detection in noisy environment*

the use of a mean mixed distribution $M$ of the healthy and faulty ones [64] $f$ and $g$, respectively.

$$JSD(f,g) = \frac{1}{2}I(f||M) + \frac{1}{2}I(g||M) \tag{3.8}$$

whereas $M = \frac{1}{2}(f+g)$ and $I$ is defined as the Kullback–Leibler information (see (3.1)).

Its efficiency has been presented recently in [65] for incipient crack detection problems similar to those described in Section 3.2 and compared to KLD. Figure 3.40 displays the detection performance comparison for cracks with severity $l_c = 0.1$ mm and $d_c = 0.1$ mm in several noise environments.

The results highlight an interesting benefit of this new proposal in case of incipient fault. This new trend is then a promising way for obtaining performed incipient fault detection and estimation in complex systems.

## 3.7 Conclusion

In this chapter, we have proposed a new approach for incipient FDD by drawing a parallel with information hiding paradigm. We have shown that a fault can be viewed as unknown embedded information in a signal pattern. So, we have proposed to adopt a similar approach as an attacker in the information-hiding domain to detect and diagnose a fault. Through the analysis, we have also shown that the FDD performances can be presented as the trade-off of three criteria (robustness, sensitivity, and simplicity). After their definition, we have described and explained how to evaluate them with the inspiration from the information-hiding domain. After the introduction of SNR, FNR, and SFR, we have rigorously defined an incipient fault as a deviation in the particular case of high SNR and $FNR \leq 0$ dB. Therefore, incipient fault detection methods should be developed while taking into account the environmental nuisances. Because incipient fault effect may be concealed in the noise, classical features usually fail to detect the fault. Based on the paradigm between data security and fault diagnosis, dissimilarity measure of PDFs has proved its efficiency. The computation of the PFA and the probability of (missed) detection has assessed the detection capability of the KLD.

Several machine learning techniques [66] (SVM, PCA, neural network, etc.) have already been evaluated for FDD. Currently, deep learning techniques [67,68] are promoted for automatic feature extraction and analysis. Other techniques developed in the signal processing and telecommunication areas like source separation, optimization techniques, or non-cooperative game theory are also potential candidates in the hot topic health-monitoring domain.

## References

[1] Isermann R. Fault Diagnosis Applications. Springer-Verlag Berlin and Heidelberg; 2011.

[2] Isermann R. Fault-Diagnosis Systems: An Introduction from Fault Detection to Fault Tolerance. Springer-Verlag Berlin and Heidelberg; 2005.

[3] Dai X and Gao Z. From model, signal to knowledge: a data-driven perspective of fault detection and diagnosis. IEEE Transactions on Industrial Informatics. 2013;9(4):2226–2238.

[4] Gao Z, Cecati C, and Ding SX. A survey of fault diagnosis and fault-tolerant techniques–Part I: Fault diagnosis with model-based and signal-based approaches. IEEE Transactions on Industrial Electronics. 2015;62(6):3757–3767.

[5] Gao Z, Cecati C, and Ding SX. A survey of fault diagnosis and fault-tolerant techniques–Part II: Fault diagnosis with knowledge-based and hybrid/active approaches. IEEE Transactions on Industrial Electronics. 2015;62(6):3768–3774.

[6] Soualhi A and Razik H. Electrical Systems 1: From Diagnosis to Prognosis. ISTE Ltd and John Wiley and Sons, Inc., Hoboken, USA; 2020.

[7] Soualhi A and Razik H. Electrical Systems 2: From Diagnosis to Prognosis. ISTE Ltd and John Wiley and Sons, Inc., Hoboken, USA; 2020.

[8] Venkatasubramanian V, Rengaswamy R, Yin K, and Kavari SN. A review of process fault detection and diagnosis, Part I: Quantitative model-based methods. Elsevier Journal on Computer and Chemical Engineering. 2003;27:293–311.

[9] Venkatasubramanian V, Rengaswamy R, and Kavuri SN. A review of process fault detection and diagnosis, Part II: Qualitative models and search strategies. Elsevier Journal on Computer and Chemical Engineering. 2003;27:313–326.

[10] Benbouzid MEH. Bibliography on induction motors faults detection and diagnosis. IEEE Transaction on Energy Conversion. 1999;14(4):1065–1074.
[11] Venkatasubramanian V, Rengaswamy R, Kavuri SN, *et al.* A review of process fault detection and diagnosis, Part III: Process history based methods. Elsevier Journal on Computer and Chemical Engineering. 2003;27:327–346.
[12] Soualhi A, Clerc G, and Razik H. Detection and diagnosis of faults in induction motor using an improved artificial ant clustering technique. IEEE Transactions on Industrial Electronics. 2013;60(9):4053–4062.
[13] Dou C and Lin J. Extraction of fault features of machinery based on Fourier decomposition method. IEEE Access. 2019;7:183468–183478.
[14] Su N, Li X, and Zhang Q. Fault diagnosis of rotating machinery based on wavelet domain denoising and metric distance. IEEE Access. 2019;7:73262–73270.
[15] Abdelkader R, Kaddour A, Bendiabdellah A, *et al.* Rolling bearing fault diagnosis based on an improved denoising method using the complete ensemble empirical mode decomposition and the optimized thresholding operation. IEEE Sensors Journal. 2018;18(17):7166–7172.
[16] Darong H, Lanyan K, Bo M, *et al.* A new incipient fault diagnosis method combining improved RLS and LMD algorithm for rolling bearings with strong background noise. IEEE Access. 2018;6:26001–26010.
[17] Jiang F, Zhu Z, and Li W. An improved VMD with empirical mode decomposition and its application in incipient fault detection of rolling bearing. IEEE Access. 2018;6:44483–44493.
[18] Li Z, Wang T, Wang Y, *et al.* A wavelet threshold denoising-based imbalance fault detection method for marine current turbines. IEEE Access. 2020;8:29815–29825.
[19] Nandi AK and Hosameldin A. Artificial neural networks (ANNs). In: Nandi AK and Hosameldin A, editors. Condition Monitoring with Vibration Signals: Compressive Sampling and Learning Algorithms for Rotating Machines. Wiley-IEEE Press, New York; 2019. pp. 239–258.
[20] Mezni Z, Delpha C, Diallo D, *et al.* Intrinsic mode function selection and statistical information analysis for bearing ball fault detection. In: Derbel N, Ghommam J, Zhu Q. editors. Diagnostic, Fault Detection and Tolerant Control. Springer, New York; 2020. 350 pp. Published in Book Series Studies in Systems, Decision and Control, Chapter 6.
[21] Harmouche J, Delpha C, and Diallo D. Incipient fault detection and diagnosis based on Kullback–Leibler divergence using principal component analysis: Part II. Elsevier Journal on Signal Processing. 2015;109:334–344.
[22] Harmouche J, Delpha C, and Diallo D. Incipient fault detection and diagnosis based on Kullback–Leibler divergence using principal component analysis: Part I. Elsevier Journal on Signal Processing. 2014;94(1):278–287.
[23] Harmouche J, Delpha C, and Diallo D. Improved fault diagnosis of ball bearings based on the global spectrum of vibration signals. IEEE Transaction on Energy Conversion. 2015;30(1):376–383.

[24] Mbo'o CP and Hameyer K. Fault diagnosis of bearing damage by means of the linear discriminant analysis of stator current features from the frequency selection. IEEE Transactions on Industry Applications. 2016;52(5):3861–3868.

[25] Haddad RZ and Strangas EG. On the accuracy of fault detection and separation in permanent magnet synchronous machines using MCSA/MVSA and LDA. IEEE Transactions on Energy Conversion. 2016;31(3):924–934.

[26] Fadhel S, Delpha C, Diallo D, *et al.* PV shading fault detection and classification based on I-V curve using principal component analysis: application to isolated PV system. Solar Energy, Elsevier Journal. 2019;179:1–10.

[27] Kruger U and Xie L. Advances in Statistical Monitoring of Complex Multivariate Processes. Wiley, New York, USA; 2012.

[28] Le Bihan Y, Pavo J, and Marchand C. Characterization of small cracks in eddy current testing. The European Physical Journal Applied Physics. 2008;43(2):231–237.

[29] Sencar HT, Ramkumar M, and Akansu AN. Data Hiding Fundamentals and Applications: Content Security in Digital Media. Elsevier Academic Press, Cambridge, Massachusetts; 2004.

[30] Cox IJ, Miller ML, Bloom JA, *et al.* Digital Watermarking and Steganography. 2nd ed. Morgan Kaufmann; 2008.

[31] Fridrich J. Steganography in Digital Media: Principles, Algorithms and Applications. Cambridge University Press; 2010.

[32] Gupta MD. Watermarking. vol. 1. Intech, IntechOpen, London; 2012.

[33] Braci S, Boyer R, and Delpha C. On the tradeoff between security and robustness of the Trellis coded quantization scheme. IEEE International Conference on Acoustics, Speech and Signal Processing (ICASSP); April 2008.

[34] Delpha C, Hijazi S, and Boyer R. A compressive sensing based quantized watermarking scheme with statistical transparency constraint. In: International Workshop on Digital-Forensics and Watermarking IWDW 2013. Auckland, New Zealand: LNCS, Springer; 2013.

[35] Cover T and Thomas J. Elements of Information Theory. 2nd ed. Wiley, New Jersey; 2006.

[36] Bohme R. Advanced Statistical Steganalysis. Springer-Verlag, Berlin-Heidelberg; 2010.

[37] Ferentinos K and Papaiopannou T. New parametric measures of information. Information and Control. 1981;51:193–208.

[38] Basseville M. Distance measures for signal processing and pattern recognition. Signal Processing, Elsevier Science. 1989;18:349–369.

[39] Toussaint GT. Some inequalities between distance measures for feature evaluation. IEEE Transaction on Computers. 1972;C-21(4):409–410.

[40] Silva J and Narayanan S. Average divergence distance as a statistical discrimination measure for hidden Markov models. IEEE Transactions on Audio Speech and Language Processing. 2006;14(4):890–906.

[41] Afgani M, Sinanovic S, and Haas H. Anomaly detection using the Kullback–Leibler divergence metric. In: 1st International Symposium on Applied Science

on Biomedical and Computer Technology, ISABEL'08. Aalborg, Denmark; 2008.
[42] Anderson A and Haas H. Kullback–Leibler divergence (KLD) based anomaly detection and monotonic sequence analysis. In: IEEE Vehicular Technology Conference (VTCFall). Budapest, Hungary; 2011.
[43] Chai Y, Tao S, Mao W, *et al.* Online incipient fault diagnosis based on Kullback–Leibler divergence and recursive principle component analysis. Canadian Journal Chemical Engineering. 2018;96(4):426–433.
[44] Chen H, Jiang B, and Lu N. An improved incipient fault detection method based on Kullback-Leibler divergence. ISA Transactions. 2018;79:127–136.
[45] Blanke M, Kinnaert M, Lunze J, and Staroswiecki M. Diagnosis and Fault-tolerant Control. 2nd ed. Springer, Berlin-Heidelberg; 2006. Chapter 6, pp. 238–263.
[46] Diallo D, Benbouzid MEH, and Masrur MA. Condition monitoring and fault accommodation in electric and hybrid propulsion systems. IEEE Transaction on Vehicular Technology, Special Section. 2013;62(3):962–964.
[47] Seshadrinath J, Singh B, and Panigrahi BK. Vibration analysis based inter-turn fault diagnosis in induction machines. IEEE Transactions on Industrial Informatics. 2014;10(1):340–350.
[48] Amirat Y, Choqueuse V, and Benbouzid M. EEMD-based wind turbine bearing failure detection using the generator stator current homopolar component. Mechanical Systems and Signal Processing. 2013;41(1–2):667–678.
[49] Faiz J, Ghorbanian V, and Ebrahimi BM. EMD-based analysis of industrial induction motors with broken rotor bars for identification of operating point at different supply modes. IEEE Transactions on Industrial Informatics. 2014;10(2):957–966.
[50] Harmouche J, Delpha C, Diallo D, *et al.* Statistical approach for non-destructive incipient damage detection and characterisation using Kullback-Leibler divergence. IEEE Transaction on Reliability. 2016;65(3):1360–1368.
[51] Wu B and Narimani M. Diode-clamped multilevel inverters. In: High-Power Converters and AC Drives. Wiley-IEEE Press, New York; 2017. pp. 143–183.
[52] Errabelli RR and Mutschler P. Fault-tolerant voltage source inverter for permanent magnet drives. IEEE Transactions on Power Electronics. 2012;27(2):500–508.
[53] Baghli M, Delpha C, Diallo D, *et al.* Three-level NPC inverter incipient fault detection and classification using output current statistical analysis. Energies. 2019;12(7):1372.
[54] Green DM and Swets JM. Signal Detection Theory and Psychophysics. John Wiley and Sons Inc., New York (NY), USA; 1966.
[55] Youssef A, Delpha C, and Diallo D. Performances Theoretical Model-based Optimization for Incipient Fault Detection with KL Divergence. In: EUSIPCO 2014. Lisbon, Portugal; 2014. pp. 466–470.
[56] Youssef A, Delpha C, and Diallo D. An optimal fault detection threshold for early detection using Kullback-Leibler divergence for unknown distribution data. Signal Processing. 2016;120C:266–279.

[57] Scott DW. Multivariate Density Estimation. 2nd ed. Wiley, New Jersey, USA; 2015.

[58] Delpha C, Diallo D, and Youssef A. Kullback-Leibler divergence for fault estimation and isolation: application to gamma distributed data. Mechanical Systems and Signal Processing Journal, Elsevier. 2017;93(C):118–135.

[59] Youssef A, Delpha C, and Diallo D. Enhancement of incipient fault detection and estimation using the multivariate Kullback-Leibler divergence. In: European Signal Processing Conference (EUSIPCO 2016). Budapest, Hungary: IEEE; 2016. pp. 1408–1412.

[60] Wang Q, Liu YB, He X, *et al.* Fault diagnosis of bearing based on KPCA and KNN method. In: Advanced Materials Research. vol. 986. Trans Tech Publ; 2014. pp. 1491–1496.

[61] Zhang X and Delpha C. Improved incipient fault detection using Jensen-Shannon divergence and KPCA. In: 2020 Prognostics and Health Management Conference (PHM 2020). Besancon, France: IEEE; 2020. pp. 241–246.

[62] Zhang X, Delpha C, and Diallo D. Nondestructive incipient crack detection based on wavelet and Jensen-Shannon divergence in the NICA framework. In: IEEE International Conference on Industrial Electronics (IEEE IECON 2019). Lisbon, Portugal: IEEE; 2019. pp. 3685–3690.

[63] Zhang X, Delpha C, and Diallo D. Incipient fault detection and estimation based on Jensen-Shannon divergence in a data-driven approach. Elsevier Journal on Signal Processing. 2020;169C(107410):1–12. DOI: 10.1016/j.sigpro.2019.107410.

[64] Lin J. Divergence measures based on the Shannon entropy. IEEE Transactions on Information Theory. 1991;37(1):145–151.

[65] Zhang X, Delpha C, and Diallo D. Jensen-Shannon divergence for non-destructive incipient crack detection and estimation. IEEE Access. 2020;8:116148–116162.

[66] Munikoti S, Das L, Natarajan B, *et al.* Data-driven approaches for diagnosis of incipient faults in DC motors. IEEE Transactions on Industrial Informatics. 2019;15(9):5299–5308.

[67] Lei Y, Yang B, Jiang X, *et al.* Applications of machine learning to machine fault diagnosis: a review and roadmap. Mechanical Systems and Signal Processing. 2020;138:106587.

[68] Pecht MG and Kang M. Machine learning: diagnostics and prognostics. In: Prognostics and Health Management of Electronics; 2019. pp. 163–191.

# Chapter 4
# Higher-order spectra

*Lotfi Saidi*[1,2,3]

## 4.1 Introduction

Over the past decades, higher-order spectra (HOS), also called polyspectra, have established a status as a suitable mathematical and signal processing tool for nonlinear system analysis. However, a major problem with this kind of signal processing tool application is the interpretation of the obtained results, and much uncertainty still exists about the relation between HOS contribution compared with the second-order statistics. This chapter provided an important opportunity to advance the understanding advantages of HOS.

The classical power spectrum (PS), which is defined as the Fourier transform (FT) of the autocorrelation sequence (the second-order cumulant), does not give any information about the phase of system frequency response; therefore, it is unable to give any indication about system nonlinearity. However, the HOS [1–11] are defined as the multidimensional FT of higher-order cumulants of a stationary random process and can overcome the inability of PS to detect these nonlinearities.

From the structure of HOS, it is possible to deduce various properties of signals that do not appear when using the PS. For example, many different signals can have the same correlation function or the same PS, but they can be distinguished by using HOS. Furthermore, there are various methods of signal processing using HOS that can solve problems that cannot be addressed using only second-order statistics [1–5].

Often there are situations in which the interaction between two harmonic components causes a contribution to the power at their sum and/or difference frequencies. For example, an important class of nonlinear interaction, called quadratic phase coupling (QPC), involves frequency triplet, $F_0$, $F_1$, and $F_2$. QPC means that the sum of the phases at $F_0(\theta_0)$ and $F_1(\theta_1)$ is the phase at frequency $F_0 + F_1$ (i.e. $\theta_0 + \theta_1$), which is often an indication of second-order nonlinearities. In certain applications, it

[1]ENSIT – Laboratory of Signal Image and Energy Mastery (SIME), Université de Tunis, Tunis, Tunisia
[2]ESSTHS – Department of Electronics and Computer Engineering, University of Sousse, Sousse, Tunisia
[3]Institut de Recherche Dupuy de Lôme, CNRS, University of Brest, Brest, France

is necessary to determine if peaks at harmonically related positions in the PS are phase coupled. Since the PS suppresses all phase relations, it cannot provide the answer. Third-order statistics called the bispectrum is a powerful tool to detect QPC and has been applied successfully to evaluate QPC types of nonlinear effects [1–24]. This is best illustrated in the examples given in Figures 4.5–4.11.

The bispectrum is the third-order spectrum and it results in a frequency–frequency–amplitude relationship that shows coupling effects between signals at different frequencies. Therefore, the bispectrum is sensitive to the non-Gaussianity of signals and can effectively extract information due to deviations from Gaussianity. If a signal is Gaussian, its bispectrum would be identically zero, for a non-Gaussian signal, the bispectrum can be nonzero [1–5].

As a result, the bispectrum has been widely used in many detection applications including biomedical engineering [13,24], plasma physics [14], engineering structures [10,15,22], mechanical systems [16,17,25–27], the coupling assessment between modes in a power generation systems [18], condition monitoring of electrical machines and drives [11,19–21], etc.

There are very few references dedicating to the bispectrum for nonlinear system outputs. This is because deriving the bispectrum for nonlinear system outputs is much more difficult as compared to linear systems. Without any additional specification, the only general structure can be used in this case, as the nonlinearity does not introduce any precise definition of the statistics of a signal. Therefore, if we require explicit analytic expressions of HOS, it is necessary to introduce statistical models of signals that can represent physical phenomena and have a structure leading to possible explicit calculations. This is one of the main purposes of this chapter.

This chapter has considered the potential advantages of using the bispectrum to analyze data from nonlinear systems driven with deterministic harmonic signals. The bispectrum detects signals that are quadratically phase coupled and suppresses those that are not, which is why it is of interest for identifying system nonlinearity.

Results of simulation and experimental studies have been used to verify the theoretical analysis and to demonstrate the effectiveness of the derived relationship. The established analytical relationship between the bispectrum and the nonlinear characteristic parameters could pave a way for quantifying the nonlinearity degree of systems. Also, it was shown that bispectral signal processing can be used to enhance the signal-to-noise ratio (SNR) of the nonlinear effects, principally by reducing the noise in the signal.

The overall structure of this chapter takes the form of four sections, including this introductory section. Section 2 begins by laying out the theoretical dimensions of the research, to motivate applications of HOS in signal analysis. The third section is concerned with the applications of the HOS used for this study application as a signal processing for electromechanical systems' fault detection and diagnosis. Finally, the conclusion gives a summary and critique of the findings.

The following section summarizes the main definitions of HOS analysis.

## 4.2 Higher-order statistics analysis: definitions and properties

### 4.2.1 Higher-order moments

In probability theory and statistics, $n$th-order central moment of a random variable $X$ is calculated as the expected value of integer power, $n$, of the random variable $X$ around its mean, as follows:

$$m_x^{(n)} = E\{(X - E\{X\})^n\} = \int_{-\infty}^{+\infty} (x - E\{X\})^n f_X(x)dx \tag{4.1}$$

where $E\{\cdot\}$ denotes the expected value operator, superscript $^{(n)}$ describes the order of the central moment, and $f_X(x)$ is the probability density function (pdf) of the random variable $X$. Thus, $m_x^{(1)} = 0 =$ the mean value, $m_x^{(2)} =$ mean square value, $m_x^{(3)} =$ mean cube value, and so on. Central moments are used in preference to ordinary moments, computed in terms of deviations from zero instead of the mean, because the higher-order central moments relate only to the spread and shape of the distribution, rather than also to its location [1–7].

HOS signal processing involves a generalization of various order moments in the case of a random variable to moment functions (i.e. correlation functions) in the case of a random process. Therefore, it is mathematically desired to assume that the random process has zero mean for computation convenience, which what we adopt throughout the rest of this dissertation. In the practical cases when we will be dealing with real vibration data from monitored mechanical components, the mean of the signal is first computed and subtracted from the signal.

Based on the mathematical foundations of higher-order statistical signal processing in [1–9], various order correlation functions can be calculated for the random process as follows:

$$\mu_x = E\{x(t)\} = 0 \quad \text{(or, a constant)} \tag{4.2}$$

$$R_{xx}(\tau) = E\{x^*(t)x(t+\tau)\} \tag{4.3}$$

$$R_{xxx}(\tau_1, \tau_2) = E\{x^*(t)x(t+\tau_1)x(t+\tau_2)\} \tag{4.4}$$

$$R_{xxx}(\tau_1, \tau_2, \ldots, \tau_n) = E\{x^*(t)x(t+\tau_1)x(t+\tau_2)\cdots x(t+\tau_n)\} \tag{4.5}$$

where $E\{\cdot\}$ denotes the expected value operator and the superscript asterisk (*) denotes the complex conjugate. It is worthwhile to note here that the second-order correlation function, $R_{xx}(\tau)$, in (4.3) is the familiar autocorrelation function. The third-order correlation function, $R_{xxx}(\tau_1, \tau_2)$, is often called bicorrelation function, presumably because it is a function of two-time variables. The fourth-order correlation function, $R_{xxxx}(\tau_1, \tau_2, \tau_3)$, is often called tricorrelation, and so on.

In the case of analyzing linear signals and systems, it is enough to have only (4.2) and (4.3) satisfied. This case is called a weakly stationary (wide-sense stationary) signal. For three-wave interaction in the quadratically nonlinear system as will be discussed later, a random signal is assumed to be stationary to third-order ((4.4) and (4.5)).

### 4.2.2 Power spectrum

The physical interpretation of the classical PS is a one-dimensional (1D) function of frequency and has demonstrated very powerfully in modeling linear phenomena. The discrete PS is the FT of the autocorrelation $R_{xx}(\tau)$, and can be estimated by

$$P_{xx}(f) = E\{X(f)X^*(f)\} = E\{|X(f)|^2\} \tag{4.6}$$

where $X^*$ denotes the complex conjugate of $X$, $X(f)$ is the discrete FT of $x(n)$, and $E$ is the expectation operator a zero-mean stationary random signal, given by

$$X(f) = \sum_{n=0}^{N-1} x(n)e^{-j\frac{2\pi fn}{N}} \tag{4.7}$$

where $N$ is the number of samples present in the signal.

The PS in (4.61) is defined regardless of whether the signal is zero mean. It can also be defined in a short-time form for nonstationary signals if the discrete version over finite records as in (4.62) is used. Assumptions about stationarity are only made when applied to random signals.

Because all phase information is destroyed in computing the PS, it is unable to detect phase coupling signatures.

### 4.2.3 Bispectrum and bicoherence

The next higher-order spectrum is the bispectrum, a 2D FT of the third-order autocorrelation function $R_{xxx}(\tau_1, \tau_2)$, which is very powerful in detecting and quantifying quadratic effects in a time-series [10,15,17,18,22,23].

Meanwhile, the bispectrum describes the statistical relationships between the signal frequency components. These relationships are, in particular, the nonlinear indicator in the signal.

The bispectrum is defined as

$$B(f_1,f_2) = E\{X(f_1)X(f_2)X^*(f_1+f_2)\} \tag{4.8}$$

The bispectrum is all about statistics (HOS). It is the FT of the third cumulant or moment. Nonlinearity affects this cumulant that is captured by the bispectrum. The expectation operation is very important in this context and cannot be ignored especially in the detection and quantification of phase coupling. It involves "ensemble averaging" for an estimate such that if phases are random, the bispectrum goes to zero and if phases are coupled it does not.

As shown in Figure 4.1, it can be noted that the bispectrum presents 12 symmetry regions [1,2]. Hence, the analysis can take into consideration only a single nonredundant region. Hereafter, $B(f_1,f_2)$ will denote the bispectrum in the triangular region $\Im$ shown in Figure 4.2 and defined by

$$\Im = \{(f_1,f_2): 0 \le f_2 \le f_1 \le f_e/2,\ f_1+f_2 \le f_e/2\}$$

where $f_e$ is the sampling frequency. Regions of computation are discussed in references [1,2,5,6].

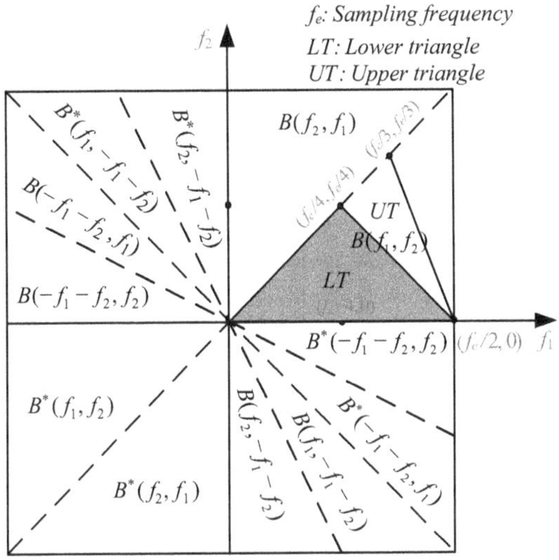


Figure 4.1 Symmetry regions of the bispectrum

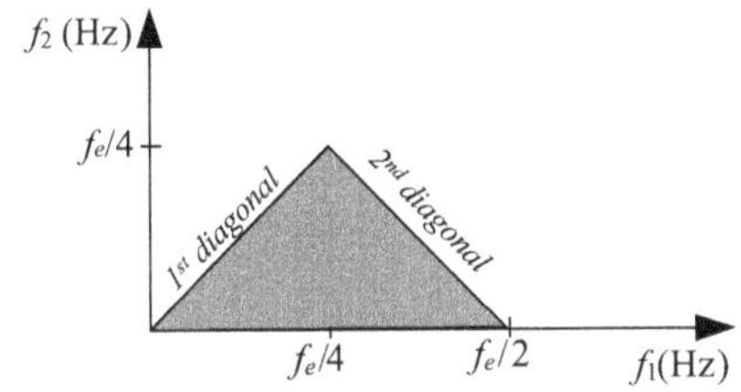


Figure 4.2 A nonredundant region of the bispectrum

For the bispectrum to be nonzero at $(f_1, f_2)$, the FTs at $f_1$, $f_2$, and $f_1 + f_2$ must be nonzero.

As a result, the bispectrum can be used to solve several practical problems effectively. Examples are expressed as follows [10–24]:

- Gaussian processes: If $x(n)$ is a stationary zero-mean Gaussian process, its bispectrum $B(f_1, f_2)$ is identically zero.
- Linear phase shifts: While the PS suppresses all phase information, the bispectrum does not.
- Non-Gaussian white noise: Its bispectrum is flat, only if the third-order autocorrelation is an impulse at the origin.

The bispectrum is a quantified measure of HOS. It is the FT of the third cumulant or moment. Nonlinearity affects these cumulants that are captured by the bispectrum.

For the interested reader, a theoretical introduction of HOS is given in [1,2,5,6].

The bicoherence or the normalized bispectrum presented in (4.9) is a measure of the amount of phase coupling that occurs in a signal or between two signals.

As mentioned above, phase coupling is the estimate of the proportion of energy in every possible pair of frequency components, $f_1,f_2$, which satisfies the definition of QPC (phase of the component at $f_3$, which is $f_1+f_2$, equals phase of $f_1$ + phase of $f_2$):

$$bic(f_1,f_2) = \frac{|B(f_1,f_2)|^2}{X(f_1)X(f_2)X(f_1+f_2)} \tag{4.9}$$

When the analyzed signal reveals the structure of any kind whatsoever, it might be expected that some phase coupling arises. Bicoherence analysis can detect coherent signals in very noisy data, provided that the coherency remains constant for sufficiently long times since the noise contribution falls off rapidly with increasing $N$.

### 4.2.4 Estimation

In general, the expected values coming from (4.6) and (4.8) need to be estimated from a finite quantity of available data.

Nonparametric (conventional) approaches for bispectrum estimation may be either direct or indirect. The usual, indirect method requires estimation of the third-order cumulant and computation of the 2D-FT. Instead, we employ the direct method, which may be achieved by dividing a signal into $M$ overlapping segments with $k$ as a subscript, $k = 1,2,3,\ldots,M$. A windowing function has been applied to each segment and the FTs of all segments are averaged. The aim is to reduce the variance of the estimate by increasing the number of records $M$ [1–6].

The estimated bispectrum $\hat{B}(f_1,f_2)$ is given by

$$\hat{B}(f_1,f_2) = \frac{1}{M}\sum_{k=1}^{M} X_k(f_1)X_k(f_2)X_k^*(f_1+f_2) \approx E\{X(f_1)X(f_2)X^*(f_1+f_2)\} \tag{4.10}$$

The expectation operation is very important in this context and cannot be ignored especially in the detection and quantification of phase coupling. It involves "ensemble averaging" for an estimate: if phases are random, the bispectrum goes to zero and if phases are coupled, it does not.

In all numerical simulations presented here, the bispectrum was computed using the direct method, which is an approximation of formula (4.10).

Using bispectrum for harmonic signal nonlinearities detection will be dealt with in more detail in the next subsection.

## 4.3 Bispectrum use for harmonic signals' nonlinearity detection

Preliminary work on bispectrum was undertaken by Kim and Powers [14]. Let first consider the academic example of [14]. Kim and Powers showed that the interaction of two waves of frequency $f_1$ and $f_2$ can generate two waves of the interaction of frequencies $f_1+f_2$ and $f_1-f_2$. The frequency components of a signal may, therefore, interact between themselves and produce other frequency components including the wave number, and frequencies are obtained from the sum or the difference of these primary components.

The frequencies $f_1, f_2$, and $f_3$ can be linked by a quadratic nonlinear interaction if the following condition is satisfied $f_1 \pm f_2 \pm f_3 = 0$. Whatever the indices derived from the PS, they remain independent of relationships that may exist between different frequencies, particularly in terms of phase. If the spectrum of a signal shows different peaks in frequency, the bispectrum allows searching the couplings, called "quadratic," existing between them. The resulting signal is described by (4.11).

$$x(t) = \cos(2\pi f_1 t + \varphi_1) + \cos(2\pi f_2 t + \varphi_2) + \frac{1}{2}\cos(2\pi f_3 t + \varphi_3) \tag{4.11}$$

Consider the three harmonics ($S_1$, $S_2$, and $S_3$) with frequencies $f_1 = 1,100$ Hz, $f_2 = 1875$ Hz, and $f_3 = 2975$ Hz and phases $\varphi_1$, $\varphi_2$, and $\varphi_3$ such that $\varphi_3 = \varphi_1 + \varphi_2$. These three harmonics are quadratically coupled. Figure 4.3 shows that the PS contains the three frequencies of the signal $x(t)$, whereas the bispectrum result (represented in 3D in the figure) contains only the peak at bifrequencies $(f_1, f_2)$, reflecting their coupling. This is because the sinusoids $S_3$ has for frequency and phase the respective sum of those of the harmonics $S_1$ and $S_2$.

This signal is sampled at the frequency 10 kHz, and the spectral and bispectral calculations are performed with Hanning-type filtering. In Figure 4.4, the figure at the center represents the bispectrum of the previous signal represented in the 2D plane $(f_1, f_2)$, the spectrum of the same signal is represented on the left and below. This example illustrates the nonlinear interaction between frequencies $f_1, f_2$ (discontinuous arrows) and $f_3$ (continuous arrow). Consequently, the bispectrum of

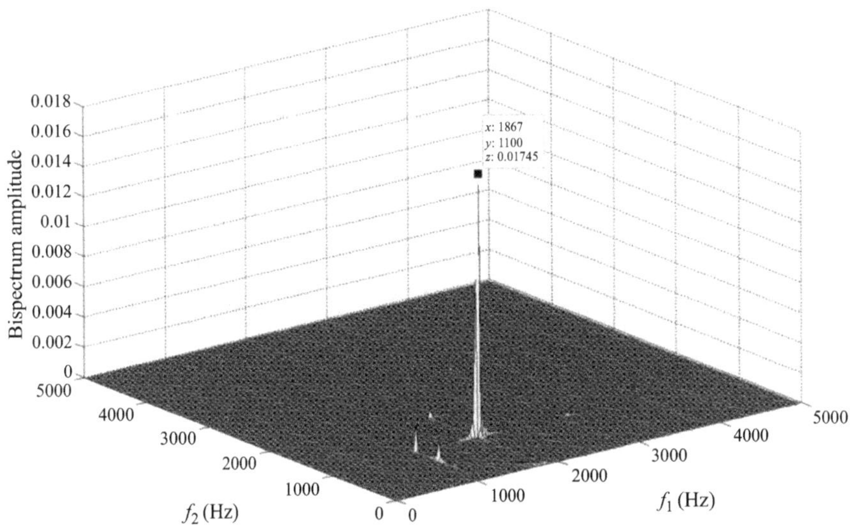


*Figure 4.3 Example of QPC presented by Kim and Powers [14], in a 3D presentation*

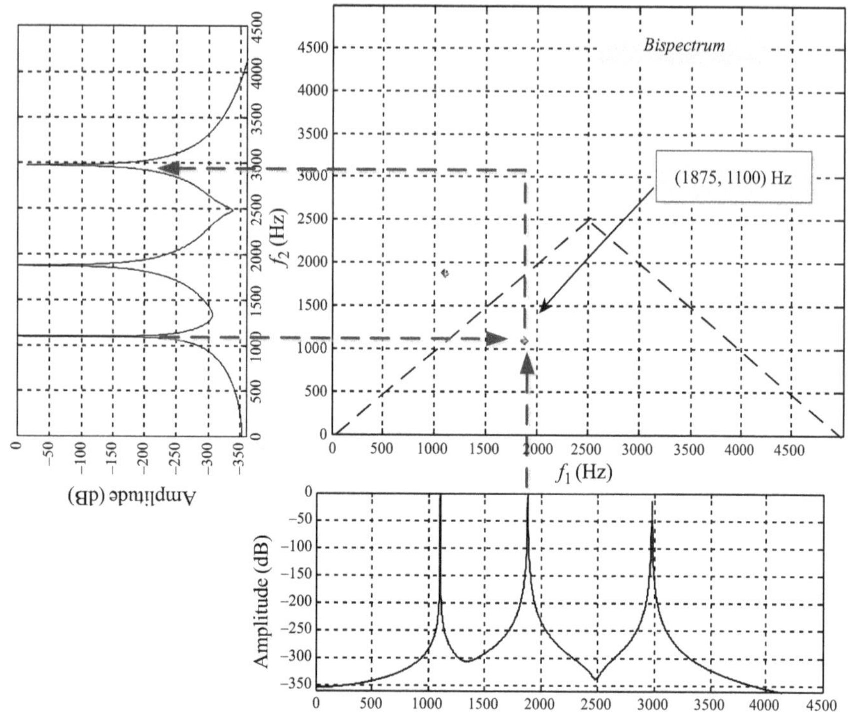


*Figure 4.4 Example of QPC presented by Kim and Powers [14], in a 2D presentation*

the signal shows a peak at (1100, 1875) Hz and by symmetry at (1875, 1100) Hz. The bispectrum harmonic signals nonlinearities detection can be coded in MATLAB as follow.

This example shows the efficiency of the bispectrum to detect the degree of coupling between the different frequencies of the signal.

It should be noted that the bispectrum only shows quadratically coupled frequencies (sometimes with themselves) and leads to the suppression of the information related to the "uncoupled" frequencies.

Typical simple examples, using cosine waves, are given to demonstrate some of the possible frequency interactions that can occur in the bispectrum. Cosine waves are used as an example because they produce easily understood results although they do not conform to an assumption of being stationary random signals.

Note that all the following considerations were used to estimate the example bispectrum. The diagonal line in the central plot defines the two regions of symmetry of the bispectrum, while the magnitude of the $Z$-axis (indicated by gray levels bar) was generated at 512 Hz sampling rate and contained 10240 data points. This corresponds to the trial duration of 20 s. For all test signals, 40 segments each contain 256 data points, with no overlapping.

*Table 4.1 A MATLAB® code of the bispectrum harmonic signals nonlinearities detection given in Section 4.3*

```
%%Direct estimation of the bispectrum, using the triple % product
B(f1, f2)=1/N% sum(X(f1)X(f2)X*(f1+f2))
clear all
clc
fe=10e3; % sampling frequency
Tmax=10; % acquisition time
N=fe*Tmax; % number of samples
t=0:1/fe:Tmax-1/fe; % time axis
f1=1100;f2=1875;f3=f1+f2;
teta1=pi/5;teta2=5*pi/18;teta3=teta1+teta2;
y=cos(2*pi*f1*t+teta1)+cos(2*pi*f2*t+teta2)+0.5*cos(2*pi*f3*t+teta3);
K=50; % number of segment
M=300; % length of each segment
w=hamming(1,M); % Hanning window
B=zeros(K,M/2,M/2); % Bispectrum initialization
%%compute the bispectrum
for x0=fix(rand*(N-M))
x=y(x0+1:x0+M); % dived a signal into K overlapping segments
x1=x-mean(x); % compute the mean of each segment
x2=x1.*w; % windowing function
X=fft(x2,M)/M; % performing the fast FT of each segment
i=1:K;
  for f1=1:M/2;
    for f2=1:M/2;
B(i,f1,f2)=X(f1)*X(f2)*conj(X(f1+f2)); % the bispectrum triple product
  end
    end
end
B0(1:M/2,1:M/2)=mean(B); % bispectrum averaging
B0=abs(triu(B0)); % non-redundant region of the % bispectrum
mesh((0:fe/M:fe/2-fe/M),(0:fe/M:fe/2 fe/M),triu(B0)),grid
xlabel('f_1[Hz]'),ylabel('f_2[Hz]'),zlabel('Bispectrum amplitude')
```

### 4.3.1 Case 1: a simple harmonic wave at frequency $F_0$

In (4.12), consider a complex cosine wave of frequency $F_0$ and phase angle $\theta_0$ randomly generated between $-\pi$ and $\pi$ with a uniform distribution. A cosine wave is used to suppress unwanted cross terms between the positive and negative frequency components.

The first numerical simulation presented in Figure 4.5 was done taking $F_0 =$ 50 Hz.

$$x_0(t) = \cos(2\pi F_0 t - \theta_0) \tag{4.12}$$

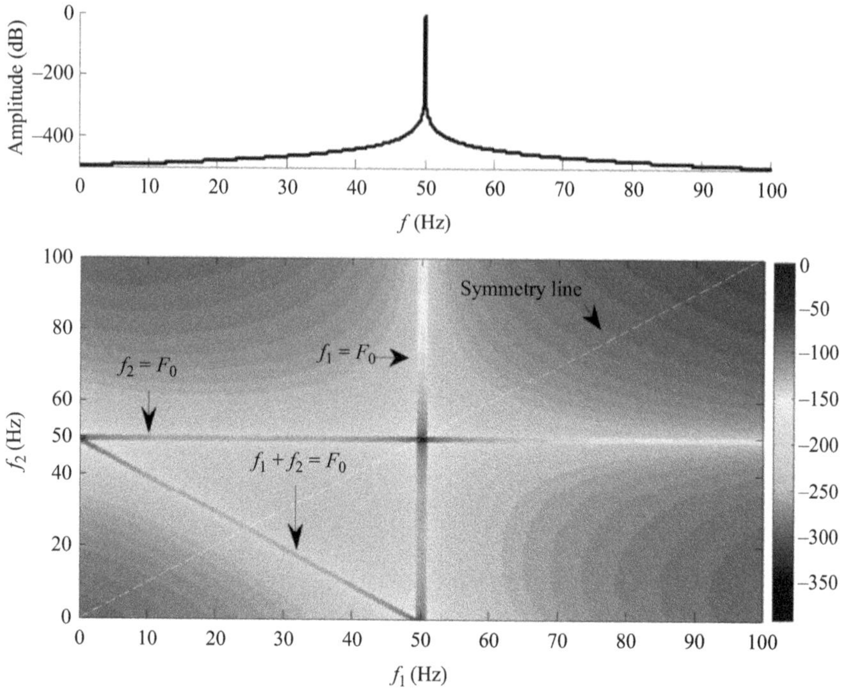


*Figure 4.5 PS and bispectrum of a cosine wave at frequency $F_0$*

The FT of the harmonic signal $x_0$ is

$$X_0(f) = \frac{1}{2}(\delta(f - F_0)e^{j\theta_0} + \delta(f + F_0)e^{-j\theta_0}) \tag{4.13}$$

where $\delta(\cdot)$ represents the Kronecker delta function.

By ignoring contributions at negative frequencies which fall outside the useful region of the PS and bispectrum, (4.13) can be written as

$$X_0(f) = \frac{1}{2}\delta(f - F_0)e^{j\theta_0} \tag{4.14}$$

Thus each factor, $X_0(f_1)$, $X_0(f_2)$, $X_0{}^*(f_1 + f_2)$, produces one line in $(f_1, f_2)$ plane. For instance, $X_0(f_1)$ will be represented by one line at frequency $f_1 = F_0$.

If $X_0(f)$ is substituted from (4.14) into (4.10), the associated bispectrum is equal to

$$\hat{B}_{x0}(f_1, f_2) \approx \frac{1}{8}\delta(f_1 - F_0)\delta(f_2 - F_0)\delta(f_1 + f_2 - F_0)e^{j\theta_0} \tag{4.15}$$

Note that in (4.15), with an expectation over the uniform random distribution of phase, the right-hand side will be zero regardless of whether the three frequencies are related or not.

Because the mesh-type plot, which shows the magnitude of the bispectrum as a 3D surface, is difficult to interpret, a simple contour map is used, which allows interpreting the fine detail with more accuracy as a 2D surface.

This bispectrum contains the triple product: there will only be a nonzero point in bispectrum when all three terms in the above product are nonzero. Plotting the three terms in the $(f_1, f_2)$ plane leads to the three lines, $f_1 = F_0, f_2 = F_0$, and $f_1 + f_2 = F_0$, as shown in Figure 4.5. There is no point of intersection of all three lines and hence the bispectrum of a cosine wave is zero. However, spectral leakage due to finite data sets seems to be the reason of a nonzero bispectrum.

### 4.3.2 *Case 2: sum of two harmonic waves at independent frequencies $F_0$,$F_1$; and with $F_1 = 2F_0$*

Next consider a signal $x_1(t)$ consisting of two cosine waves at independent frequencies, $F_0 = 50$ Hz and $F_1 = 70$ Hz. The FT of this signal is

$$X_1(f) = \frac{1}{2}[\delta(f - F_0)e^{j\theta_0} + \delta(f - F_1)e^{j\theta_1}] \tag{4.16}$$

This is shown in Figure 4.6. The deterministic bispectrum is now equal to

$$\begin{aligned}\hat{B}_{x_1}(f_1,f_2) \approx \frac{1}{8}&(\delta(f_1 - F_0)e^{j\theta_0} + \delta(f_1 - F_1)e^{j\theta_1}) \\ &\times (\delta(f_2 - F_0)e^{j\theta_0} + \delta(f_2 - F_1)e^{j\theta_1}) \\ &\times (\delta(f_1 + f_2 - F_0)e^{-j\theta_0} + \delta(f_1 + f_2 - F_1)e^{-j\theta_1})\end{aligned} \tag{4.17}$$

This can be shown to consist of eight terms, each of which is a triple product. If these are plotted in the $(f_l,f_2)$ plane, they appear as the six possible lines $f_1 = F_0$, $f_1 = F_1$, $f_2 = F_0$, $f_2 = F_1$, $f_l + f_2 = F_0$, and $f_l + f_2 = F_1$ as shown in Figure 4.6. Because there is no interaction between $F_0$ and $F_1$, the bispectrum is identically zero.

In another hand, in Figure 4.7, there will be an intersection of the three terms if $F_1 = 2F_0$. The intersection will then occur at $(F_0, F_0)$ as shown by the dashed circle in Figure 4.7, where it can be seen that there is a peak at (50, 50) Hz.

### 4.3.3 *Case 3: sum of three harmonic waves at coupled frequencies, $F_2 = F_0 + F_1$*

To illustrate the usefulness of the bispectrum estimation with statistical analysis, a simple simulation example is provided. The simulation consists of two test signals, both involving three frequencies, as shown next:

$$x_2(t) = \cos(2\pi F_0 t - \theta_0) + \cos(2\pi F_1 t - \theta_1) + \cos(2\pi F_2 t - \theta_2) \tag{4.18}$$

where $F_0$ and $F_1$ are set to 50 and 70 Hz, respectively. For both test signals, the third frequency $F_2$ is set to $F_0 + F_1 = 120$ Hz to achieve the frequency coupling phases associated with the first two frequencies ($\theta_0$ and $\theta_1$) that are randomly generated between $-\pi$ and $\pi$ with a uniform distribution.

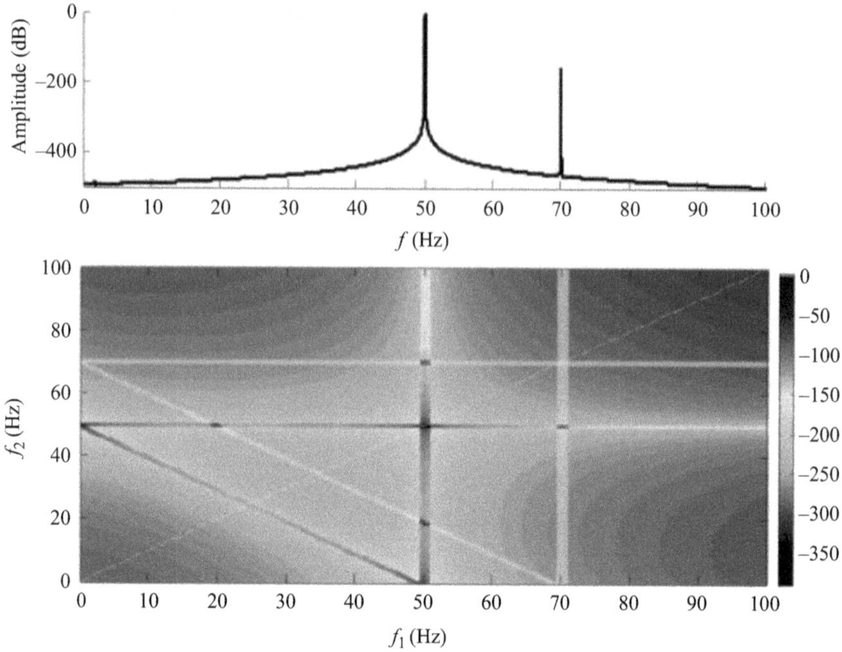


*Figure 4.6 PS and bispectrum of the sum of two cosine waves at independent frequencies $F_0$ and $F_1$*

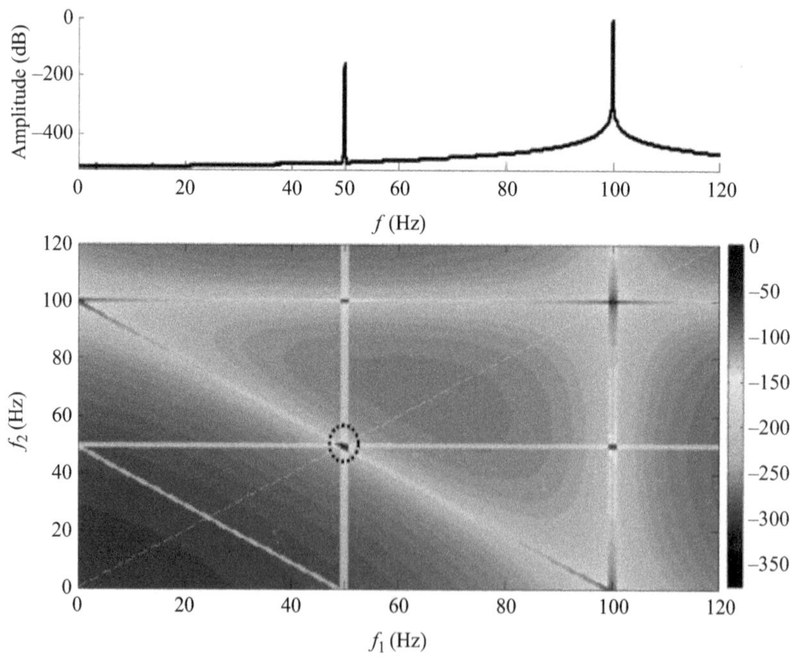


*Figure 4.7 PS and bispectrum of the sum of two cosine waves ($F_1 = 2F_0$)*

This could represent the output from a nonlinear system excited by input signals at frequencies $F_0$ and $F_1$ (i.e. the signal component at the frequency $F_2$ is generated as a result of the nonlinear system response, as discussed later in the section).

The FT of this signal is given by

$$X_2(f) = \frac{1}{2}\sum_{i=0}^{2} \delta(f - F_i)e^{j\theta_i} \tag{4.19}$$

Two sharp peaks in Figure 4.8 are located at frequency coordinates (70, 50) Hz and (50, 70) Hz. One peak is a symmetric image of the other peak due to the frequency domain symmetry property of the bispectrum [1,2]. These two peaks are generated from QPC relationship between 50 and 70 Hz components.

The signal contains three frequency components 50, 70, and 120 Hz. The 120 Hz component comes from the coupling between 50 and 70 Hz. Therefore, the phase at 120 Hz is equal to the sum of the phases at 50 and 70 Hz (i.e. there is QPC between the frequency triplet 50, 70, and 120 Hz). However, in Figure 4.9, the signal also has three frequency components at 50, 70, and 90 Hz; however, the 90 Hz component is not from the coupling. Hence, there is no QPC. This plot gives no indication of the coupling between these frequencies that is evident in the bispectral plots.

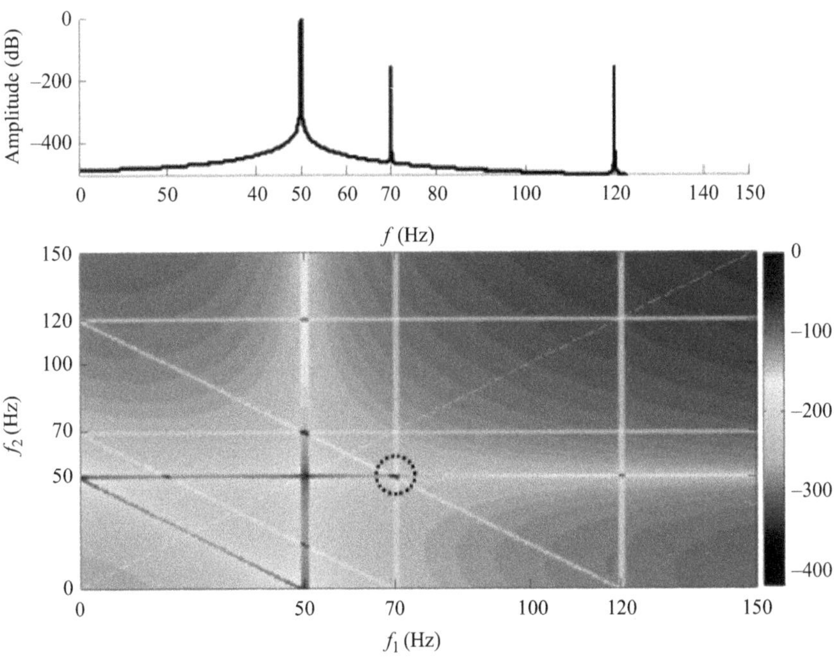


*Figure 4.8 PS and bispectrum of three cosine waves with QPC ($F_2 = F_1 + F_0$)*

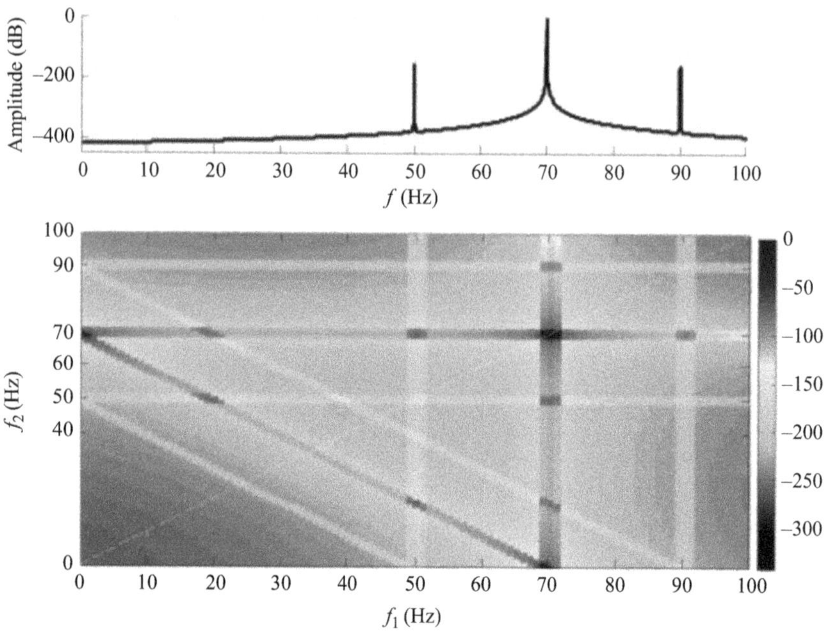


*Figure 4.9 PS and bispectrum of three cosine waves without QPC*

A model is given in Figure 4.9 has no QPC, whereas the model is given in Figure 4.8 has a QPC. It is trivial to show that $E[x_{2\text{QPC}}] = E[x_{2\text{noQPC}}] = 0$. The autocorrelation function is defined by

$$\begin{aligned} r_{x2\text{QPC}}(\tau) &= E(x_{2\text{QPC}}(t)x_{2\text{QPC}}(t+\tau)) \\ &= r_{x2\text{noQPC}}(\tau) = \frac{1}{2}[\cos(F_0\tau) + \cos(F_1\tau) + \cos(F_2\tau)] \end{aligned} \tag{4.20}$$

where $\tau$ is a discrete-time delay between successive observations.

It is the way; the power spectra of the QPC and no-QPC signals are similar. Therefore, the second-order statistics (PS) are incapable of detecting the QPC.

Note that the third-order moment (or third-order cumulant) is defined by

$$m(\tau_1, \tau_2) = E(x_2(t)x_2(t+\tau_1)x_2(t+\tau_2)) \tag{4.21}$$

where $\tau_1$ and $\tau_2$ are discrete-time delays.

For the model with QPC, we show that:

$$m(\tau_1, \tau_2) = 0 \tag{4.22}$$

While for a model without QPC, we have

$$m(\tau_1, \tau_2) = \frac{1}{4}[\cos(F_1\tau_1 + F_0\tau_2) + \cos(F_2\tau_1 - F_0\tau_2) + \cos(F_0\tau_1 + F_1\tau_2) + \cos(F_2\tau_1 - F_1\tau_2) + \cos(F_0\tau_1 - F_2\tau_2) + \cos(F_1\tau_1 - F_2\tau_2) \quad (4.23)$$

Because the bispectrum $B(f_1, f_2)$ is defined as 2D FT of $m(\tau_1, \tau_2)$, the no-QPC signals $B(f_1, f_2) = 0$. In contrast to $B(f_1, f_2)$ in the QPC case, we have a peak at bifrequency $(f_1, f_2)$.

Clearly, the bispectrum can distinguish the QPC from the no-QPC signals. Thus, in the case where the frequencies and their phases $F_0$, $F_1$, and $F_2$ are random and statistically independent, the bispectrum will be zero after the expectation operation is carried out. However, if $F_2$ is phase coupled to $F_0$ and $F_1$, the bispectrum will not be zero.

## 4.3.4 The use of bispectrum to detect and characterize nonlinearity

### 4.3.4.1 QPC detection

A signal from a nonlinear system often provides information about the type of nonlinearity [1–4]. A characteristic of all nonlinear phenomena is the generation of "*new*" frequencies corresponding to harmonics, and sum and difference combinations of the "original" nonlinear interacting frequencies. Both the new and original frequencies must satisfy a particular frequency selection rule that depends on the order of the nonlinearity. A diagram of a general quadratic nonlinear system is given in Figure 4.10.

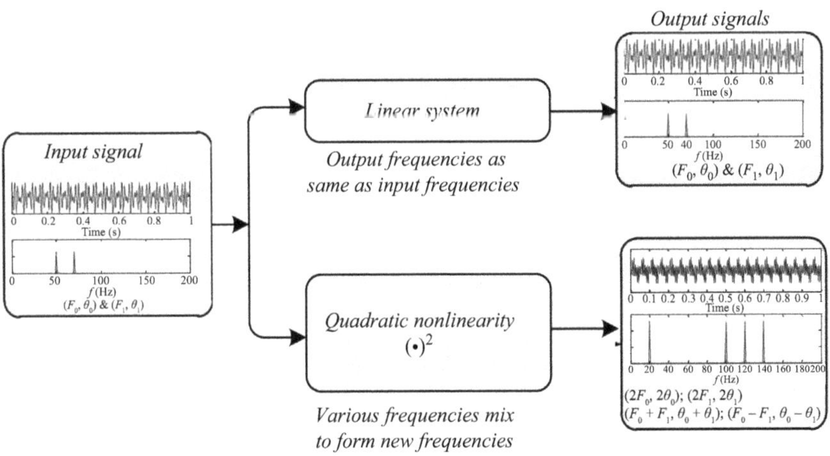


*Figure 4.10 The diagram indicating the quadratically nonlinear interaction between the frequency modes $F_0$ and $F_1$*

The example given in Figure 4.9 doesn't give any QPC information. Now, this signal is passed throughout a simple quadratic nonlinear system as given in Figure 4.10.

Assume that $y(t)$ is the output of this nonlinear system

$$y(t) = x_1(t) + \varepsilon x_1^2(t) \tag{4.24}$$

where the parameter $\varepsilon$ represents the coefficient of nonlinearity.

The quadratic interaction involves the multiplication of two spectral components, and the multiplication can be described using trigonometric identities (ignoring terms that are constant in time). One can now be expanding $y(t)$ and be rewritten in terms of the harmonics using basic trigonometric identities. These are demonstrated in Table 4.2. These relationships are called phase coupling and are considered to be a "true" signature of quadratic nonlinear systems [1–24].

In the output of this system, the signal will include the harmonic components with frequencies and phases that are correlated (Table 4.2). Such a phenomenon, which produces a formation of these phase relations, is called QPC, which the bispectrum shows. The PS is the frequency domain decomposition of signal power. When this concept is extended to higher orders, the bispectrum provides information regarding such signal features as phase coherence, which is absent in the second-order PS. Indeed, peaks in the bispectrum occur when the interaction between two harmonic components causes a contribution to the power at their sum and/or difference frequencies.

The form of the bispectrum in (4.10) suggests that, if the energy at the sum or difference of frequency is generated by a nonlinear process, phase coherence among the bifrequency components ($F_0$, $F_1$, $F_0 + F_1$) exists, and therefore the statistical average will lead to a nonzero value of the bispectrum.

Figure 4.11 illustrates the use of the bispectrum for QPC detection of the process (4.24) with $F_0 = 50$ Hz, $F_1 = 70$ Hz, $\varepsilon = 0.1$.

Note the generation of second harmonics at 100 and 140 Hz, and phase-coupled intermodulation components at $F_0 + F_1 = 120$ Hz and $F_1 - F_0 = 20$ Hz in the PS.

The corresponding bispectrum estimate resulting from (4.10) shows a peak at the bifrequency (70, 50) Hz.

In practical applications, spurious peaks may appear in the bispectrum at locations without significant QPC due to various aspects such as finite data length.

In the following subsection, we demonstrate the robustness of the bispectrum against the Gaussian noise reduction.

*Table 4.2 A simple nonlinearity-introduced new harmonics with higher-order correlations*

| Signal | Frequency | Phase |
|---|---|---|
| $x_1(t)$ | $F_0, F_1$ | $\theta_0, \theta_1$ |
| $y(t)$ | $F_0, F_1, 2F_0, 2F_1, F_0 + F_1, F_1 - F_0$ | $\theta_0, \theta_1, 2\theta_0, 2\theta_1, \theta_0 + \theta_1, \theta_1 - \theta_0$ |

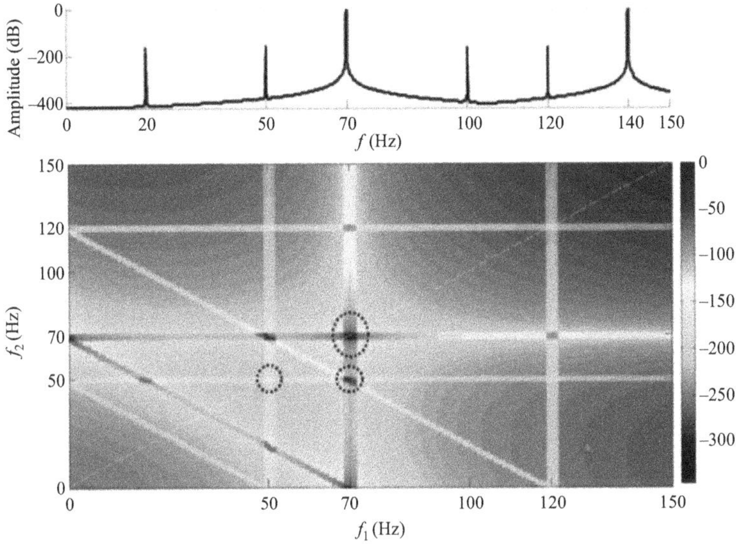


*Figure 4.11 PS and bispectrum of the process given in (4.24)*

#### 4.3.4.2 Robustness against the presence of additive Gaussian noise

In this subsection, we study the effect of noise on the bispectrum. To determine the statistical behavior of the bispectrum under the effect of white Gaussian noise, we start with recalling that white Gaussian noise affects all the frequency components with the same amount of noise power which is normally distributed ($N(f) \sim \aleph(0, \sigma_n^2/L)$), where $L$ is the number discrete FT points. Thus, the bispectrum becomes as follows:

$$\hat{B}(f_1, f_2) = E\{(X(f_1) + N(f_1))(X(f_2) + N(f_2))(X(f_3 = f_1 + f_2) + N(f_3)^*)\} \tag{4.25}$$

Since we are dealing with harmonics, our signal of interest mainly consists of sinusoidal components. Thus, the signal spectral component at certain frequencies can be represented as magnitude and phase; for example, $X(f_1) = A_1/2e^{j\theta}$. Thus, bispectrum in (4.25) can be rewritten as follows:

$$\hat{B}(f_1, f_2) = E\left\{\left(\frac{A_1}{2}e^{j\theta_1} + N(f_1)\right)\left(\frac{A_2}{2}e^{j\theta_2} + N(f_2)\right)\left(\frac{A_3}{2}e^{j(\theta_1+\theta_2)} + N(f_3)^*\right)\right\} \tag{4.26}$$

Note that magnitude $A$ and phase $\theta$ of each sinusoidal frequency components are deterministic. Therefore, the $E\{\cdot\}$ works only on the noise components which we will refer to as $N_1$, $N_2$, and $N_3$ for simplicity. Hence, (4.26) can be simplified as follows:

$$\begin{aligned}\hat{B}(f_1, f_2) &= \left\{\left(\frac{A_1A_2A_3}{8}\right) + E\{N_1N_2N_3^*\}\right\} \\ &+ \frac{A_1}{2}e^{j\theta_1}\frac{A_2}{2}e^{j\theta_2}E\{N_3^*\} + \frac{A_1A_3}{4}e^{j\theta_2}E\{N_2\} + \frac{A_2A_3}{4}e^{j\theta_1}E\{N_1\} \\ &+ \underset{\substack{\downarrow \\ E_1}}{\frac{A_1}{2}e^{j\theta_1}E\{N_2N_3^*\}} + \underset{\substack{\downarrow \\ E_2}}{\frac{A_2}{2}e^{j\theta_2}E\{N_1N_3^*\}} + \underset{\substack{\downarrow \\ E3}}{\frac{A_3}{2}e^{-j(\theta_1+\theta_2)}E\{N_1N_2\}} \\ &= \frac{A_1A_2A_3}{8} + 0 + 0 + 0 + 0 + E_1 + E_2 + E_3 \end{aligned} \tag{4.27}$$

The noise is assumed to be zero-mean white Gaussian noise. In (4.63), first-order moments and third-order moments are then equal to zero: $E\{N_1\} = E\{N_2\} = E\{N_3\} = E\{N_1N_2N_3^*\} = 0$.

To quantitatively compare the performances between the PS and bispectrum, let us consider the signals given in (4.28) in the presence of additive white Gaussian noise (AWGN):

$$x_n(t) = x_2(t) + n(t) \tag{4.28}$$

where $n(t)$ is AWGN with $\sigma_n^2 = 0.15$, equivalent to 10 dB SNR, in decibels, defined by

$$SNR \triangleq 20\log_{10}\left(\frac{\sigma_x^2}{\sigma_n^2}\right) \tag{4.29}$$

where $\sigma_x^2$ the variance of the signal $x_v(t)$.

The effect of additive noise on the magnitude spectrum of a signal is to increase the mean and the variance of the PS as illustrated in Figures 4.12 and 4.13. The increase in the variance of the signal spectrum results from the random fluctuations of the noise, and cannot be canceled out.

Comparing power spectra results of Figures 4.12 and 4.13, it can be seen that PS analysis is influenced by noise, and the signal frequency components are buried in the noise. Note that bispectrum is less sensitive to noise when compared to the PS.

Harmonic random signals (Fourier series expansion of periodic signals) such as those expected from rotating machinery are not strictly stationary and are cyclostationary. Their autocorrelation and higher-order correlations will be periodic. At such, another direction of the research can be extended to machine condition monitoring. Because, the stator current signals of rotating machinery present a strongly nonlinear and non-Gaussian behavior, and bispectrum is well suitable to analyze this kind of signals. This is the next phase of our research work.

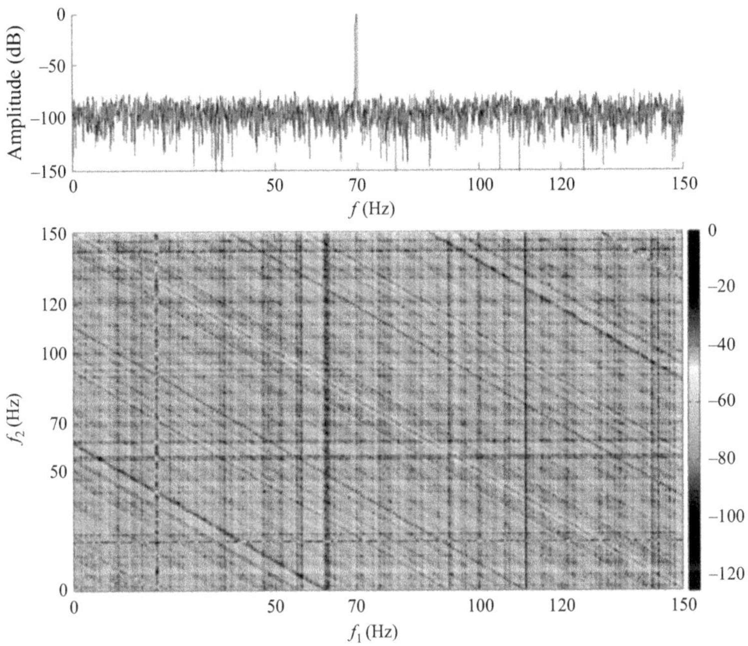


*Figure 4.12 PS and bispectrum of the simulated signal given in (4.28), with AWGN (SNR = 10 dB)*

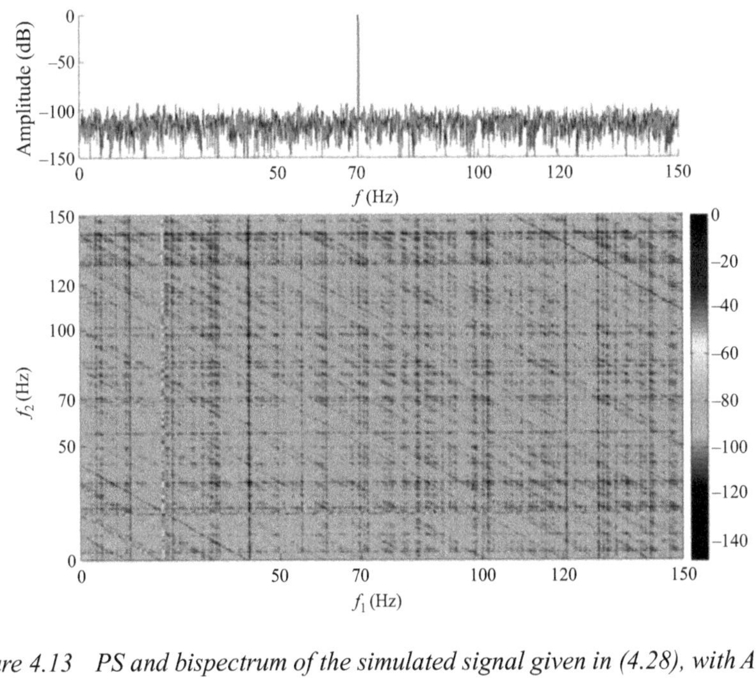


*Figure 4.13 PS and bispectrum of the simulated signal given in (4.28), with AWGN (SNR = 20 dB)*

## 4.4 Practical applications of bispectrum-based fault diagnosis

Since a damaged or abnormal state machine often generates highly nonlinear signals, it is advantageous to use a tool that can effectively detect and analyze nonlinear signatures [17]. The bispectrum has been proposed for such nonlinear analysis since it is a measure of the phase coupling between interacting frequency components.

### *4.4.1 BRB fault detection*

Rotor failures account for 5%–10% of the total induction motor (IM) failures [19–21,25,26,28–43]. The detection of broken rotor bar (BRB) faults can be done by the inspection of the frequency components ($f_{\mathrm{BRBs}}$) in the current spectrum as a fault indicator,

$$f_{\mathrm{BRBs}} = \left[\left(\frac{k}{p}\right)(1-s) \pm s\right] f_s \tag{4.30}$$

where $f_s$ is the supply frequency, $p$ is the number of pole pairs, $k$ is an integer, and $s$ is the rotor slip.

By considering the speed ripple effects, it has been reported that other frequency components may be observed in the stator current spectrum and may be determined by the following additional equation [25,26,28–43]:

$$f_{\mathrm{BRBs}} = (1 \pm 2ks) f_s \tag{4.31}$$

While the lower sideband is fault-related, the upper sideband is due to consequent speed oscillations. It has been shown that the sum of magnitudes of these two sideband frequency components is a good diagnostic index given by

$$I_{\mathrm{dB}} = \left[20\log_{10}\left(\frac{I_l}{I}\right) + 20\log_{10}\left(\frac{I_r}{I}\right)\right] \Big/ 2 \tag{4.32}$$

where $I_l$, $I_r$, and $I$ are the amplitude of the lower, upper sideband frequency components ($k = 1$) and the fundamental frequency of the stator current, respectively.

Figure 4.14 presents the theoretical spectral content given by (4.31), of the current signal, as well as the variation of the amplitude of right (higher) and left (lower) sideband frequency components (HSB and LSB, respectively).

#### 4.4.1.1 Simulation and experimental tests for BRB fault

In this section, the bispectrum stator current signal is derived theoretically. In particular, the cases regarding current signals with BRB fault is considered. This theoretical analysis will be also confirmed by some experimental results.

The energy distribution in the bispectrum domain is validated experimentally by the analysis of real stator current IM with the following characteristics 18.5kW-220V/380V-50Hz-2 poles. The experimental test bed is presented in Figure 4.15. One IM is undamaged (0BRB) and it is defined as the reference condition whereas another one is a synthetic rotor fault with one broken rotor bar (1BRB) or three broken rotor bars (3BRB) at different load levels. The synthetic rotor faults were obtained by

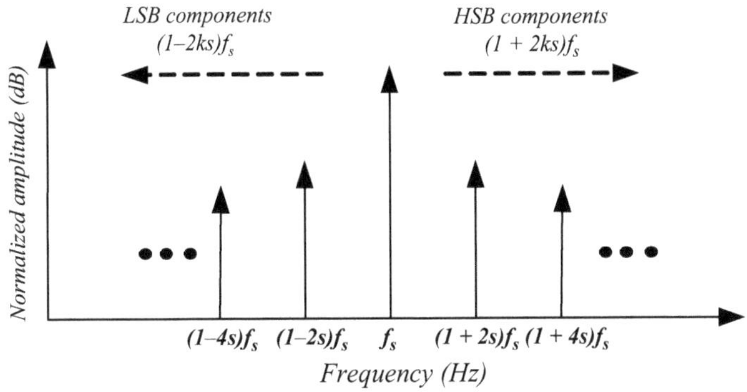


*Figure 4.14 Frequency patterns included in the PS of an IM with broken bars fault*

drilling a small hole of 3-mm diameter in all the rotor bar depth, without harming the rotor shaft.

The experiments used one current sensor with a 20 kHz frequency bandwidth. The analog signals are passed through a low-pass anti-aliasing filter with a cut-off frequency of 2 kHz. The current signals are collected at a sampling frequency of 16,384 Hz for all the experiments using a 12-bit A/D converter. The measurements were carried out for different load conditions starting from 0% up to 100% of the rated torque, and at different numbers of broken bars, a healthy rotor (0BRB), with 1BRB and with 3BRB.

To minimize the fast FT (FFT) leakage effect, the Hanning window has been applied. The signal processing is performed by using the MATLAB® environment to generate power spectra and bispectra and LabView™ software for the data acquisition.

### 4.4.1.2 Model of the BRB stator current

A simple model which characterizes the IM stator current, with electrical rotor asymmetries, includes the so-called sideband frequencies and it can be expressed by the following [19,20,25–28,33–47]:

$$i_a(t) = i_f \cos(\omega t - \varphi) + \sum_k i_{l,k} \cos((\omega - \omega_{f,k})t - \varphi_{l,k}) + \sum_k i_{r,k} \cos((\omega + \omega_{f,k})t - \varphi_{r,k}) \tag{4.33}$$

with
$f_s$ : as the fundamental frequency of the power grid
$i_f$ : is the fundamental value of the stator current amplitude (index $f$),
$\omega$ : is its angular frequency
$\varphi$ : phase angle at the supply frequency
$\omega_{f,k}$: is its main phase shift angle

$i_{l,k}, \varphi_{l,k}, i_{r,k}$, and $\varphi_{r,k}$ $(k = 1, 2, 3, \ldots)$ are the magnitude and the phase of the left (index $l$) and right (index $r$) sideband components, respectively.

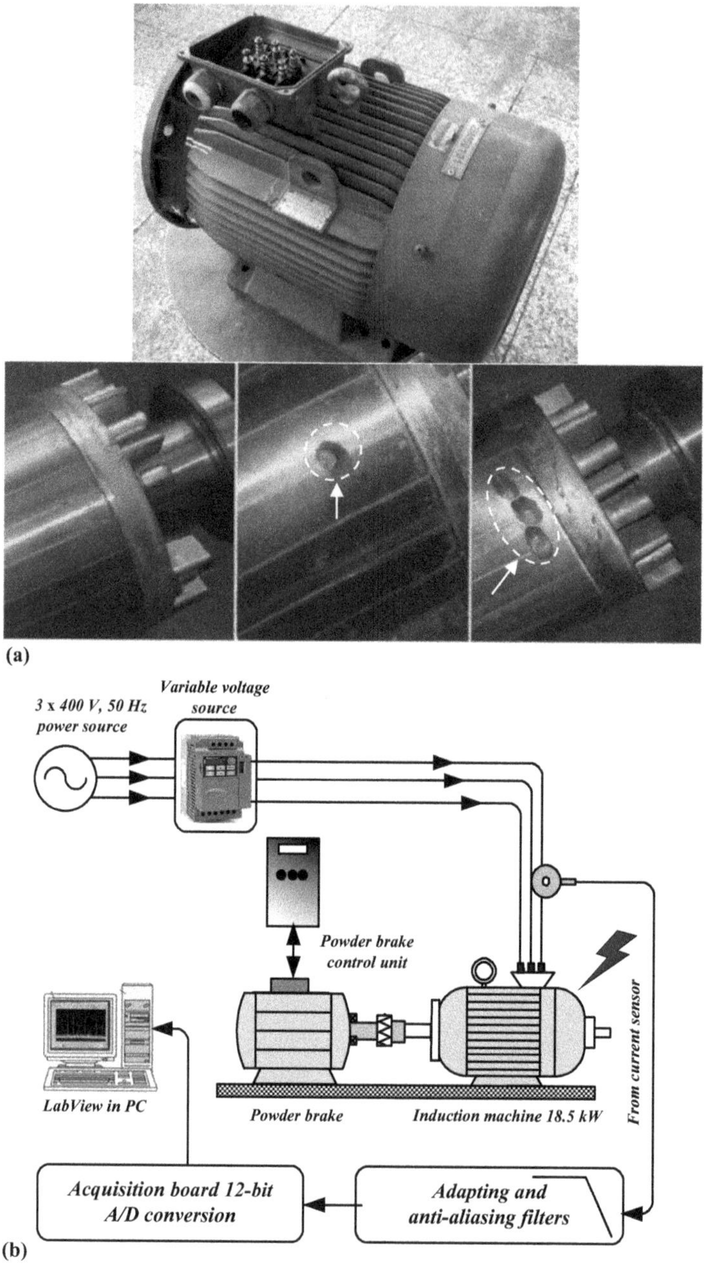


*Figure 4.15 Configuration of the experimental setup for BRB detection: (a) a real healthy rotor and rotors with one and three broken bars, (b) block diagram of the laboratory setup*

As $\omega_{f,k} = 2ks\omega$ and $\omega = 2\pi f_s$ the expression (4.33) can be rewritten as

$$i_a(t) = i_f \cos(2\pi f_s t - \varphi) + \sum_k i_{l,k} \cos(2\pi f_{l,k} t - \varphi_{l,k}) + \sum_k i_{r,k} \cos(2\pi f_{r,k} t - \varphi_{r,k}) \tag{4.34}$$

where $f_{l,k} = (1 - 2ks)f_s$ and $f_{r,k} = (1 + 2ks)f_s$, which represents the frequency sidebands of bar breakages.

This simplified expression of the stator current signal has been used extensively for IM fault condition monitoring. By checking the magnitude of the sideband frequency components through the spectrum computation, various faults such as rotor bar breakage can be detected with a high level of accuracy [25–28,33–47].

From (4.34), the FT can be computed as follows:

$$I_a(f) = \frac{i_f}{2}\delta(f \pm f_s)e^{\mp j\varphi} + \frac{1}{2}\sum_k i_{l,k}\delta(f \pm f_{l,k})e^{\mp j\varphi_{l,k}} + \frac{1}{2}\sum_k i_{r,k}\delta(f \pm f_{r,k})e^{\mp j\varphi_{r,k}} \tag{4.35}$$

where $\delta(\cdot)$ represents the Dirac delta function.

By ignoring contributions at negative frequencies which fall outside the useful region of the bispectrum given in Figure 4.2, (4.35) can be written as

$$I_a(f) = \frac{i_f}{2}\delta(f - f_s)e^{j\varphi} + \frac{1}{2}\sum_k i_{l,k}\delta(f - f_{l,k})e^{j\varphi_{l,k}} + \frac{1}{2}\sum_k i_{r,k}\delta(f - f_{r,k})e^{j\varphi_{r,k}} \tag{4.36}$$

If the expression (4.36) is substituted in (4.10), the stator current bispectrum becomes (Appendix 1):

$$\begin{aligned}\hat{B}(f_1, f_2) &\approx I_a(f_1)\, I_a(f_2)\, I_a^*(f_3 = f_1 + f_2) \\ &\approx \frac{1}{8}\left(\begin{array}{l} i_f\delta(f_1 - f_s)e^{j\varphi} + \sum_{k_1} i_{l,k_1}\delta(f_1 - f_{l,k_1})e^{j\varphi_{l,k_1}} \\ \qquad + \sum_{k_1} i_{r,k_1}\delta(f_1 - f_{r,k_1})e^{j\varphi_{r,k_1}} \end{array}\right) \\ &\quad \times \left(\begin{array}{l} i_f\delta(f_2 - f_s)e^{j\varphi} + \sum_{k_2} i_{l,k_2}\delta(f_2 - f_{l,k_2})e^{j\varphi_{l,k_2}} \\ \qquad + \sum_{k_2} i_{r,k_2}\delta(f_2 - f_{r,k_2})e^{j\varphi_{r,k_2}} \end{array}\right) \\ &\quad \times \left(\begin{array}{l} i_f\delta(f_3 - f_s)e^{-j\varphi} + \sum_{k_3} i_{l,k_3}\delta(f_3 - f_{l,k_3})e^{-j\varphi_{l,k_3}} \\ \qquad + \sum_{k_3} i_{r,k_3}\delta(f_3 - f_{r,k_3})e^{-j\varphi_{r,k_3}} \end{array}\right)\end{aligned} \tag{4.37}$$

#### 4.4.1.3 Numerical simulation

To evaluate the bispectrum performance, numerical simulations have been performed. Let consider that the stator current given by (4.38) with AWGN $n(t)$. This signal is similar to the measured current for one broken bar at rated load, and it is generated at a sampling rate of 512 Hz. Assume the simplest case of a stator current signal, where only the first sideband frequencies associated with the BRBs ($f_{l,1} = 48.87$ Hz and $f_{r,1} = 51.13$ Hz) are considered.

$$i_a(t) = i_f \cos(2\pi f_s t - \varphi) + i_{l,1} \cos(2\pi f_{l,1} t - \varphi_{l,1}) + i_{r,1} \cos(2\pi f_{r,1} t - \varphi_{r,1}) + n(t) \tag{4.38}$$

with

$i_f$ : RMS value of the supply phase current

$I_{l,1}, \varphi_{l,1}$ : RMS value of the lower current component at $(1 - 2s)f_s$, and its phase angle, respectively

$I_{r,1}, \varphi_{r,1}$: RMS value of the upper current component at $(1 + 2s)f_s$, and its phase angle, respectively

$\varphi$ : phase angle at the supply frequency

By plotting $B(f_1, f_2)$ in the bispectrum domain as shown in Figures 4.16–4.18, it can be seen that each of three factors consists of three parallel delta function lines ($f_i = f_j$, where $i = 1, 2, 3$ and $j = l, s, r$). Therefore, the nonredundant region of the

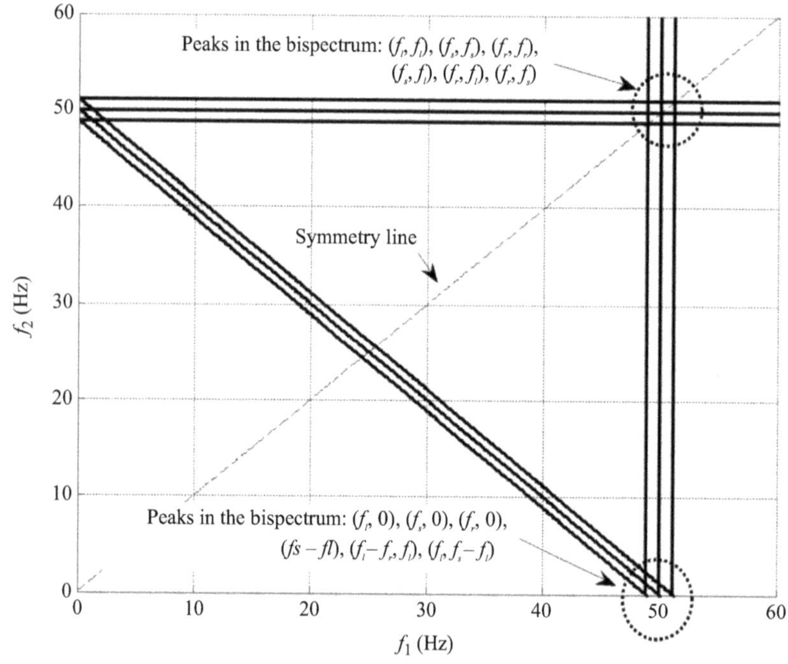


*Figure 4.16 Graphical view of stator current bispectrum in (4.38)*

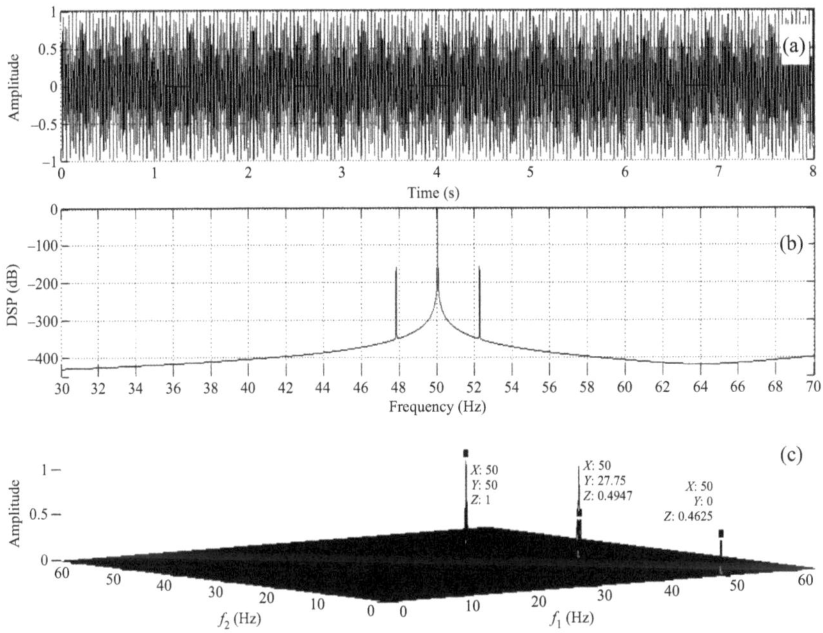


*Figure 4.17 A simulated signal given by (4.38) with noise effect (a), and its normalized PS (b) and bispectrum (c)*

bispectrum comutation is nonzero are the 12 points (peaks in the bispectrum) shown in Figure 4.16. Note that in the remaining parts of this chapter, all the bispectrum figures, the line crossing points (the sharpness peaks) represent possible frequencies of failures.

Insight from (4.37) shows that every stator current signal generates a bidimensional bispectral pattern characterized by peaks positions shown in Figure 4.16. Additionally, Figure 4.18(b) shows the bispectrum of a real stator current with one BRBs at rated load sampled at $f_e = 16384$ Hz, in the frequency bandwidth [0, 60 Hz]. The power spectra and bispectra are performed by tools of MATLAB while data acquisition uses LabView™ software.

Figures 4.19 and 4.20 present, respectively, the stator current bispectrum results for a healthy rotor and 1BRB fault, under different load conditions (no-load and full load), in 3D and 2D representations. All the bifrequency component magnitudes have been normalized and expressed in dB scale.

Two symmetrical pulses are detected at ($f_s$, 28 Hz) and (28 Hz, $f_s$), that is at the intersection between the line $f_3 = f_1 + f_2 = f_s + 28$ (here called decimation line) and the lines $f_1 = f_s$ and $f_2 = f_s$. By changing the sampling frequency, the decimation line translates and the peak at ($f_s$, 28 Hz) disappears and can be relocated at other frequencies without disturbing the region of interest. It can be concluded that these peaks are only an artifact of the numerical technique. More details about these results are given in [28,29].

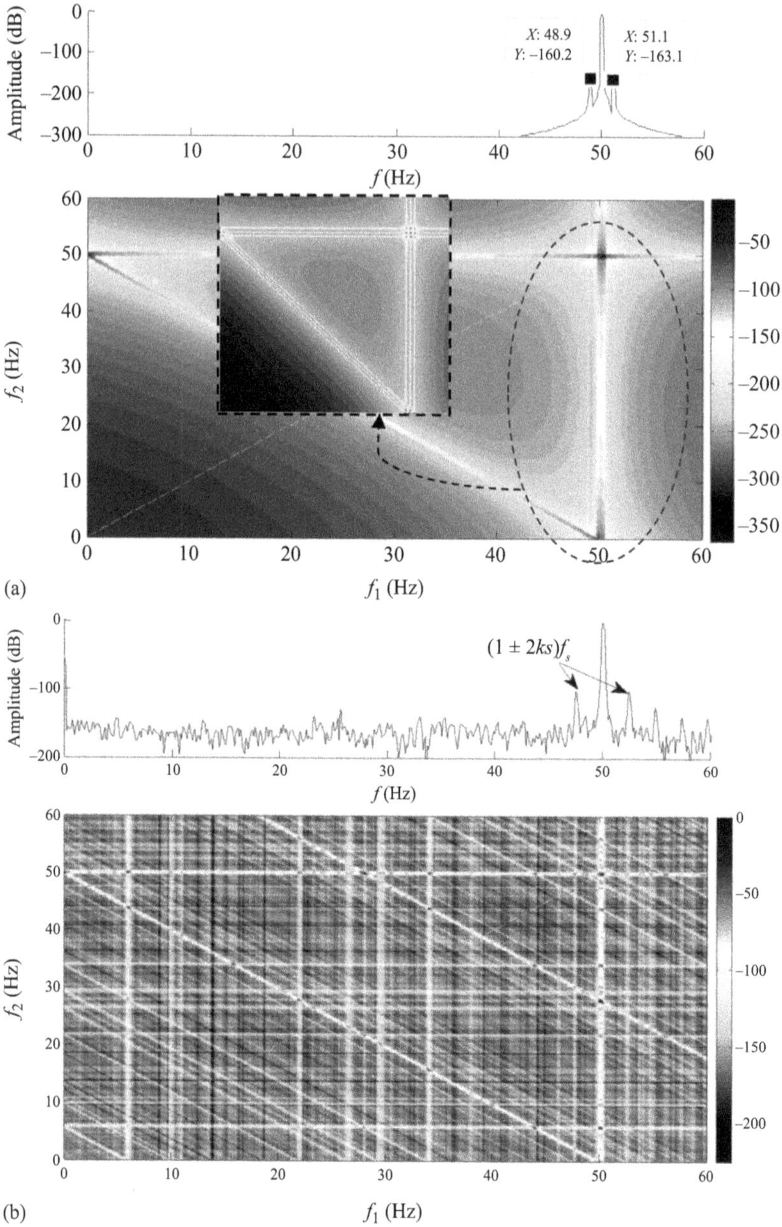


*Figure 4.18 Normalized stator current PS and bispectrum image plot, generated by (a) synthetic current signal given in (4.38) without noise effect and with 1BRB, (b) real stator current at full load (s = 2.4%)*

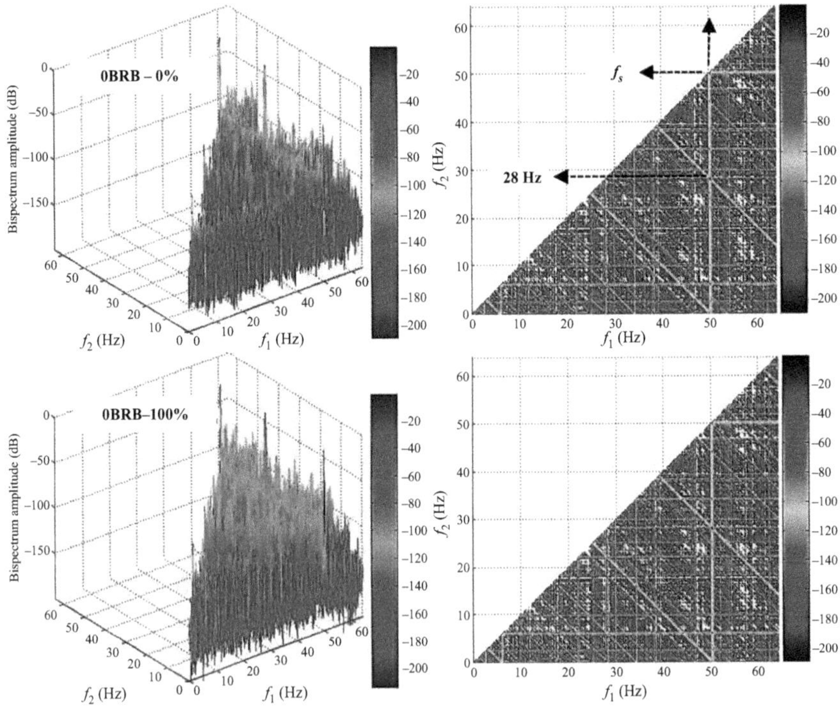


*Figure 4.19 Normalized bispectrum amplitude, stator current bispectrum of a healthy rotor (0BRB) under different load conditions*

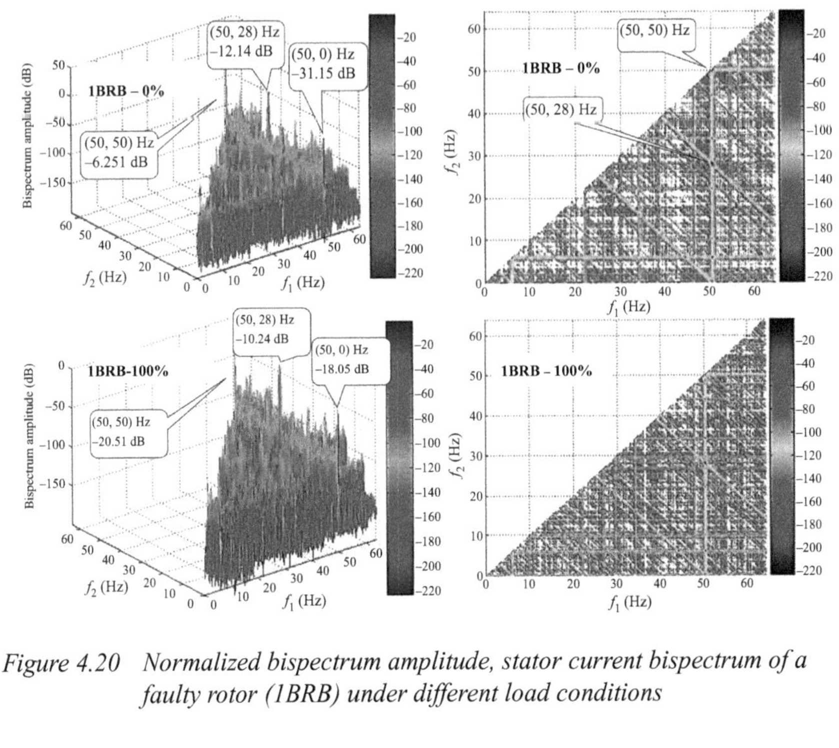


*Figure 4.20 Normalized bispectrum amplitude, stator current bispectrum of a faulty rotor (1BRB) under different load conditions*

Let us denote "$p$" the number of significant harmonics. If this number is beforehand known, the positions and the number "$N$" of peaks in the bispectrum domain is given by; $N = 2\sum_{i=1}^{p} i$ Then the proposed method can be directly computed in a 1D bispectrum diagonal slice (BDS) which condenses the 2D distribution into a 1D curve. It requires less computation and it can be used in a real-time system. The BDS of a process $x(n)$ can be obtained by setting $f_1 = f_2 = f$ and it is defined as follows:

$$\hat{B}(f_1, f_2)|_{f_1=f_2=f} = \hat{D}(f) \approx E\{X^2(f)X^*(2f)\}. \tag{4.39}$$

Figure 4.16 shows only the peaks which do not depend on the studied fault. Nevertheless, other peaks giving information about BRBs fault can be observed in the 1D BDS with $f_1 = f_2 = f$ and given as follows [11,28,29] (Appendix 2):

$$\hat{D}(f) = \underbrace{\frac{i_f^3}{8}\delta(f - f_s)e^{j\varphi}}_{\text{Peak at } f_s} + \sum_k \underbrace{\frac{i_{l,k}^3}{8}\delta(f - f_{l,k})e^{j\varphi_{l,k}}}_{\text{Peak at } f_{l,k}} + \sum_k \underbrace{\frac{i_{r,k}^3}{8}\delta(f - f_{r,k})e^{j\varphi_{r,k}}}_{\text{Peak at } f_{r,k}} \tag{4.40}$$

It has been shown (Figure 4.21) that the BDS method gives similar results compared to the PS computed for the same cases at different shaft load levels. However, the detection is even smoother than in the case of the PS and it is much more visible. This is since the bispectrum can effectively filter out the Gaussian noise.

The different magnitudes of frequency components related to BRBs have been computed in the three previous cases (healthy, 1BRB, 3BRB) by using the formula (4.32) for the average value $I_{dB}$ (Figure 4.22). It can be seen that even when the machine is operating at no load, the healthy rotor can be clearly distinguished from the faulty case (this difference is indicated by dashed circles in Figure 4.22).

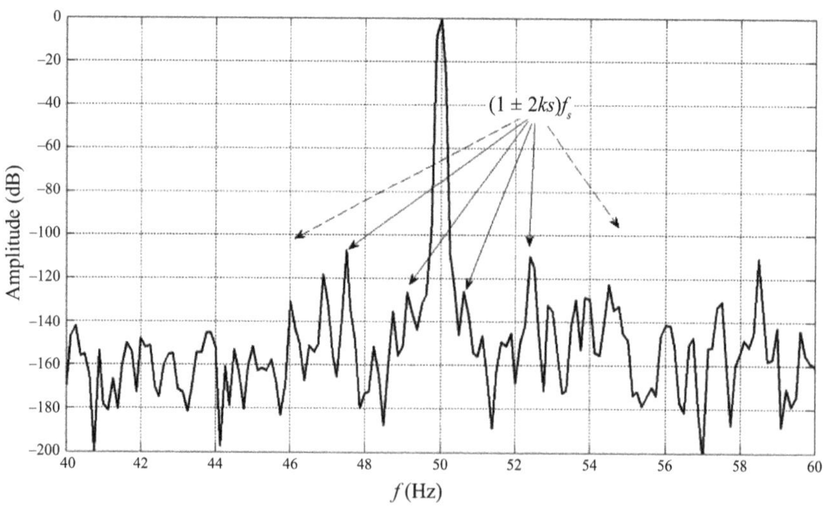


*Figure 4.21 Zoomed stator current BDS with 1BRB at full load condition*

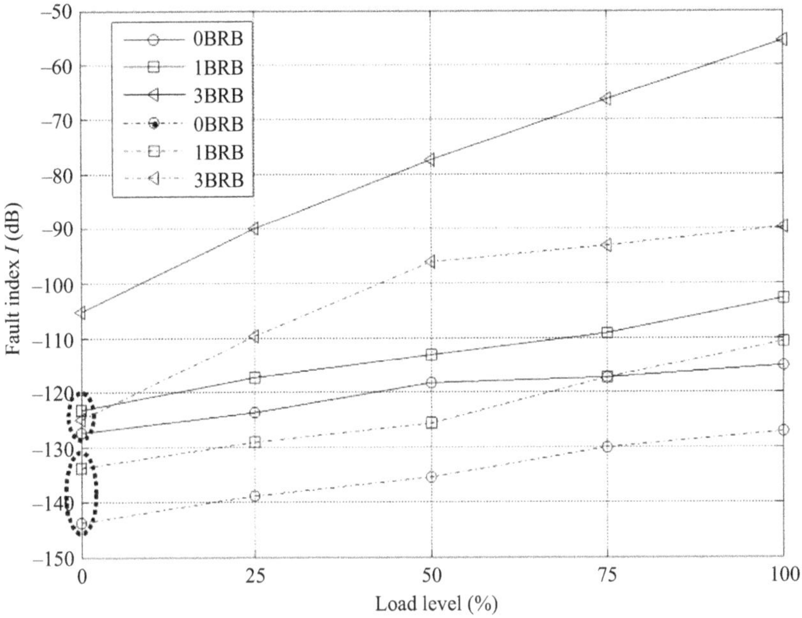


*Figure 4.22 Fault index using the* $(1 \pm 2s)f_s$ *components for different shaft load levels for both PS (solid line) and BDS (dashed line)*

The BDS technique discussed is characterized by a higher sensitivity than the PS, especially when the machine is operating at no load.

The use of nonlinear features motivated by the HOS has been reported to be a promising approach to analyze the nonlinear and non-Gaussian characteristics of the stator current signals. This problem will be further discussed in the next subsection when a novel pattern classification approach for bearings diagnostics which combine the HOS analysis features and support vector machine (SVM) classifier will be proposed.

## *4.4.2 Bearing multi-fault diagnosis based on stator current HOS features and SVMs*

### 4.4.2.1 Bearing defect signatures

Condition monitoring and fault diagnosis of rolling element bearings (REBs) timely and accurately are very important to ensure the reliability of rotating machinery. This subsection presents a novel pattern classification approach for bearings diagnostics, which combine the HOS analysis features and SVM classifier. The use of nonlinear features motivated by the HOS has been reported to be a promising approach to analyze the nonlinear and non-Gaussian characteristics of the stator current signals. To deal with the frequency analysis, a mathematical model of the stator current has been derived and used in the bispectrum formulas. The stator current bispectrum patterns are extracted as feature vectors presenting different faults of the bearings. The extracted bispectrum features are subjected to principal component analysis (PCA) for dimensionality reduction. These *principal components* (PCs) were fed to SVM to distinguish six kinds of fault bearing signals which were measured in the experimental test bench running under different working conditions. To find the optimal parameters for the multi-class SVM model, a grid-search method in combination with tenfold cross-validation has been used. The results indicated that the proposed method can reliably identify different fault patterns of REBs based on the stator current signals.

Table 4.3 provides a summary of the studies on automated identification of IMs faults.

Failure surveys by the Electric Power Research Institute indicate that IMs bearing-related faults are about 40% among the most frequent faults in IMs. As shown in Figure 4.23, the bearings consist mainly of the outer and inner raceways, the balls, and the cage. Bearing defects (BDs) can be classified into two classes: single-point defects and generalized roughness. Single-point defects are localized and classified into the following [48–67]:

- outer raceway defect;
- inner raceway defect;
- ball defect.

Generalized roughness is a type of fault where the condition of a bearing surface has degraded considerably over a large area and become rough, irregular, or deformed.

These faults may enhance the vibration and noise level. Moreover, there are internal operating stresses caused by vibration, eccentricity, and bearing current. Additionally, bearings can also be affected by other external causes such as the following [48–67]:

- contamination and corrosion;
- lack of lubrication causing heating and abrasion;
- defect of bearing's mounting, by improperly forcing the bearing onto the shaft or in the IM's stand.

The single-point defect may be seen by fault frequencies appearing in the machine vibration spectrum record. The frequencies at which these components occur are predictable and depend on the surface on which the bearing contains the fault. Therefore, there are different fault frequency characteristics associated with each component among the four parts of the bearing [48].

These frequencies are the following: $f_O$: outer race fault (ORF) frequency; $f_I$: inner race fault (IRF) frequency; $f_B$: ball fault (BF) frequency; and $f_C$: cage fault frequency; their mathematical equations are as follows:

$$f_O = \frac{f_r}{2} N_b \left( 1 - \frac{D_b \cos \beta}{D_c} \right) \tag{4.41}$$

$$f_I = \frac{f_r}{2} N_b \left( 1 + \frac{D_b \cos \beta}{D_c} \right) \tag{4.42}$$

*Table 4.3 Some previous research on automatic identification of IM faults*

| Literature | Features | Classifier | Number of classes | Accuracy (%) |
|---|---|---|---|---|
| [49] | Histogram features (mean, standard deviation, skewness, kurtosis, energy, entropy) derived from thermal images + generalized discriminant analysis (GDA) | SVM-OAO, SVM-OAA, adaptive neuro-fuzzy inference system (ANFIS), and relevance vector machine (RVM) | **4**: Normal, misalignment, bearing fault, mass unbalance | 97.5, 95, 82.5, and 97.5, respectively |
| [50] | Frequency domain-based vibration energy features | SVM-OAA | **3**: Normal, inner race and outer race BDs | 100 |
| [51] | The start-up transient current and discrete WT with nonlinear feature reduction using kernel PCA and kernel independent component analysis (ICA) | SVM-OAA | **7**: Powder rotor, BRBs, faulty bearing, eccentricity, phase unbalance, normal condition, and mass unbalance | 78.75 using PCA<br>80.95 using ICA<br>76.19 using kernel PCA<br>83.33 using kernel ICA |
| [52] | Time-domain, and frequency domain features derived from vibration and three phases of current signals | ICA and SVM (OAO, OAA) | **7**: BRB, bowed rotor, faulty bearing, rotor unbalance, eccentricity, phase unbalance, and normal condition | 99.97 |
| [60] | Hilbert modulus current space vector (HMCSV) and Hilbert phase current space vector (HPCSV) | SVM-OAA | **4**: Normal condition, electrical fault, air-gap eccentricity fault, outer raceway bearing fault | 97.5 |
| [63] | Time-domain and frequency-domain vibration signals + PCA | Artificial immunization algorithm SVM (AIA-SVM) | **14**: Gear damage, structure resonance, rotor radial touch friction, rotor axial touch friction, shaft crack, bearing damage, etc. | 97 |
| [68] | Features derived from HOSA of vibration | PCA + SVM-OAA | **6**: HB, IRF, ORF, IORF, and BF | 98 |

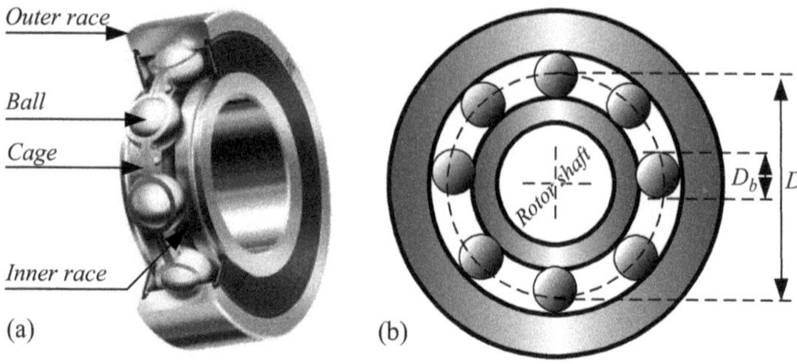


*Figure 4.23 (a) Exploded view and (b) geometry of REB*

$$f_B = \frac{f_r}{2}\frac{D_c}{D_b}\left[1 - \left(\frac{D_b\cos\beta}{D_c}\right)^2\right] \tag{4.43}$$

$$f_C = \frac{f_r}{2}\left(1 - \frac{D_b\cos\beta}{D_c}\right) \tag{4.44}$$

where
$f_r$ : rotor shaft frequency
$N_b$: number of rolling elements
$D_b$: ball diameter
$D_c$: pitch diameter
$\beta$ : ball contact angle

However, these characteristic race frequencies in (4.41) and (4.42) can be approximated for most bearings with between 6 and 12 balls by the following [48,69,70]:

$$f_O = 0.4N_bf_r \tag{4.45}$$

$$f_I = 0.6N_bf_r \tag{4.46}$$

The torque oscillations generate stator current components at predictable fault bearing frequencies. The bearing fault frequencies $f_{\text{Bng}}$ are related to the oscillations and electrical supply frequency by

$$f_{\text{Bng}} = |f_s \pm mf_v| \tag{4.47}$$

where $f_s$ is the power supply frequency, $f_v$ is one of the characteristic vibration frequencies ($f_C, f_O, f_I, f_B$), and $m = 1, 2, 3, \ldots$.

#### 4.4.2.2 BDs stator current bispectrum: a theoretical approach

In this section, the BDs' bispectrum stator current signal is theoretically presented. This theoretical analysis will be also confirmed by some experimental results.

A simulated signal is built to testify the proposed method. Let consider that the stator current is given by (4.48). This signal is similar to the measured current for an IRF at rated speed (2780 rpm), and rated torque, thus accordingly to (4.42), $f_v = 251.12$ Hz. This signal is generated at a sampling rate of 4,096 Hz. Assume the simplest case of a stator current signal, where only the first sideband frequencies associated with the BDs ($f_l = |f_s - f_v| = 201.12$ Hz and $f_r = |f_s + f_v| = 301.12$ Hz) are considered. Here index $l$ for left, and index $r$ for right, in absolute values:

$$i_a(t) = i_f \cos(2\pi f_s t - \varphi) + i_l \cos(2\pi f_l t - \varphi_l) + i_r \cos(2\pi f_r t - \varphi_r) \tag{4.48}$$

with

$i_f$ : RMS value of the supply phase current

$i_l,\ \varphi_l$ : RMS value of the lower current component at $f_l = |f_s - f_v|$, and its phase angle, respectively

$i_r,\ \varphi_r$: RMS value of the upper current component at $f_r = |f_s + f_v|$, and its phase angle, respectively

$\varphi$ : phase angle at the supply frequency

The bispectrum of the stator current signal generated by BDs is theoretically calculated using the FT of expression (4.48) and substituted in the bispectrum formula given by (4.10), and is given by (4.49):

$$\begin{aligned} B(f_1,f_2) &= I_a(f_1)I_a(f_2)I_a^*(f_3 = f_1 + f_2) \\ &= \frac{1}{8}\left(i_f\delta(f_1 - f_s)e^{j\varphi} + i_l\delta(f_1 - f_l)e^{j\varphi_l} + i_r\delta(f_1 - f_r)e^{j\varphi_r}\right) \\ &\quad \times \left(i_f\delta(f_2 - f_s)e^{j\varphi} + i_l\delta(f_2 - f_l)e^{j\varphi_l} + i_r\delta(f_2 - f_r)e^{j\varphi_r}\right) \\ &\quad \times \left(i_f\delta(f_3 - f_s)e^{-j\varphi} + i_l\delta(f_3 - f_l)e^{-j\varphi_l} + i_r\delta(f_3 - f_r)e^{-j\varphi_r}\right) \end{aligned} \tag{4.49}$$

By plotting B($f_1$,$f_2$) in the bispectrum domain as shown in Figures 4.24 and 4.25, it can be seen that each of the three factors consists of three parallel delta function lines ($f_i = f_j$, where $i = 1, 2, 3$ and $j = l, s, r$). Therefore, the nonredundant region of computation of the bispectrum $B(f_1,f_2)$ is nonzero at 12 points (peaks in the bispectrum) shown in Figure 4.24. Equation (4.49) shows that every stator current signal generates a bi-dimensional bispectral pattern characterized by peaks positions shown in Figures 4.24 and 4.25. The energy distribution in the bispectrum domain is validated experimentally by the analysis of real stator current IM recorded using the test bench shown in Figure 4.26. Figure 4.26(b) shows the bispectrum of stator current with IRF at rated speed sampled at $f_e = 10$ kHz, in the frequency bandwidth [0, 350 Hz]. All the bifrequency component magnitudes have been normalized and expressed in dB scale.

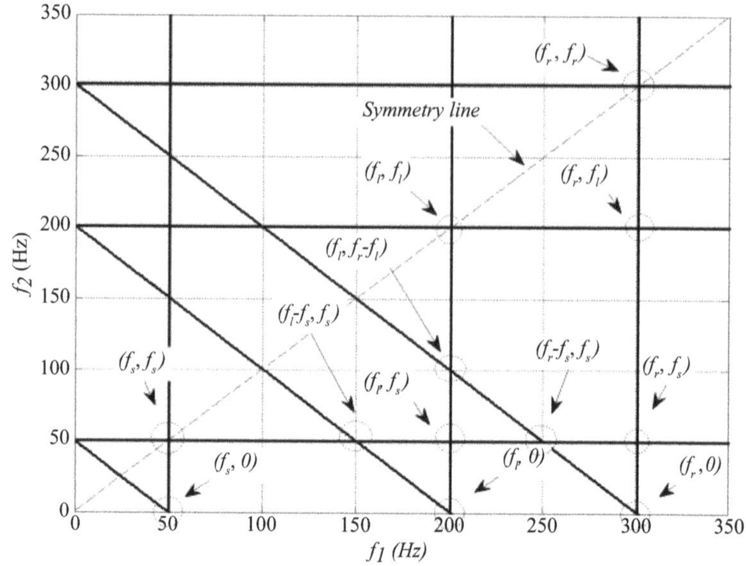


*Figure 4.24 Graphical illustration of stator current bispectrum given in (4.49), where the bispectrum peaks are indicated by arrows*

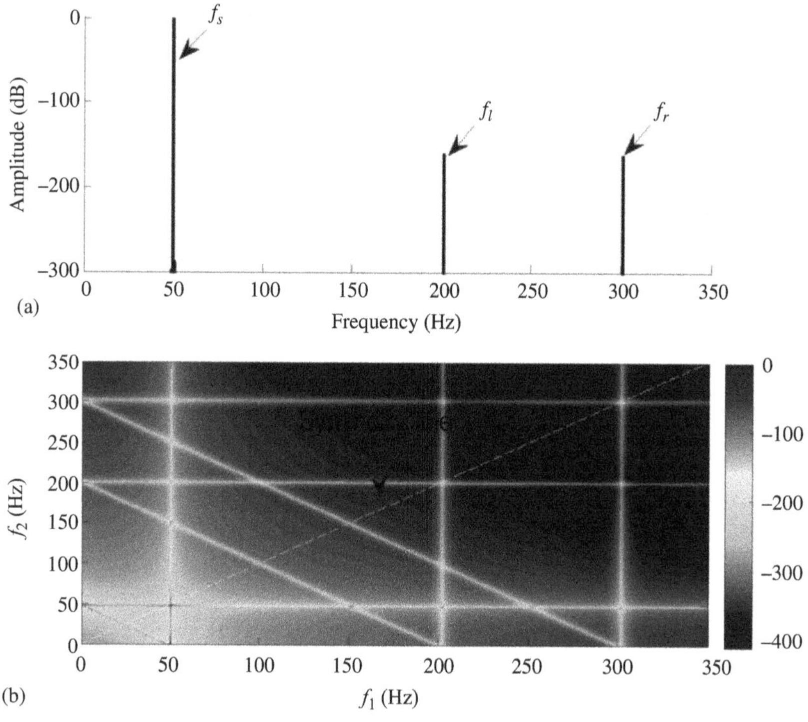


*Figure 4.25 (a) Stator current PS, and (b) its bispectrum displayed as color images, associated with the current signal given in (4.48)*

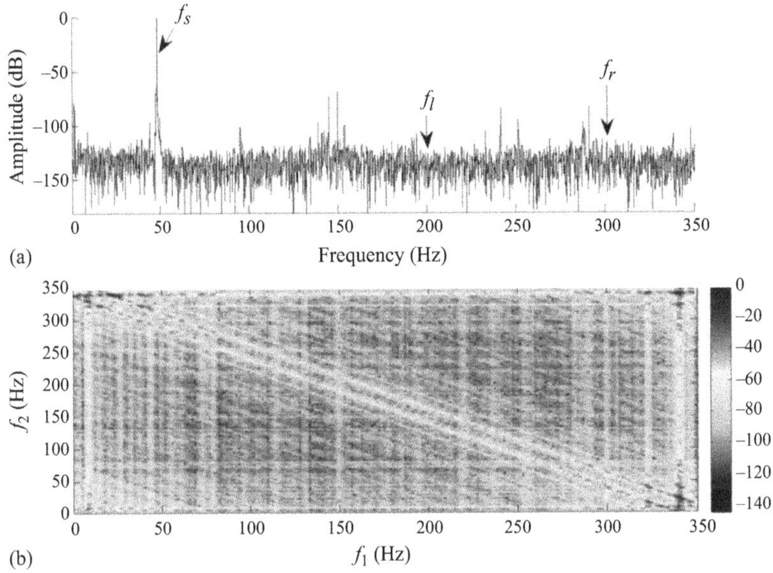


*Figure 4.26 (a) Zoomed stator current PS and (b) its bispectrum displayed as color images, IM running at rated speed and rated torque condition; this is to be compared with the results in Figures 4.24 and 4.25*

### 4.4.2.3 Features extraction and reduction

*Features extraction*

To characterize the frequency information within REB data, this subsection proposes to use the following derived bispectrum features from the stator current bispectrum, which are presented in Table 4.4 [13,68]:

Then the features vector is expressed as

$$T = [F_1, F_2, F_3, P_1, P_2, P_e, WCOB_1, WCOB_2] \tag{4.50}$$

In the next step, the PCA is introduced to eliminate correlations between features and reduce the dimensionality of the original feature vectors.

*Features reduction*

Feature reduction means transforming the original features into a lower-dimensional space [69]. Most of the feature extraction techniques have based on linear techniques such as PCA. PCA is a quantitatively rigorous method for achieving data dimensionality reduction. The method generates a new set of variables, called PCs, which maximizes the variance of the projected vectors. Each PC is a linear combination of the original variables. All the PCs are orthogonal to each other, so there is no redundant information. The PCs as a whole form an orthogonal basis for the space of the data. Thus, the first PC consists of the highest variability, the second PC consist of the next highest variability and so on for other directions. The first few components are kept and others with less variability are discarded.

*Table 4.4 Derived bispectrum features from the stator current signals*

| Features | Expression | Notes |
|---|---|---|
| Sum of logarithmic amplitudes of the bispectrum | $F_1 = \sum_{f_1, f_2 \in \Im} \log(\lvert B(f_1, f_2)\rvert)$ | |
| Sum of logarithmic amplitudes of diagonal elements in the bispectrum | $F_2 = \sum_{f_k \in \Im} \log(\lvert B(f_k, f_k)\rvert)$ | |
| First-order spectral moment of amplitudes of diagonal elements in the bispectrum | $F_3 = \sum_{f_k \in \Im} k \cdot \log(\lvert B(f_k, f_k)\rvert)$ | |
| Normalized bispectral entropy | $P_1 = -\sum_n p_n \log(p_n)$ | where $p_n = \frac{\lvert B(f_1, f_2)\rvert}{\sum_{f_1, f_2 \in \Im} \lvert B(f_1, f_2)\rvert}$ |
| Normalized bispectral squared entropy | $P_2 = -\sum_n q_n \log(q_n)$ | where $q_n = \frac{\lvert B(f_1, f_2)\rvert^2}{\sum_{f_1, f_2 \in \Im} \lvert B(f_1, f_2)\rvert^2}$ |
| Bispectrum phase entropy | $P_e = \sum_n p\psi_n \log(p\psi_n)$ | where $p\psi_n = \frac{1}{L}\sum_{\Im} 1(\phi(B(f_1, f_2)) \in \Psi_n$<br>and $\psi_n = \left\{\phi \mid -\pi + \frac{2\pi n}{N} \le \phi < -\pi + \frac{2\pi(n+1)}{N}\right\}$,<br>$n = 0, 1, \ldots, N-1$<br>$L$ is the number of points within the nonredundant region, $\phi$ refers to the phase angle of the bispectrum, and $1(\cdot)$ is an indicator function which gives a value of 1 when the phase angle $\phi$ is within the range of bin $\Psi_n$ |
| Weighted center of bispectrum (WCOB) | $WCOB_1 = \frac{\sum_{\Im} iB(i, j)}{\sum_{\Im} B(i, j)}$; $WCOB_2 = \frac{\sum_{\Im} jB(i, j)}{\sum_{\Im} B(i, j)}$ | where $i$ and $j$ are the frequency bin index in the non-redundant region |

#### 4.4.2.4 Bearings *multi-fault classification proposed method

*Binary SVM: formalization*

SVM is a binary classifier developed by Vapnik [71], which performs IM fault detection using its superior capacity in the classification process. Based on the input data vectors that consist of IMs fault features, SVM will identify these patterns. Usually, each fault produces special features that can be considered as patterns. Linear decision function is designed in the feature space basically to classify the input data. SVM maps the inner product of feature space using the kernel function nonlinearly. Its basic principle can be illustrated in a 2D way as Figure 4.27. A more detailed description of SVMs can be found in [71].

Figure 4.27 shows the classification of a series of points for two different classes of data, class $A$ (circles) and class $B$ (squares). The SVM tries to place a linear boundary $H$ between the two classes and orients it in such a way that the margin is maximized, namely, the distance between the boundary and the nearest data point in each class is maximal. The nearest data points are used to define the margin and are known as support vectors.

Suppose there is a given training sample set $G = \{(x_i, y_i), i = 1, \ldots, l\}$, where $l$ is the number of training sets, each sample $x_i \in R^N$, $N$ is the input space dimension, belongs to a class by $y \in \{+1, -1\}$. The boundary can be expressed as follows:

$$\omega \cdot x + b = 0 \tag{4.51}$$

where $\omega$ is a weight vector and $b$ is a bias. So the following decision function can be used to classify any data point in either class $A$ or $B$:

$$f(x) = sgn(\omega \cdot x + b = 0) \tag{4.52}$$

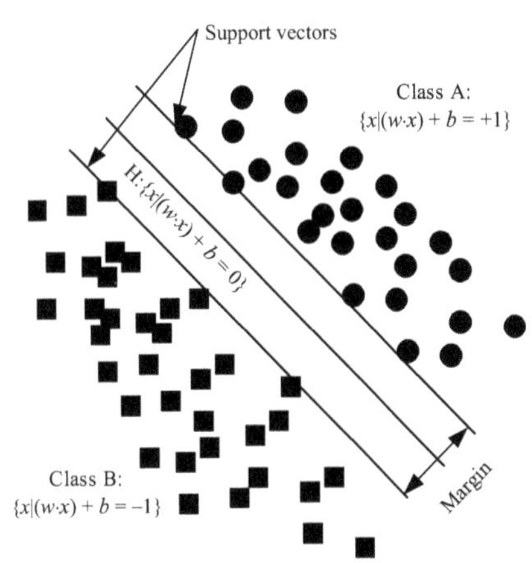


*Figure 4.27 Separation of two classes by SVM*

The optimal *hyperplane* separating the data can be obtained as a solution to the following constrained optimization problem for training the SVM model is to minimize:

$$\left(\frac{1}{2}\|\omega\|^2 + C\sum_{i=1}^{l} \xi_i\right) \tag{4.53}$$

subject to

$$y_i[\omega \cdot \Phi(x_i) + b] \geq 1 - \xi_i, \quad i = 1, \ldots, l \tag{4.54}$$

where $\xi_i$ are the slack-variables. The aim of introducing slack variables is to relax the hard margin constraints to allow misclassifications. The constant $C > 0$ is the penalty parameter used to control the trade-off between maximization of the soft margin width and minimizing the training error. $\Phi$ denotes a nonlinear mapping used by SVM to map the training pattern from the original finite-dimensional input space to a very high-dimensional feature space.

The mapping appears in terms of the kernel function $K(x_i, x_j) = \Phi(x_i) \cdot \Phi(x_j)$, where $k(x_i, x_j)$ is a symmetric positive definite kernel function.

Among the kernel functions in common use are linear functions, polynomials functions, radial basis function (RBF), and sigmoid functions. The necessary and sufficient condition for choosing a kernel is that it should satisfy Mercer's theorem [71], such kernel is the RBF, which is given by

$$K(x, y) = \exp\left(-\gamma\|x - y\|^2\right), \gamma > 0 \tag{4.55}$$

where $\gamma$ is the kernel parameter.

Furthermore, binary SVM can be extended to multi-class. This is the subject of the next subsection.

### *Multiple classes SVM*

As mentioned above, SVMs were originally designed for binary (two classes) classification [6,10–12,24]. In binary classification, the class labels can take only two values: 1 and $-1$. In the real problem, however, we deal more than two classes: for example, in condition monitoring of IMs, there are several classes such as mechanical unbalance, misalignment, different load conditions, bearing faults, and gear faults. Therefore, multi-class SVM is obtained by decomposing the multi-class problem into several numbers of binary class problems.

Two different approaches are taken into account: one-against-all (OAA) and one-against-one (OAO). In the first one, the *i*th SVM is trained with all the examples in the *j*th class with positive labels and all the other examples with negative labels, while in the latter one each classifier is trained on data from two classes. Here, SVM-OAA is chosen to classify different bearing faults.

We do not make any comparison between SVM-OAA strategy and other popular approaches like SVM-OAO in this subsection, because of the following reasons:

- Benchmark comparisons on multi-class SVM approaches already exist in the literature [51,52].
- It has been concluded [51,52] that SVM-OAA is as accurate as any other approach, assuming that all underlying binary SVMs are well-tuned.

The training procedure and choice of SVM parameters for training is very important for classification. In this work, the process of optimizing the SVM parameters with the cross-validation method is adopted.

Detailed information of this strategy has been clearly explained in [51,52,71].

For training and testing SVM-OAA, bispectrum features relative to the studied faults are extracted to develop the input vector, which is necessary for the training and the test of the BD classification. This is the next phase of our research work.

#### 4.4.2.5 BD classification based on SVM

The diagram of the fault diagnosis scheme is presented in Figure 4.28. The procedure can be summarized as follows:

- ***Step* 1**: Data acquisition is carried out through the designed test rig.
- ***Step* 2**: Signal processing is performed using bispectrum analysis.
- ***Step* 3**: Bispectrum features calculation from the stator current signals.
- ***Step* 4**: Bispectrum features reduction using the PCA method.
- ***Step* 5**: Classification process for fault diagnosis by SVM-OAA based on multi-fault classification.

Since only six bearing conditions need to be identified, just five SVM classifiers need to be designed, as indicated in Figure 4.28 and Table 4.5. For SVM1, define the IRF condition as $y = +1$, and the remaining five other conditions, as another class, identified as $-1$; thus, the IRF could be separated from other conditions by SVM1. Then define the condition with ORF as $y = +1$ and the other conditions as $y = -1$ for SVM2; thus, the ORF could be separated from other conditions by

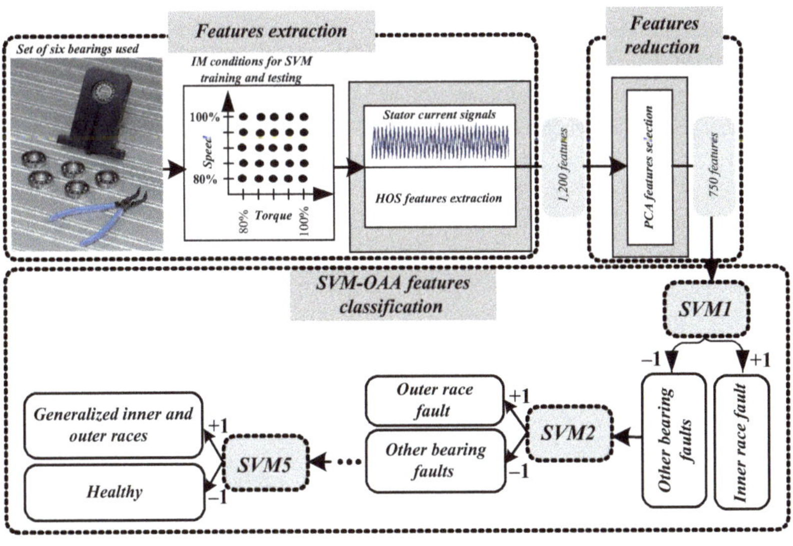


*Figure 4.28 Proposed diagnostic methodology for multi-fault diagnosis scheme based on SVM-OAA*

Table 4.5 Binary encoding for each BDs

| | SVM1 | SVM2 | SVM3 | SVM4 | SVM5 |
|---|---|---|---|---|---|
| IRF | +1 | | | | |
| ORF | −1 | +1 | | | |
| BF | −1 | −1 | +1 | | |
| IOBF | −1 | −1 | −1 | +1 | |
| GIOF | −1 | −1 | −1 | −1 | +1 |
| HB | −1 | −1 | −1 | −1 | −1 |

SVM2. Similarly, the BF could be separated from other conditions by SVM3 and so on, until the classification test completed. Then they become a multi-class fault diagnosis system as shown in Figure 4.28. Note that all five SVMs adopt RBF as their kernel function. We choose the RBF for the SVM classifier since previous studies have shown that it has the best performance in pattern recognition tasks. As well, the RBF kernel is a better choice than other kernels like polynomial kernel because it has lesser hyper-parameters and so the problem becomes less computationally intensive.

SVM-OAA algorithm is used, however, some parameters are predefined for this classifier such as the regularizing parameter $C$ and the kernel parameter $\gamma$ are set to 100 and 0.5, respectively, which were selected based on the cross-validation method.

#### 4.4.2.6 Experimental results

This section presents the experimental results for damage in REB classification. Descriptions of our test rig and how experiments and the data set used are established and the results are given.

The test rig is shown in Figure 4.29(a) was composed by a variable speed 0.37 kW IM with 2780 rpm of rated speed controlled by an inverter, driving a shaft rotor and a controlled brake, assembly through flexible couplers; shafts were rested on two ball bearings. The bearings under analysis (type SKF6004 given in Figure 4.29(b)) were placed at the load end side for ease of replacement. The BDs were carried out during the manufacture. A milling cutter was used to scratch the corresponding surfaces. The test rig was used for modeling different fault types such as eccentricities, misalignment, and different types of BDs. Stator current signals were collected using a 16-bit A/D converter at a sampling rate of 10 kS/s. The numbers of samples collected were one hundred thousand for a duration of 10 s. LabView™ software is used for data acquisition and was post-processed in a MATLAB environment.

Table 4.6 shows the parameters of the bearing used in our experimental test bench, taken from the datasheet.

To take into account different speed and torque combination, 25 measurements for each bearing conditions are considered: from 80% to 100% rated speed, and from 80% to 100% of rated torque, every variation of 5% of each parameter.

The bearing data set was obtained from the experimental test rig under six different operating conditions as presented in Figure 4.29(b); six identical bearings

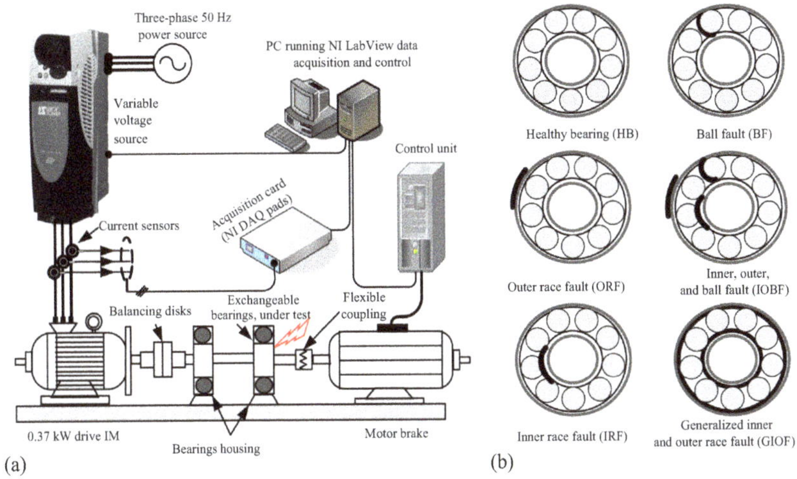


*Figure 4.29* *(a) The instrumentation of the experimental setup for BDs detection, and (b) a series of bearing components with faults induced in them indicated in bold line*

*Table 4.6* *REB parameters used in the experimental setup*

| **Type** | **Outside diameter** | **Inside diameter** | $N_b$ | $D_b$ | $D_p$ | $\cos\beta$ |
|---|---|---|---|---|---|---|
| SKF6004 | 42 mm | 20 mm | 9 | 6.35 mm | 31 mm | 1 |

(SKF6004) have been used covering the most important BDs scenarios: (a) HB (25 measurements), (b) with an ORF (25 measurements), (c) with an IRF (25 measurements), (d) with a BF, (e) with an inner and outer race as well as ball-bearing faults (IOBF) (25 measurements), and (f) with generalized inner and outer races degradation (GIOF) (25 measurements).

The SVMs training experiments are conducted on a data set (150 current signals include 25 signals for six different bearing conditions). The classification results are shown in Tables 4.7–4.9.

Stator current bispectra are presented in Figure 4.30 in the full region of bispectrum, to compare with the theoretical results given in Figures 4.24 and 4.25.

As an example, the HB stator current bispectrum at rated speed and rated torque is computed in the triangular region $\Im$, with its corresponding bispectrum features as follows: $T = [0.5765; 0.1654; 0.0993; 3.5721; 138.6543; 4.6175; -3.43 \times 10^4; -4.6175 \times 10^4]$, where the bispectrum plots are depicted in 2D space of coordinates $(f_1, f_2)$ and the amplitude will be represented in dB scale by a color bar.

To show the speed and torque effects on the distribution of the bispectrum features, we take as an example the normalized bispectrum entropy values for the 25 acquisitions in each condition evaluated, as presented in Figure 4.31. This plot shows

*Table 4.7 Confusion matrix for the multi-class SVM-OAA resulting from the evaluation of the whole data set*

| | | Predicted | | | | | |
|---|---|---|---|---|---|---|---|
| | | **HB** | **IRF** | **ORF** | **BF** | **IOBF** | **GIOF** |
| Actual | HB | 78 | 0 | 2 | 0 | 0 | 0 |
| | IRF | 1 | 79 | 0 | 0 | 0 | 0 |
| | ORF | 3 | 1 | 74 | 0 | 0 | 2 |
| | BF | 0 | 0 | 0 | 80 | 0 | 0 |
| | IOBF | 0 | 2 | 3 | 1 | 72 | 2 |
| | GIOF | 0 | 3 | 1 | 0 | 1 | 75 |

*Table 4.8 Confusion matrix for the multi-class SVM-OAA resulting from the evaluation of the reduced data set*

| | | Predicted | | | | | |
|---|---|---|---|---|---|---|---|
| | | **HB** | **IRF** | **ORF** | **BF** | **IOBF** | **GIOF** |
| Actual | HB | 49 | 0 | 0 | 0 | 0 | 1 |
| | IRF | 1 | 49 | 0 | 0 | 0 | 0 |
| | ORF | 1 | 1 | 47 | 0 | 0 | 1 |
| | BF | 0 | 0 | 0 | 50 | 0 | 0 |
| | IOBF | 0 | 2 | 1 | 1 | 46 | 0 |
| | GIOF | 0 | 1 | 1 | 0 | 1 | 47 |

*Table 4.9 The testing accuracy* for six different bearing conditions using SVM-OAA*

| **Bearing conditions** | **SVM including all features** Classification accuracy (%) | | **SVM including selected features** Classification accuracy (%) | |
|---|---|---|---|---|
| | **Training** | **Testing** | **Training** | **Testing** |
| HB | 99.16 | 97.50 | 100 | 98.00 |
| IRF | 100 | 98.75 | 100 | 98.00 |
| ORF | 100 | 92.50 | 100 | 94.00 |
| BF | 100 | 100 | 100 | 100 |
| IOBF | 98.33 | 90.00 | 97.33 | 92.00 |
| GIOF | 97.5 | 93.75 | 98.66 | 94.00 |
| Average | 99.165 | 95.416 | 99.331 | 96.00 |

*Accuracy is computed based on the confusion matrix provided in this work.

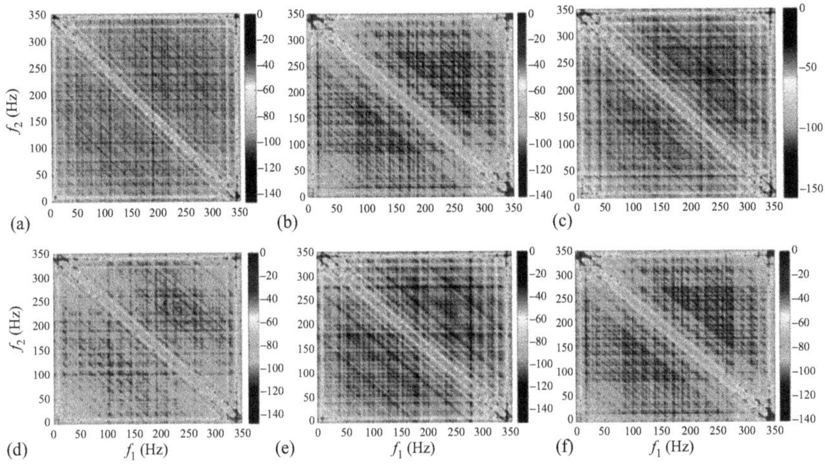


Figure 4.30 *Stator current bispectra of different BDs types: (a) HB, (b) IRF, (c) ORF, (d) BF, (e) IOBF, and (f) GIOF. The magnitude coming out of the page and indicated by gray levels-bar*

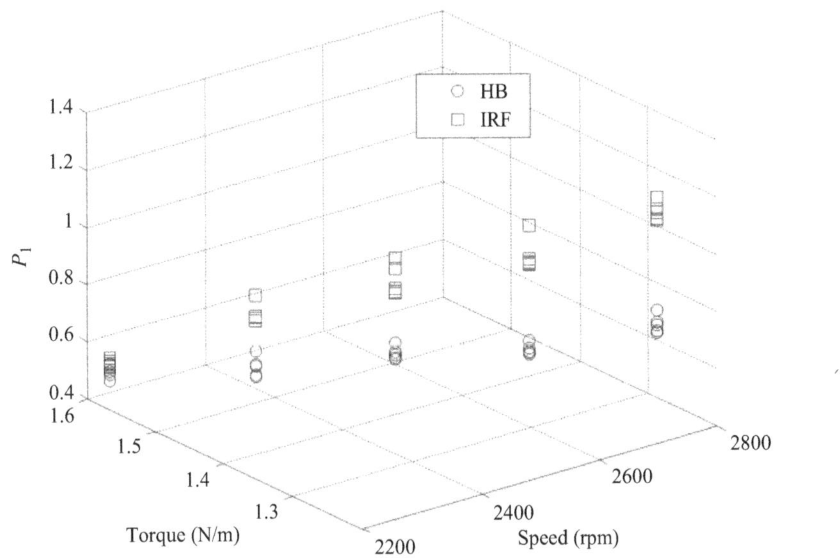


Figure 4.31 *The effect of speed and torque on statistical normalized bispectral entropy parameter distribution*

how this parameter is influenced by the speed and the torque both for a healthy and damaged case and it increases with higher speeds. Moreover, it can be noticed that in low-speed cases this parameter value for the damaged bearing is almost near to the healthy one when it reaches the highest speed.

The bispectrum is computed for each of the six types of bearing conditions including various combinations of speed and load, in the principal domain $\Im$. After feature calculation, normalized bispectral entropy ($P_1$), the normalized bispectral squared entropy ($P_2$), and bispectrum phase entropy ($P_e$) were plotted in Figure 4.32 to know the structure of the original features. Figure 4.32 represents the original features which are not well clustered and have disorder structure. Plotting original feature parameters indicates the necessity of preprocessing of the original features to make them separable and ready for classification. The disordered structure of the original features tends to decrease the performance of the classifier if it is directly processed in the classifier.

To avoid this disadvantage, PCA was proposed to extract and to reduce the feature dimensionality based on the eigenvalue of the covariance matrix. Therefore, the first five PCs have been selected to replace the original feature vector.

Figure 4.33 shows the feature reduction in the component analysis based on the eigenvalues of the covariance matrix. The projection result is illustrated in Figure 4.33. The axes of the projection plane correspond to the maximum variance directions in the initial space. As shown in Figure 4.34, the number of features is reduced from 8 to 5.

#### 4.4.2.7 Training and test vectors

The training and testing of the SVM model with real-time data sets were implemented with the help of LIBSVM software [23]. The total databases comprised 1200 (25 × 6 × 8) original features. The number of features is reduced to 750 (25 × 6 × 5);

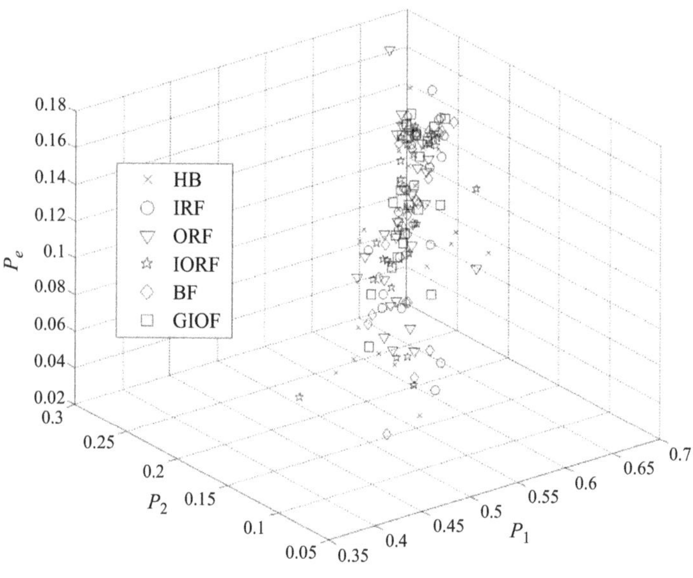


*Figure 4.32 Distribution of "$P_1$," "$P_2$," and "$P_e$" features of original data features (belonging to the six bearing classes)*

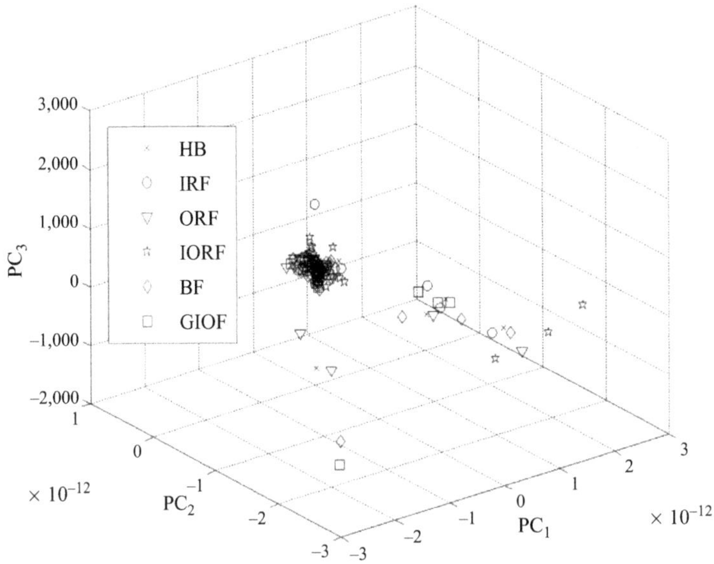


*Figure 4.33 Original features obtained from PCA*

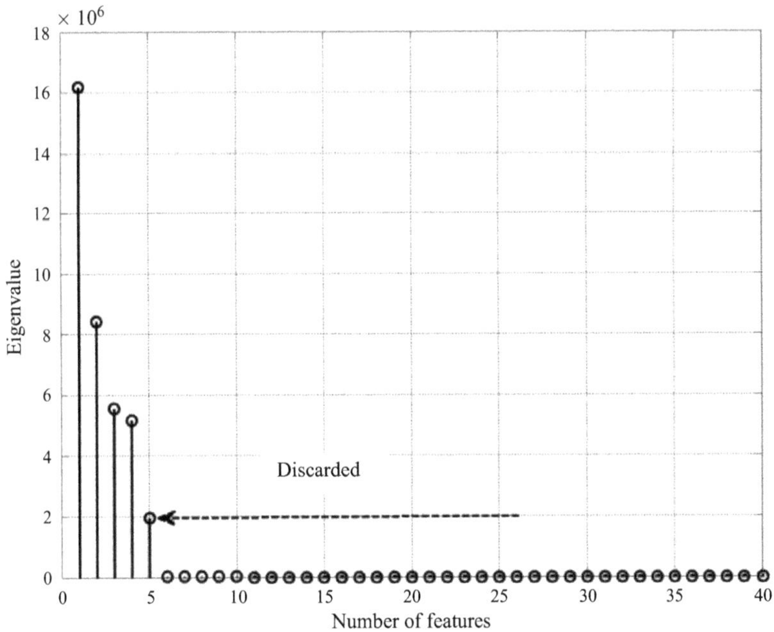


*Figure 4.34 Eigenvalues of the covariance matrix for feature reduction*

selected features were divided into two sets: one for training (containing 60% of the samples) and the other for the test (containing 40% of the samples).

After the SVM is trained with 450 features, its performance has been tested with the 300 remaining (50 features for each BD).

The performance of the SVM-OAA is validated by calculating the following performance measure for the train set and test set separately.

The classification accuracy (CA): The ratio between the total number of correctly classified test samples to the total number of test samples is given by (4.56):

$$CA\,[\%] = \frac{number\ of\ correctly\ classified\ samples}{total\ number\ of\ samples\ in\ dataset} \times 100 \tag{4.56}$$

The classification results of all classifiers in the training and testing processes are presented in Table 4.9. In the training process, all the SVM-OAA classifiers achieve an average accuracy of 99.165% and 99.331% in the whole data set and the reduced data set, respectively, and some of them without any misclassification out of 450 samples (respectively 720 samples) of training data for all bearing features. This indicates that the classifiers are well trained and can be applied for diagnosing BDs. However, in the testing process, these classifiers are validated against the test data, the average accuracy is about 96% and 95.416% for the reduced and original features, respectively. The misclassifications are due to the overlap of machine condition features.

A confusion matrix is a useful tool for analyzing how well the classifier can recognize tuples of different groups, which contains information about actual and predicted classifications done by a classification system.

From the confusion matrix of the SVM in Tables 4.7 and 4.8, one can note that SVM finds it difficult to discriminate between GIOF and IRF on the one hand, ORF and BF on the other hand. Misclassification of 4% brings down the diagnostic ability of the SVM-OAA; however, the overall classification accuracy is reasonably good. It can be seen that the BF presented has the highest accuracy of 100%.

From the test results shown in Tables 4.7–4.9, it can be seen that the SVM-OAA classifiers recognize the defect samples effectively, especially for the HB signals, IRF signals, and the ORF signals. The recognition results of SVM are ideal because of its high accuracy and a good generalization capability when the average classification efficiency is close to 96% which is reasonably good.

## *4.4.3 Bispectrum-based EMD applied to the nonstationary vibration signals for bearing fault diagnosis*

### 4.4.3.1 Nonstationary nature of defective REB vibration response

As shown in Figure 4.35, Rollers or balls rolling over a local fault in the bearing produce a series of force impacts. If the rotational speed of the races is constant, the repetition rate of the impacts is determined solely by the geometry of the bearing. The repetition rates are denoted bearing frequencies; for example, BPFO (ball passing frequency outer race), BPFI (ball passing frequency inner race), and BFF (BF frequency) are frequently used. Their mathematical equations are given in (4.41), (4.42), and (4.43), respectively.

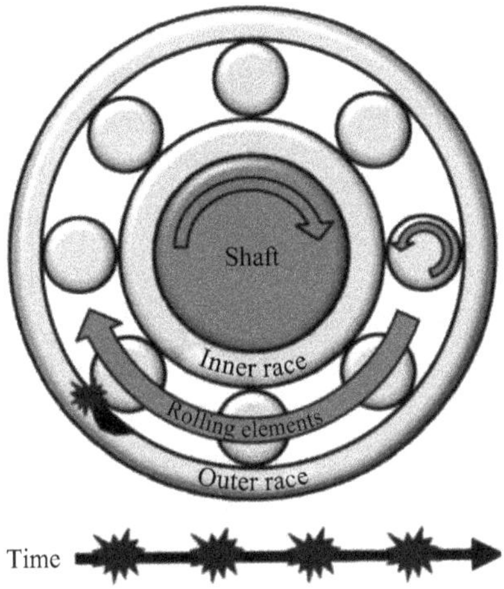


*Figure 4.35 Bearing rolling elements create impacts when the pass over damage on the bearing races, creating a periodic series of impacts through time*

A large number of models have been used to describe the dynamic behavior of REBs under different types of defects. According to the traditional approach, when rolling elements of bearing pass the defect location, wideband impulses are generated. And those impulses will then excite some of the vibrational modes of the bearing and its supporting structure. The excitation will result in the sensed vibration signals (waveforms) different in either the overall vibration level or the vibration magnitude distribution.

Measured vibration signals consist of two parts: $y(t) = v(t) + n(t)$, where $v(t)$ is the defect-induced impulse responses and $n(t)$ is the background noise, including vibration signals generated by other components, such as rotor unbalance and gear meshing. Because of the structure and the mode of operation of REBs, $v(t)$ has distinct features as follows [48,54–58]:

- Wide frequency range: BDs usually start as small pits or spalls, and give sharp impulses in the early stages covering a very wide frequency range.
- Small energy: The energy created by the BD is very small. A band has to be found where the bearing signal dominates over other components.
- Nonstationary signal: Incipient BDs produce a series of repetitive short transient forces, which in turn excite structural resonances.

Because the vibration signals generated by a defective REB have the characteristics mentioned above, it is difficult to identify their faults through simple classical frequency analysis. Empirical mode decomposition (EMD) has been widely applied to analyze vibration signal's behavior for bearing failures detection [49,58,72–74]. Vibration signals are almost always nonstationary since bearings are inherently dynamic (e.g. speed and load condition change over time). By using EMD, the complicated nonstationary vibration signal is decomposed into many stationary intrinsic mode functions (IMFs) based on the local characteristic time scale of the signal. Bispectrum, a third-order statistic, helps to identify phase coupling effects—the bispectrum is theoretically zero for Gaussian noise and flat for non-Gaussian white noise, and consequently, the bispectrum analysis is insensitive to random noise—which are useful for detecting faults in IMs. Utilizing the advantages of EMD and bispectrum, this subsection gives a joint method for detecting such faults, called bispectrum-based EMD (BSEMD). First, original vibration signals collected from accelerometers are decomposed by EMD and a set of IMFs is produced. Then, the IMF signals are analyzed via bispectrum to detect outer race BDs. The procedure is illustrated with the experimental bearing vibration data.

In this part, the effectiveness of the EMD-bispectrum method is illustrated on a synthetic signal of an REB with an ORF. This type of fault was chosen because it is a relatively simple phenomenon to simulate while being often found in rotor machinery condition monitoring. REB with ORF during operation generates a series of periodic shock pulses whose repetition rate (Figure 4.35) depends on its dimensions and rotational speed of the shaft with which the bearing is assembled. Shock pulses are generated each time rolling elements strike the defected surface of the bearing outer race and, consequently, excite resonances of the structure between the fault location and the vibration sensor. The frequency of shock occurrence is usually called the BPFO and is given in (4.41).

Assuming constant rotational speed and load of the bearing, the vibration signals generated by an REB with defected outer race were modeled by Randall *et al.* as follows [48,56,57,75,76]:

$$v(t) = \sum_i \omega\left(t - i\frac{1}{BPFO} - \tau_i\right) + n(t) \tag{4.57}$$

where $\omega(t)$ is the waveform generated by a single impact (related to resonance frequencies of the system), $\tau_i$ is an independent and identically distributed random variable, and $n(t)$ is an additive background random noise. It now has to be stated that $\tau_i$ introduces influence of the rolling elements slips into the model.

To exhibit the effect of the stochastic nonstationary nature of certain critical parameters like a slip, on the vibration spectra, a typical example is considered. The simulated signal generated by (4.57) corresponds to the typical response of a bearing with an ORF. The shaft rotation speed $f_r$ is 29.16 Hz [$f_r = rpm/60$]. The characteristic BD frequency BPFO is equal to 3.58 times the shaft rotation speed, leading to an estimation of the BPFO around 104 Hz. The excited natural frequency $f_c$ of the system is assumed to be equal to 4500 Hz, which corresponds to the largest peak in the spectrum. The signal consists of 2048 samples and the sampling rate is equal to 20 kHz. Figure 4.36 illustrates the waveform and spectra of the simulated signals with and without additive Gaussian white noise (AGWN) effect. In perfect agreement to the expected results of the model defined in (4.57), it can be found that a serious of spike pulses appear in the time domain waveforms, the interval of the

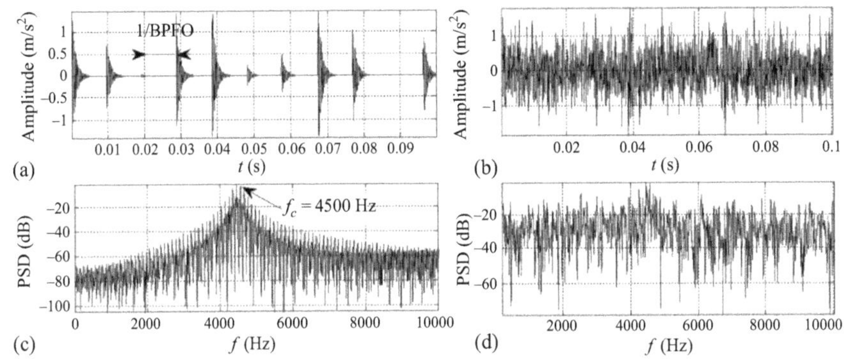


*Figure 4.36 A simulated short segment of vibration response data showing outer race impacts with (a) and without (b) noise effect, (a) and (b) waveforms, (c) and (d) their power spectral densities (PSD), respectively*

spike pulses of the defect-induced impulses (1/BPFO = 9.6 ms). Their amplitudes are maximum in the frequency band around the resonance frequency of 4500 Hz.

As seen on the spectrum in Figure 4.36(d), no distinct spectral lines can be recognized. Information that could be obtained by examination of the spectrum is only the frequency characteristic of band-pass stationary Gaussian noise used in the generated signal. For rotor-machinery vibration signals, natural resonances of the object usually manifest themselves in the same manner. However, for vibration-based condition monitoring purposes, information about high-frequency resonances have relatively limited value. We focus our interest on the phenomena that causes excitation of observed resonances. Due to this fact, the usage of an advanced statistical procedure is necessary.

#### 4.4.3.2 Brief description of EMD

*Empirical mode decomposition*

EMD is adaptable in applications where the signal is nonstationary. A recent review of EMD applications in fault diagnosis of rotating machinery can be found in [77]. In this subsection, EMD is used for nonstationary vibration bearing data. The main function of EMD is to decompose the original vibration signal into several signals which have a specific frequency called IMFs based on the enveloping technique. The results of IMFs are from high frequencies to low frequencies. In condition monitoring of rolling bearing, EMD is used to reveal the frequency content of vibration signal by decomposing the original signal into several IMFs to determine whether the bearing signal has specific frequency content corresponds to the BD frequencies or not, the selected IMF is chosen when the decomposed frequencies are identical to one of the bearing fault frequencies (for example as shown in Table 4.10). Figure 4.37 shows the first eight IMF components decomposition of the vibration signal generated by (4.57), due to space limitation.

*Table 4.10 Drive end bearing information*

| **Bearing** | | **SKF6205** |
|---|---|---|
| Geometry size (mm) | Outside diameter | 51.81 |
| | Inside diameter | 24.9 |
| | $N$ | 9 |
| | $D_c$ | 201.9 |
| | $D_b$ | 38.86 |
| | $\cos\beta$ | 0.9 |
| Defect frequencies multiple of running speed (Hz) | Inner ring | 5.4152 |
| | Outer ring | 3.5848 |
| | Cage train | 0.3983 |
| | Rolling elements | 4.7135 |

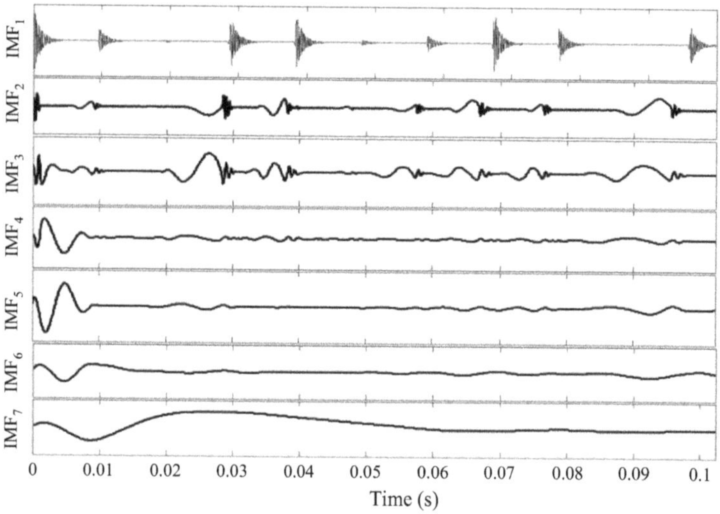


*Figure 4.37 IMF decomposition results of simulated vibration signal without noise, calculated by EMD*

The multicomponent signal (the current $v$ in our case) is then decomposed into $M$ intrinsic modes and a residue $r$.

$$v(t) = \sum_{i=1}^{M} IMF_i(t) + r(t) \tag{4.58}$$

IMFs are oscillatory signals that are locally zero mean. The residual $r(t)$ is the low-frequency mean trend. Note that all IMFs, except $r(t)$, are mean stationary. The higher-order IMF index; that is, $M$ is found by the algorithm itself and is signal-dependent. In this subsection, the standard implementation proposed by Flandrin *et al.* [9] was used.

The effective algorithm for extracting the IMFs from a signal can be summarized as follows:

---

***Step 1***: *$r_0(t) = v(t)$ (the residual), $i = 1$ (index number of IMF).*

***Step 2***: *Extract the ith IMF:*
- *a – Initialize: $h_o(t) = r_{i-1}(t)$; $j = 1$ (index number of the iteration).*
- *b – Extract the local extrema of $h_{j-1}(t)$.*
- *c – Interpolate the local maxima and the local minima by cubic splines to form upper and lower envelope $e_{max}(t)$ and $e_{min}(t)$, respectively, of $b_{j-1}(t)$.*
- *d – Calculate the mean of the upper and lower envelops: $m_{j-1}(t) = (e_{min}(t) + e_{max}(t)/2)$.*
- *e – If $b_j(t)$ is a IMF then set $c_i(t) = b_j(t)$ else go to (b) with $j = j + 1$.*

***Step 3***: *Update residual $r_i(t) = r_{i-1}(t) - c_i(t)$.*

***Step 4***: *If $r_i(t)$ still has at least two extrema then go to Step 2, with $i = i + 1$ else the decomposition is finished and $r_i(t)$ is the residue.*

---

Besides, the implementation of EMD is a data-driven process, not requiring any pre-knowledge of the signal or the machine. This particular advantage in electrical machines and drives context leads the EMD to be a promising tool to improve condition monitoring. The EMD method has however several drawbacks. The choice of a relevant stopping criterion and mode-mixing problem are the most important topics that need to be addressed to improve the EMD algorithm.

In reality, a mechanical system may have multiple natural frequencies that spread over a wide frequency range. For example, a lower one may be in the 1–2 kHz range and a higher one may be over 5 kHz [73,74,78–88]. In our case study, it just happened that the second IMF (IMF2) explained the very natural frequency that is strong and also strongly modulates (couple with) the BD frequency. Depending on how high the data acquisition frequency is (e.g. 10, 20, or 50 kHz), the IMF#2 might explain a lower or higher natural frequency that is weakly coupled with the BD frequency. Then applying the proposed reduction noise method only to the first IMFs might give the best result. Therefore, we limit our effort to vibration signal only for the first seven IMFs. The IMFs extracted from vibration signal $IMF_1$ is associated with the local highest frequency and $IMF_7$ with the lowest frequency. The component in the high-frequency band ($IMF_1$) which represents the resonance modulation (about 4500 Hz) is selected for calculation.

The flowchart of the BSEMD-based method combines the EMD and the bispectrum of the defective bearing nonstationary vibration signal is given in Figure 4.38. On the other hand, BSEMD can detect sets of frequency components that are phase-coupled. The BSEMD applied to the first IMF of simulated vibration ORF is shown in Figure 4.39, and its enlarged presentation is shown in Figure 4.40. The space between the peaks in the resonant frequency band is about 104 Hz, which is equal to the

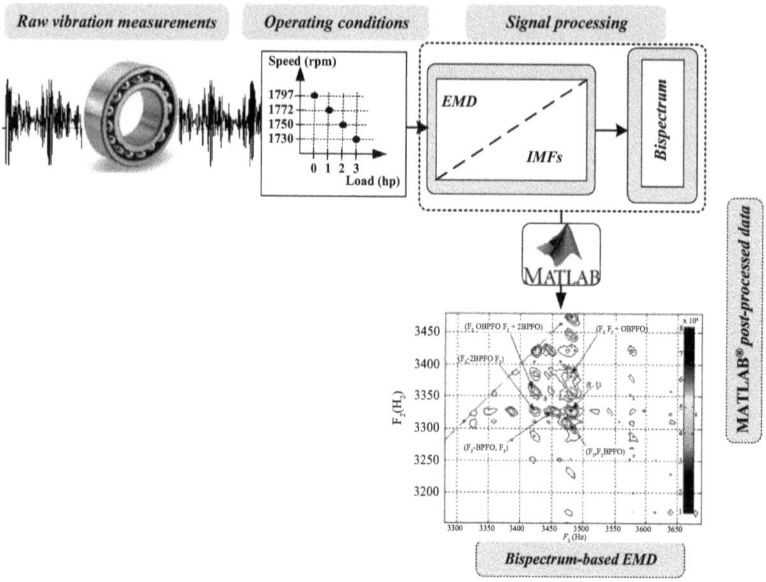


*Figure 4.38 Flowchart of the proposed BDs diagnostic methodology*

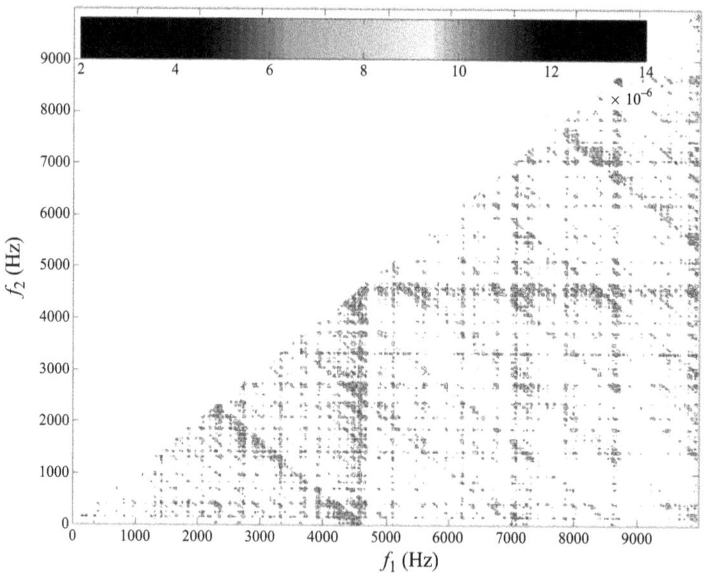


*Figure 4.39 BSEMD of the simulated outer raceway vibration signal, result obtained by applying BSEMD to IMF1*

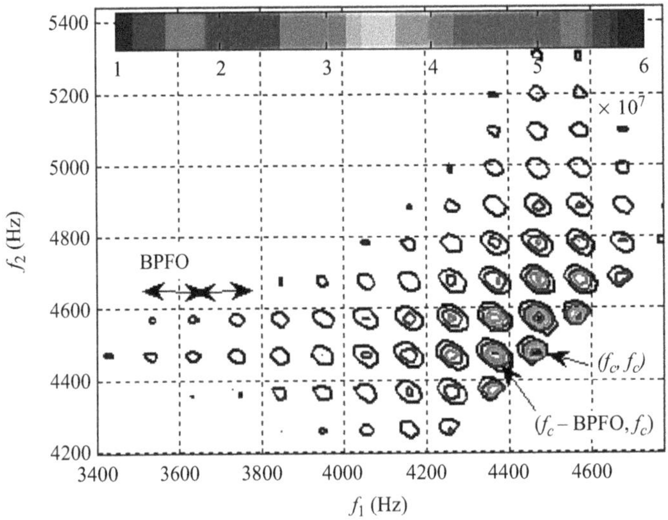


*Figure 4.40 Enlarged BSEMD of simulated outer raceway vibration signal in the bandwidth frequency between 3400 and 4800 Hz, the result obtained by applying BSEMD to $IMF_1$*

BPFO. In a higher frequency band, the peaks around the highest peak are $(f_c, f_c)$, $(f_c \pm \mathrm{BPFO}, f_c)$.

*Stationarity test*

Before applying the PS and bispectrum, signals must be stationary. Various methods exist for testing whether a given measurement signal may be regarded as a sample sequence of a stationary random sequence. A simple yet effective way to test for stationarity is to divide the signal into several (at least two) nonoverlapping segments and then test for equivalency (or compatibility) of certain statistical properties (mean, mean-square value, PS, etc.) computed from these segments. More sophisticated tests that do not require *a priori* segmentation of the signal are also available in [7–9].

The third- and fourth-order cumulants of a discrete process $x(n)$ are expressed as follows [1–2]:

$$Cum_3(x) = E[x^3] - 3E[x]E[x^2] + 2E^3[x] \tag{4.59}$$

$$Cum_4(x) = E[x^4] - 4E[x]E[x^3] - 3E^2[x^2] + 12E^2[x][x^2] - 6E^4[x] \tag{4.60}$$

Let us calculate the third- and fourth-order cumulants of the ORF vibration signal using (4.59) and (4.60). The obtained values for the two cumulants are different ($Cum_3 = 8.6736 \times 10^{-6}$, $Cum_4 = -6.5052 \times 10^{-8}$), so the nonstationarity hypothesis on ORF vibration is reinforced.

#### 4.4.3.3 Experimental results

*CWRU bearing data: description of the experimental setup and data acquisition*

The vibration data of roller bearings analyzed in this subsection comes from Case Western Reserve University (CWRU) bearing data center. The detailed description of the test rig can be found in [89]. As shown in Figures 4.41 and 4.42, the test rig consists of a 2 hp, three-phase IM (left), a torque transducer (middle), and a dynamometer load (right). The transducer is used to collect speed and horsepower data. The load is controlled so that the desired torque load levels could be achieved. The test bearing supports the motor shaft at the drive end. Single point faults with fault diameters of 0.1778, 0.3556, and 0.5334 mm, respectively, were introduced into the test bearing

*Figure 4.41 Photo of the experimental test rig from CWRU, composed of a 2 hp motor (left), a torque transducer/encoder (center), load (right). The test bearings support the motor shaft [89]*

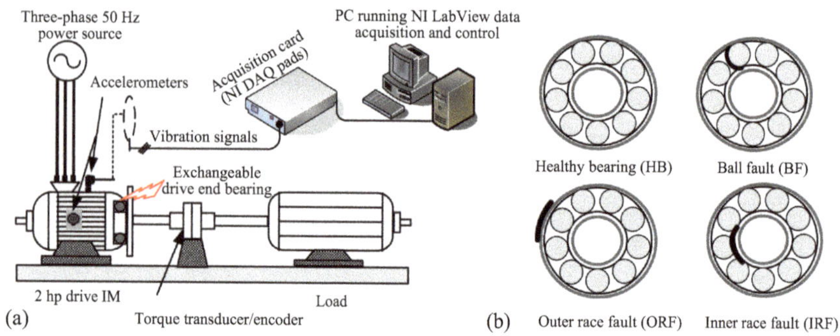


*Figure 4.42 (a) Schematic of the experimental test rig composed of a 2 hp motor (left), a torque transducer/encoder (center), load (right), and control electronics. The test bearings support the motor shaft. (b) A series of bearing components with faults induced in them indicated in bold line*

using electro-discharge machining. Vibration data are collected using an acquisition system at a sampling frequency of 12 kHz for different bearing conditions. The data recorder is equipped with low-pass filters at the input stage to avoid anti-aliasing. The geometry and defect frequencies of the two type bearings are listed in Table 4.10. Tests are carried out under different loads varying from 0 to 3 hp with 1 hp increments. The corresponding speed varies from 1797 to 1730 rpm. BDs cover inner and outer races and BFs. The deep groove ball bearing 6205-2RS JEM SKF is used in the tests. Single point defects were set on the test bearings separately at the rolling element, inner raceway, and outer raceway using electro-discharge machining. Accelerometers were placed at the 12 o'clock position when the defects were at the rolling element and inner raceway, and at the 6 o'clock position for the outer raceway defect.

An example of characteristics vibration signals under different bearing status at the same operating condition (1750 rpm speed and 1 hp load) is given in Figure 4.43.

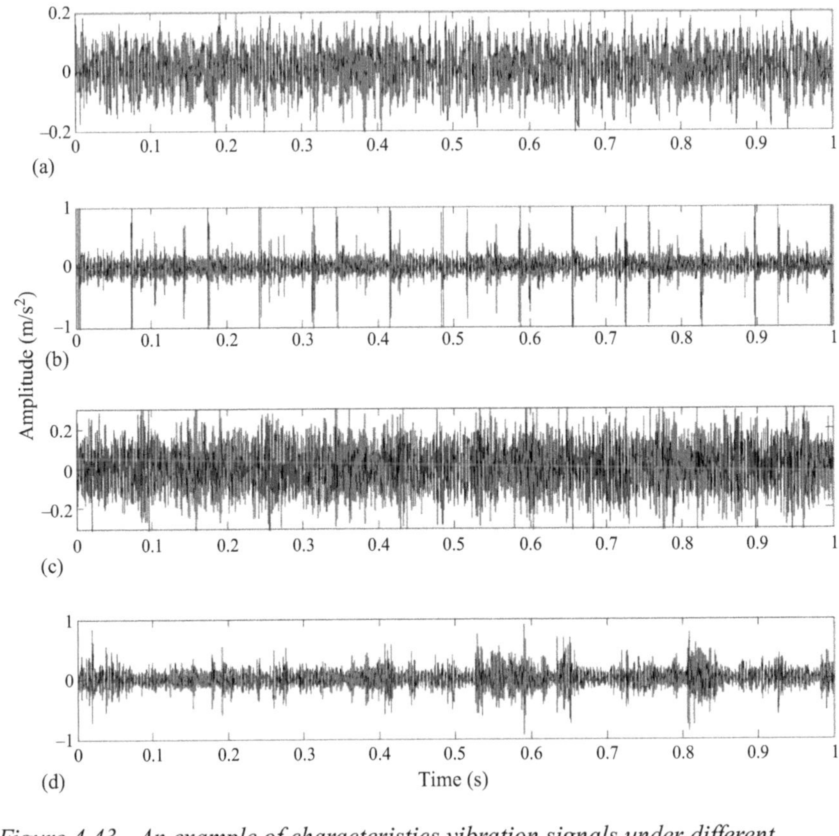


*Figure 4.43 An example of characteristics vibration signals under different bearing status at the same operating condition (1750 rpm speed and 1 hp load). (a) HB. (b) IRF; 0.1778 mm. (c) ORF; 0.1778 mm. (d) BF; 0.1778 mm*

*Analysis of the results*

Although larger FFT bins produce a bispectrum with higher resolution, the computational load will also increase significantly. As a trade-off between resolution and computational expense, we set the FFT bin size as 1024. Moreover, it is recommended that the number of averaged segments shall exceed the number of samples per segment. So each segment with the size of 1024 samples was employed to calculate segment bispectrum, and then the final bispectrum was achieved by average bispectrum of 118 segments.

For the healthy and outer raceway defective bearing signals, the raw signals and their power spectral densities (PSDs) are shown in Figure 4.44(b)–(d), where the structural defect-induced impulses cannot be seen from the waveform due to the heavy background noise interference. The central frequency $f_c$ is equal to 3480 Hz, and is got from the highest spectral line in Figure 4.44(d). Nevertheless, the fault signature is still hardly found in the PS. Moreover, it can be seen from the PS in Figure 4.44(d), BPFO characteristic frequency is buried in the background noise, making the diagnosis result be hardly convinced.

From Figures 4.45 and 4.46 we can see the EMD method is acting as a set of filters and has decomposed the original vibration signal into eight bands from high to low frequency.

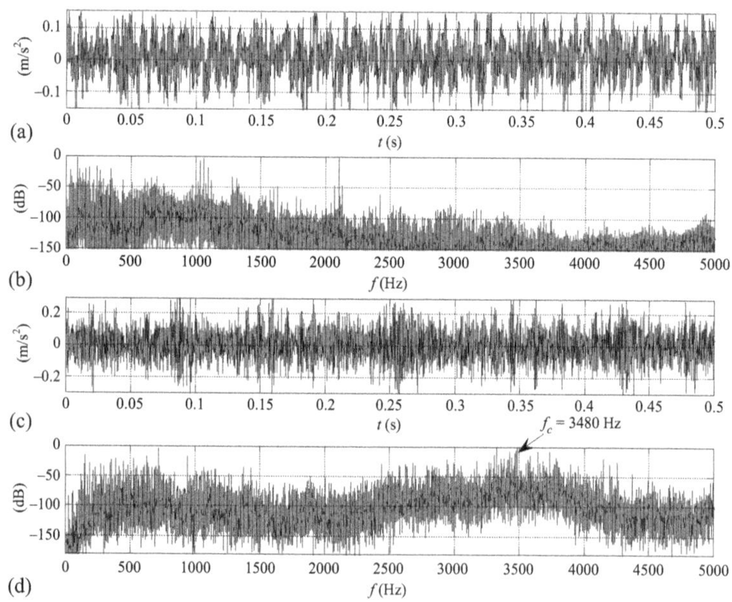


*Figure 4.44 An example of characteristics vibration signals under two different bearing statuses at the same operating condition (1750 rpm speed and 1 hp load). (a) Healthy bearing. (c) ORF; 0.1778 mm. (b) and (d) their PDS, respectively*

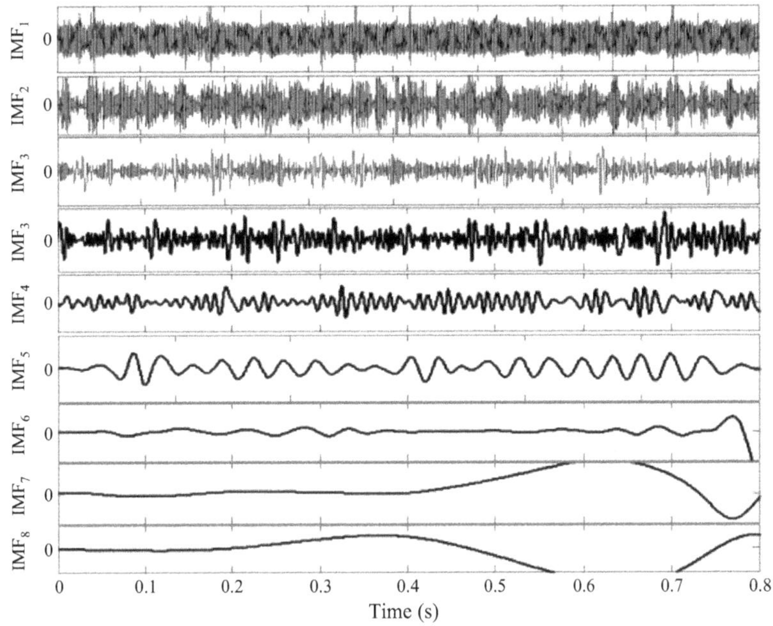


*Figure 4.45 The first eight IMF components of healthy bearing vibration signal under the operating condition: 1750 rpm speed and 1 hp load*

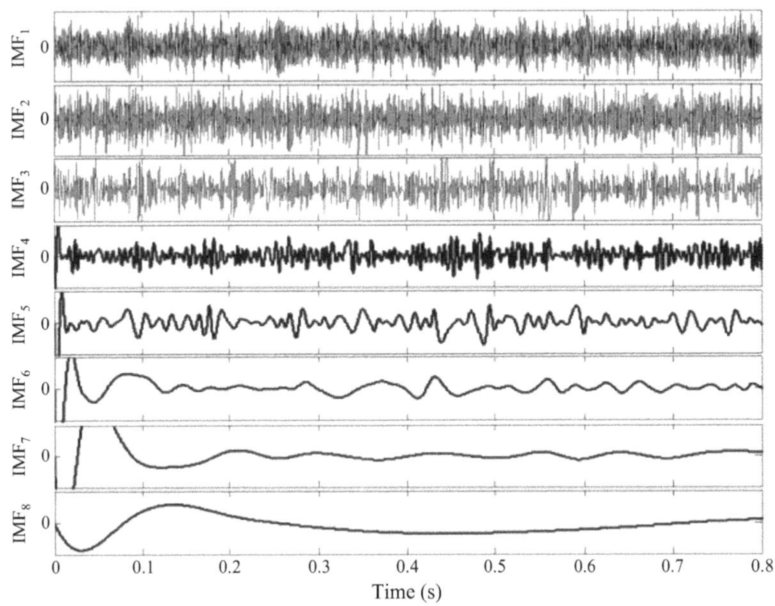


*Figure 4.46 The first eight IMF components of the ORF bearing vibration signal under the operating condition: 1750 rpm speed and 1 hp load*

As shown in Figure 4.47, the b-spectrum contour plot is dominated by the presence of several distinct peaks around the central frequency $f_c$. For better resolution, this part of the picture is enlarged in Figure 4.48. Characteristic peaks around the central frequency pair ($f_c$, $f_c$) can be observed. The peak distances are equal to the sums and the differences between the central frequency $f_c$, the BD frequency BPFO, and its sub-harmonic multiple exactly at 2 × BPFO. The presence of these peaks, which are associated with the harmonic of the BPFO, requires particular attention since it provides a further indication of the complex nonlinear mechanisms present in the vibration response of defective bearings.

*IMF energy criterion*

The energy of the $n$ IMFs obtained using the EMD method is $E_1, E_2, \ldots, E_n$. Table 4.11 shows the energy percent of the IMF components, which is designated as $P_i = E_i/E$, where $E$ is the whole signal energy ($E = \sum_{i=1}^{n} E_i$ is equal to the total energy of the original signal due to the orthogonality of the EMD decomposition). The energy percentages of the first four IMF components are named as $P_1$, $P_2$, $P_3$, and $P_4$. $P_5$ is the energy percentage of the fifth IMF components and the remaining components (note that $P_5 = (E_5 + E_6 + \cdots + E_n)/E$). From Table 4.11, we can see that the most dominant frequency and energy features of the outer race BD are mainly described

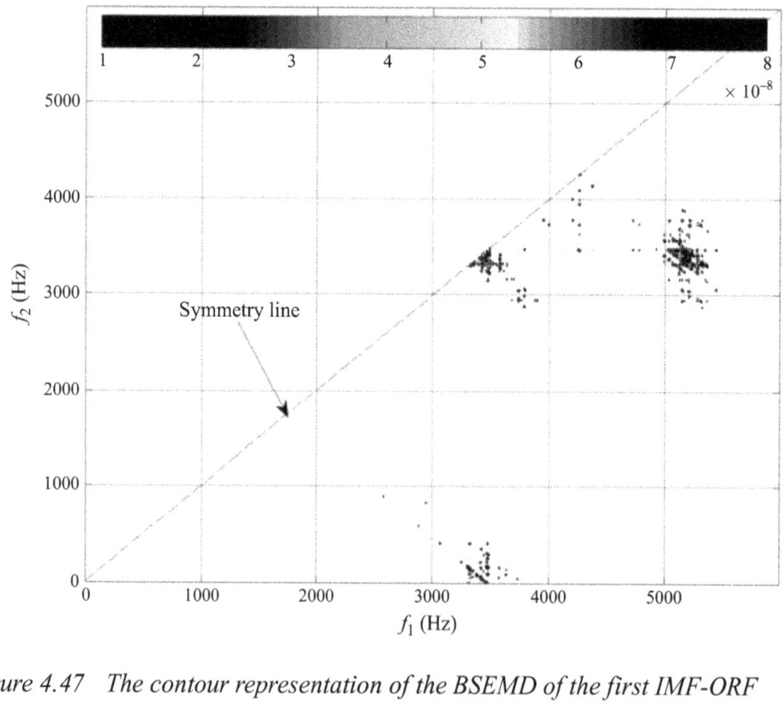


*Figure 4.47 The contour representation of the BSEMD of the first IMF-ORF bearing signal*

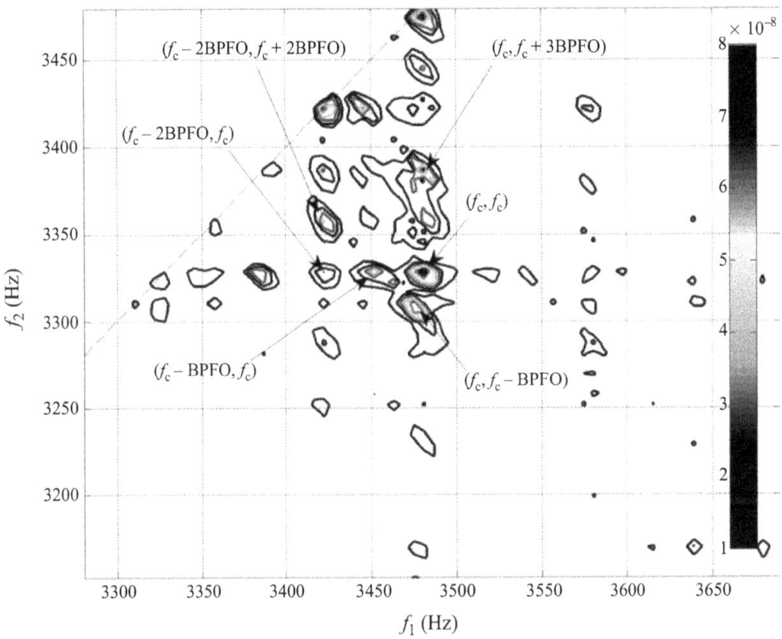


*Figure 4.48 The entire frequency band of Figure 4.9, zoom in the area between 3200 and 3700 Hz in the $f_1$–$f_2$ spectral frequency axes*

*Table 4.11 Energy percentage of IMF components*

| Vibration signals | Energy percentage | | | | |
|---|---|---|---|---|---|
| | $P_1$ | $P_2$ | $P_3$ | $P_4$ | $P_5$ |
| HB | 0.51529 | 0.26738 | 0.19391 | 0.029383 | 0.00945 |
| ORF | 0.71153 | 0.18232 | 0.14309 | 0.0334 | 0.01238 |

by the first components. When the ORF occurs, the energy percentage of the first components will increase compared to the healthy bearing.

*Statistical significance*

The results shown in the subsection seem to be from a single run of simulation/ experiment. The results generated from the data do not have statistical significance and may not be able to generalize. A series of experiments should be conducted to support the author's claims. Besides, in fault detection problems, the performance of a detection algorithm usually depends on the trade-off between robustness and sensitivity. The sensitivity and robustness of the proposed BSEMD method need to be explored by running a series of experiments. A receiver operating characteristic (ROC) curve will make the results more convincing.

To examine the BSEMD detection measure's performance, ROC curves are often the only valid method of evaluation [26]. An ROC curve is a detection performance evaluation methodology and demonstrates how effectively a certain detector can quantitatively separate two groups [90,91]. An ROC curve shows the trade-off between the probability of detection or true positives rate (*tpr*), also called sensitivity and recall versus the probability of false alarm or false positives rate (*fpr*). ROC curves are well described by Fawcett [90,91]. The *tpr* and *fpr* are mathematically expressed in (4.61) and (4.62), respectively:

$$tpr = \frac{true\,positives}{true\,positives + false\,negatives} \tag{4.61}$$

$$fpr = \frac{false\,positives}{false\,positives + true\,negatives} \tag{4.62}$$

For each case (healthy, ORF, IRF, and BF) and each load and speed combinations ((0 hp, 1797 rpm), (1 hp, 1772 rpm), (2 hp, 1750 rpm), and (3 hp, 1730 rpm)), and under different BD severities (as shown in Table 4.12), a series of 70 independent Monte-Carlo experiments are conducted. For each experiment, the probability of false alarm and the probability of detection are obtained by counting detection results out of 3360 independent Monte-Carlo experiments by the BSEMD-based method. The resultant ROC curve is shown in Figure 4.49. Thus, when applied to experimental data from real bearings, the BSEMD method successfully identified more than 98.9% of the bearing data available with less than 1.1% error.

*Table 4.12 Description of bearing data set analyzed under rated conditions*

| **Bearing condition** | | **Fault specifications** | |
|---|---|---|---|
| | | **Diameter (mm)** | **Depth (mm)** |
| HB | HB | 0 | 0 |
| IRF | $IRF_{17}$ | 0.1778 | 0.279 |
| | $IRF_{35}$ | 0.3556 | 0.279 |
| | $IRF_{53}$ | 0.5334 | 0.279 |
| | $IRF_{71}$ | 0.7112 | 0.279 |
| BF | $BF_{17}$ | 0.1778 | 0.279 |
| | $BF_{35}$ | 0.3556 | 0.279 |
| | $BF_{53}$ | 0.5334 | 0.279 |
| | $BF_{71}$ | 0.7112 | 0.279 |
| ORF | $ORF_{17}$ | 0.1778 | 0.279 |
| | $ORF_{35}$ | 0.7112 | 0.279 |
| | $ORF_{53}$ | 0.5334 | 0.279 |

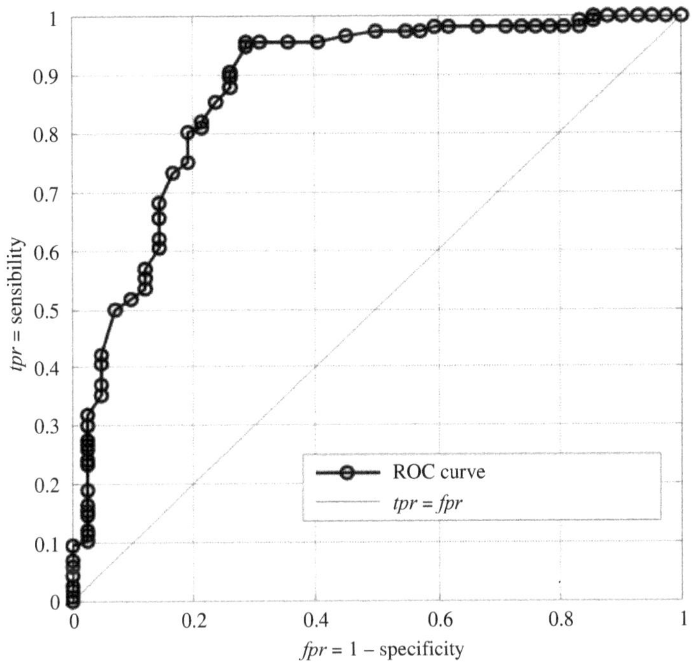


*Figure 4.49 ROC curve for BSEMD performance evaluation method*

Note that a critical work of bearing fault diagnosis is locating the optimum frequency band that contains a faulty bearing signal, which is usually buried in the noise background. Now, envelope analysis is commonly used to obtain the BD harmonics from the envelope signal spectrum analysis and has shown good results in identifying incipient failures occurring in the different parts of a bearing. However, the main step in implementing envelope analysis is to determine a frequency band that contains a faulty bearing signal component with the highest signal noise level. Conventionally, the choice of the band is made by manual spectrum comparison via identifying the resonance frequency where the largest change occurred. In the next subsection, we will present a squared envelope-based spectral kurtosis (SK) method to determine optimum envelope analysis parameters including the filtering band and center frequency through a short-time FT (STFT).

## *4.4.4 The use of SK for bearing fault diagnosis*

### 4.4.4.1 SK and its application for bearing fault diagnosis

*Definition and physical interpretation*

Dwyer [92] originally proposed SK only for stationary signals as the normalized fourth-order moment of the real part of the FT. However, in real practical applications bearing vibration signals are nonstationary. Recently, Antoni *et al.* [40,48,54–57,70,76,83] proposed the formalization for both stationary and nonstationary signals and introduced the SK technique into mechanical fault diagnosis.

SK provides a robust way of detecting incipient faults that produce impulse-like signals, even in the presence of strong noise. SK also offers a way of designing optimal filters for filtering the fault signature using the kurtogram or the fast kurtogram (ways to compute the SK) [70,75,76].

SK has been proved suitable in detecting premature faults from the strong noise and is widely used in fault diagnosis of REBs. A bearing signal is a train of impulses and an impulse has much higher kurtosis value than Gaussian type signals. Kurtosis is a statistical parameter, defined as

$$Kurtosis = \frac{\frac{1}{N}\sum_{i=1}^{N}(x_i - \bar{x})^4}{\left(\frac{1}{N}\sum_{i=1}^{N}(x_i - \bar{x})^2\right)^2} \tag{4.63}$$

where $x$ is the sampled time signal, $i$ is the sample index, $N$ is the number of samples, and $\bar{x}$ is the sample mean. This normalized fourth moment is designed to reflect the "peakedness" of the signal. The SK, of a signal, is defined as the kurtosis of its spectral components. The SK of a signal $x(t)$ can be defined as the normalized fourth-order spectral moment, i.e.:

$$SK_x(f) = \frac{\langle |X^4(t,f)| \rangle}{\langle X^2(t,f) \rangle^2} - 2 \tag{4.64}$$

where $\langle \cdot \rangle$ represents the time-frequency averaging operator, $X^4(t,f)$ and $X^2(t,f)$ are the fourth-order and the second-order cumulants, respectively, of a band-pass filtered signal of $x(t)$ around $f$. The constant 2 is used since $X(t,f)$ is the complex envelope of $x(t)$ at frequency $f$.

The most important properties of this definition are as follows [48,54–57,70,75,76,83,93]:

- *The* SK *of a stationary process is a constant function of frequency.*
- *The* SK *of a stationary Gaussian process is identical.*

The SK overcomes some limits of the global kurtosis in distinguishing a high-frequency train of shocks from noise. It can be shown that the SK of a nonstationary process $x(n)$ affected by stationary noise $b(t)$ is

$$SK_{(x+b)}(f) = \frac{SK_x(f)}{(1+\rho(f))^2} + \frac{\rho(f)^2\, SK_b}{(1+\rho(f))^2} \tag{4.65}$$

where $f \neq 0$; $\rho(f)$ is the SNR as a function of frequency. If $b(t)$ is an additive stationary Gaussian noise independent of $x(t)$, then SK becomes

$$SK_{(x+b)}(f) = \frac{SK_x(f)}{(1+\rho(f))^2} \tag{4.66}$$

The aforementioned properties clarify how the SK is capable of detecting, characterizing, and locating in frequency the presence of hidden nonstationarities. Indeed, from (4.66), we can see the value of $SK_{(x+b)}(f)$ is similar to $SK_x(f)$ at frequencies with high SNR. If the SNR is very low, it is close to zero. Therefore, SK value directly indicates the SNR of the defective signal at each frequency and can find the RFB of vibration signal automatically while designing a band-pass filter. It has been shown in [54–57] that SK is a complement of the classical PSD to detect nonstationary components of a signal and that it can be applied as well on the real part, the imaginary part of the modulus of the signal's spectrum. However, a minimum number of spectra are required to correctly estimate the SK of a signal. In practice, this number is reached using the STFT. The principle of STFT is to split a signal into *k* segments and compute the FFT on each segment. This technique is designed to find the frequency band by high kurtosis value. STFT coefficients of each time window of the vibration signal are calculated for kurtosis individually, all of which are averaged to result in the SK, as shown in Figure 4.50. This method is similar to the Welch method for power spectral estimation. The pivot of this method is that the time window must encompass only one impulse; otherwise, the bridge between impulses will smooth kurtosis out. However, it is very difficult to determine the time window length.

Technically speaking, many signal time-frequency decomposition methods have ever been adopted to perform the different multi-rate filter-bank structures used in the SK technique.

The first task is to design a filter bank that decomposes the signal through a series of sub-bands. Various architectures are possible as demonstrated in the open literature [54–56,75] such as multi-rate filters, wavelet transform (WT), wavelet packets, dual-tree wavelets, etc. In this subsection, an implementation based on the STFT is used due to its simplicity and high flexibility.

The kurtogram was proposed in [75] as a tool for blind identification of detection filters for diagnostics. As a result, a 2D map (called the kurtogram) is obtained (Figure 4.51), which presents values of SK calculated for various parameters of frequency and bandwidth, in short, a high value of the kurtogram indicates high

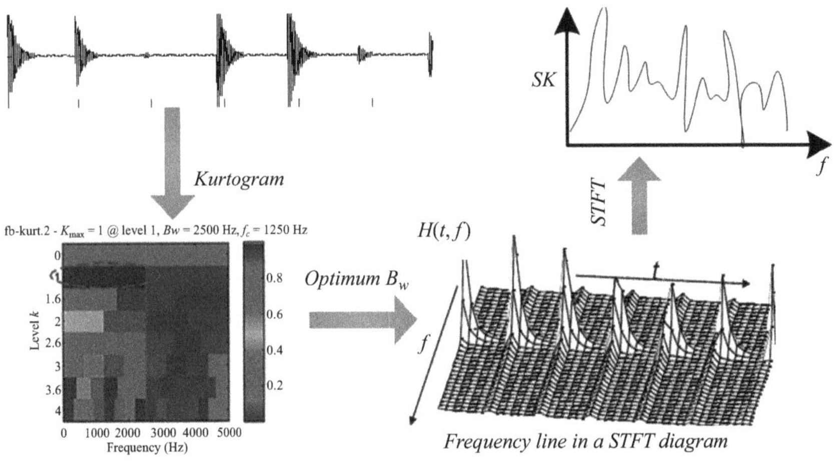


*Figure 4.50 Calculation of SK from the STFT; SK is an algorithm that indicates how kurtosis varies with frequency*

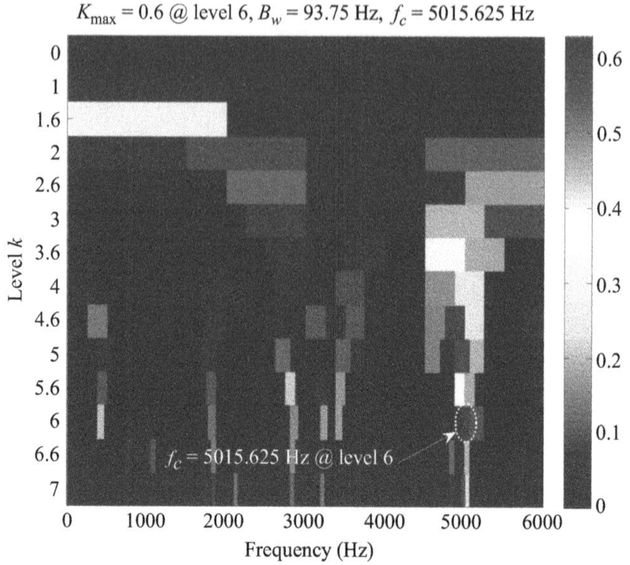


*Figure 4.51 The fast kurtogram of SK of an outer race vibration signal. The optimal filtering band is highlighted by a white dashed circle*

impulsiveness in the corresponding frequency band. The original kurtogram was based on STFT calculation. A faster version of the kurtogram is the fast kurtogram, based on the filterbank approach.

Nevertheless, the kurtosis value depends on both central frequency "$f_c$" and bandwidth "$B_w$" of each frequency band, so it is hard to determine the decomposition mode. In practice, many combinations of different central frequency and bandwidth have to be tried to find a suitable frequency band for envelope analysis, which needs considerable computation.

The principle of the kurtogram algorithm is based on an arborescent multi-rate filter-bank structure. A 1/2-binary tree kurtogram estimator is shown in Figure 4.52, where center frequency and bandwidth can be automatically determined. Those gray levels are shown in different squares and indicate the values of SK. Therefore, the maximum value can be easily found by some simple searching techniques.

As shown in Figure 4.51, 1–7 levels of filter bank decomposition are tried and level 6 is turned out to be the best in this case ($k_{max} = 0.6$ (maximum kurtosis value) at level 6, a band-pass filter of center frequency at $f_c = 50153.625$ Hz, and a bandwidth of $B_w = 93.75$ Hz was used to filter the vibration signal). The gray-level scale in Figure 4.51 denotes kurtosis value. For a comprehensive derivation of SK together with and its entire properties, one should refer to Refs. [54–57,75].

*The characteristics of rolling bearing vibration signals*

This subsection focuses on REBs supporting radial loads. Figures 4.23–4.35 show the structure of an REB. Commonly, if the vibration spectrum of a healthy bearing

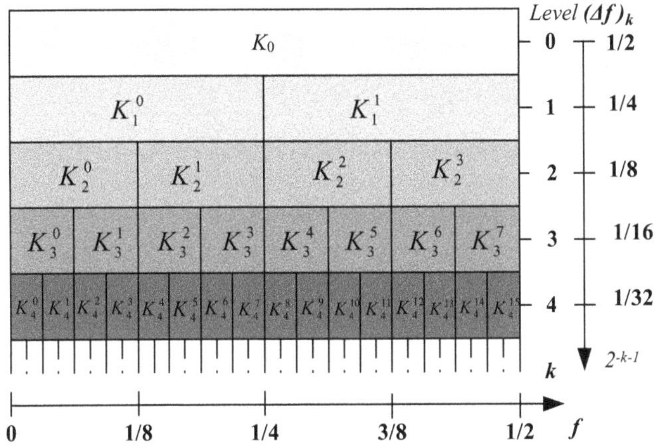


Figure 4.52 Combinations of center frequency and bandwidth for the 1/2-binary tree kurtogram estimator

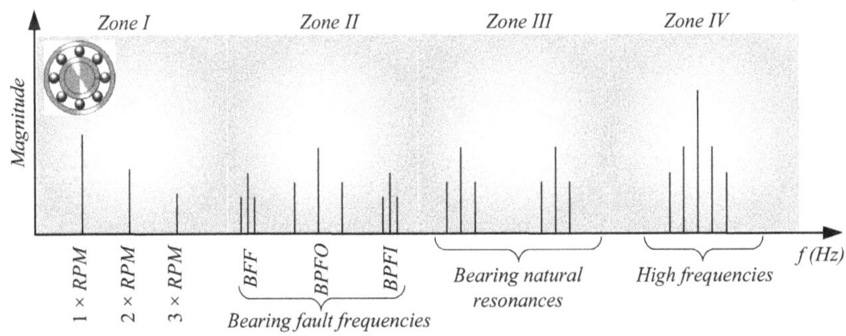


Figure 4.53 The frequency content of a vibration signal of a damaged REB

contains any information at all, then it is information related to the shaft rotation speed and its harmonics, which is shown as *Zone I* in Figure 4.53. Any other frequencies might indicate noise or frequencies related to other rotating parts operating at the same time with the bearing under test [53–60,63–67]. During its early stages, the damage on the surface is mostly only localized, e.g. pits or spalls. As shown in Figure 4.35, the vibration signal, in this case, includes repetitive impacts of the moving components on the defect. These impacts could create "repetition" frequencies that depend on whether the defect is on the inner or the outer race or the rolling element.

The repetition rates are denoted bearing frequencies, for example, BPFO, BPFI, and BFF are frequently used.

Apart from *Zone I*, in this case, *Zones II*, *III*, and/or *IV* might appear in the frequency spectrum of the vibration (Figure 4.35). Most of the time, only the vibration spectra of bearings with early faults contain information of damage since with time these faults can eventually be smoothed and not give as sharp impulses. So for early faults, the repetition impulses might create initially an increase in frequencies in the high-frequency range, *Zone IV*, and maybe excite the resonant frequencies of the bearing parts, later on, *Zone III*, as well as the repetition frequencies of *Zone II* (BFF, BPFO, BPFI). It has been observed in previous studies though that many times the vibration of a damaged bearing might not carry the desired information, and that in this case, SK analysis might be of more use for damage detection.

In this subsection, we present a squared envelope-based SK, called SESK, method to determine optimum envelope analysis parameters including the filtering band and center frequency through an STFT, as it seems better suited to analyze the nonstationarity of the random impact process caused by these faults. In short, the detection of a bearing fault in the outer race is proposed through the application of SK-based algorithms to improve the squared envelope-based spectral (SES) analysis of the vibration signals.

#### 4.4.4.2 SESK proposed method

*Methodology*

Figure 4.54 summarizes the proposed methodology used in this study for bearing fault detection based on SESK method of the raw vibration signal. The SESK application process is shown in Figure 4.54. Once the vibration signal is acquired, three main processing steps are conducted as follows:

- SK-based algorithm (fast kurtogram) indicates in a gray-level colormap, the kurtosis values for several combinations of the center frequency ($f_c$) and bandwidth ($B_w$) in a predetermined way (Figure 4.52). Then, the optimum filter, defined by $f_c$ and $B_w$ (with highest kurtosis value), is selected to be used in the envelope computation.
- SES can be viewed as a development and improvement for envelope analysis. Usually, it consists of four steps: (1) determination of the analysis frequency band; (2) design of a band-pass filter; (3) calculation of the squared band-passed signal; (4) derivation of the Fourier spectrum for the envelope signal.
- Finally, SESK is performed, if it does not contain the fault characteristic frequency; it means that the bearing under test is healthy. Otherwise, the bearing is faulty, such that, in a case of an ORF, its fault characteristic frequency and harmonics can be identified by a peak around values predicted by (4.41).

In this way, some advantages can be outlined from the analysis of envelope using Hilbert transform (as an approach to the amplitude demodulation); in this case, an optimum filter is used to extract a frequency band to be demodulated, removing adjacent components that may interfere with the analysis. In this case, according to [54–57,67], the signal envelope can be described as the analytical signal module, which is obtained by the inverse transform of the extracted one-sided band frequency.

On the other hand, the envelope signal analysis is limited by SNR. Besides, the envelope of a signal is the squared root of the squared envelope. This square root

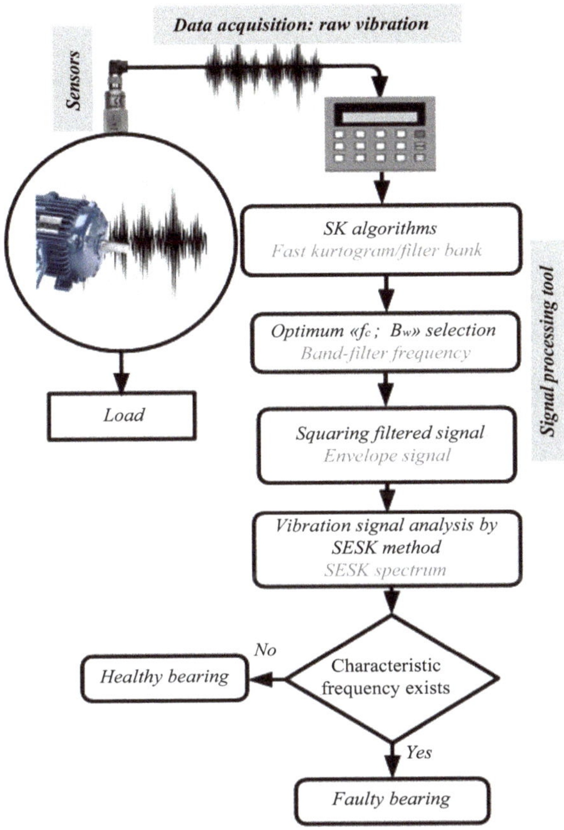


*Figure 4.54 Flowchart of the proposed SESK bearing diagnosis method*

operation inserts high-frequency components, and some of them might be aliased if their frequencies are higher than the Nyquist limit [67].

This process might mask the fault information. Thus, an SES analysis is performed rather than the envelope analysis.

Considering an analytic filtered signal $x[n]$, its SES is calculated using discrete FT (DFT) as given by

$$SES_x = \left|DFT\left(x[n]^2\right)\right|^2 \tag{4.67}$$

SES has been widely used in industrial applications, mainly due to its low complexity and efficiency. Thus, this subsection aims to use SESK to analyze vibrations to detect localized faults in REBs. However, the greatest difficulty in using envelope analysis is to define the frequency bandwidth to be used for amplitude demodulation. This difficulty has been mitigated through the advancement of SK and SK-based algorithms.

*REB signals model*

Assuming constant rotational speed and the load of the bearing, the vibration signals generated by an REB with defected outer race were modeled by Randall *et al.* as given in (4.57).

The simulated signal generated by (4.57) corresponds to the typical response of a bearing with an ORF. The shaft rotation speed $f_r$ is 29.16 Hz [$f_r = rpm/60$]. The characteristic BD frequency BPFO is equal to 3.58 times the shaft rotation speed, leading to an estimation of the BPFO around 104 Hz (Figure 4.42(c)).

To extract the fault feature, an SK analysis method based on fast kurtogram is applied to the stimulation signal. This signal is decomposed into four frequency levels, with a 1/3-binary tree structure. The corresponding kurtogram is presented in Figure 4.55, from which a kurtosis dominant frequency band with center frequency $f_c$ of 8300 Hz and bandwidth of 3333.33 Hz is identified. With this information, an optimal band-pass filter is further designed to extract the impulses from the raw vibration signal. Figure 4.56(b) illustrates the filtered signal.

In the next section, artificially produced defects are introduced in the bearing outer race to simulate a localized fault, allowing the evaluation of the SESK applied methodology.

### 4.4.4.3 Experimental results

The bearing data center of CWRU published bearing signal data online for researchers to validate new theories and techniques. All data are annotated with bearing geometric

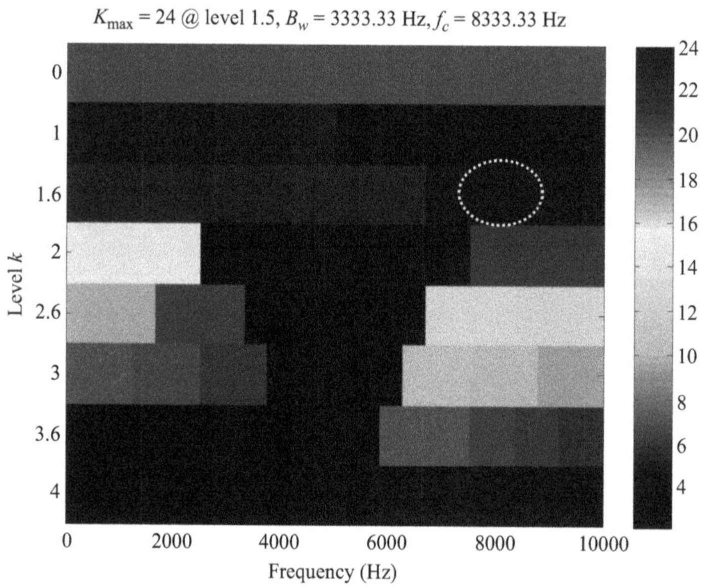


*Figure 4.55 Fast kurtogram of simulated ORF vibration signal. The optimal filtering band is highlighted by a dashed white circle ($f_c$ = 8300 Hz, $B_w$ = 3300 Hz)*

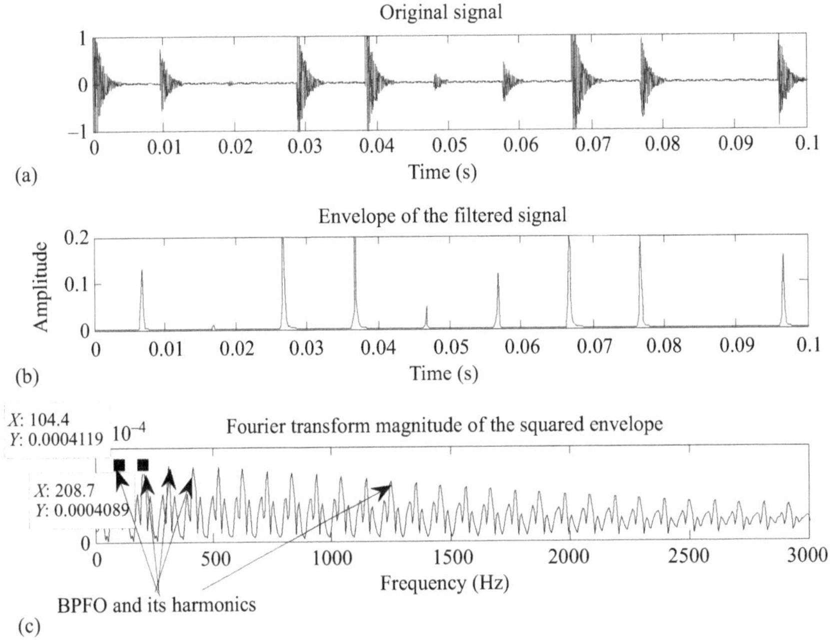


*Figure 4.56 (a) A simulated short segment of vibration response data showing outer race impacts and fast kurtogram results. (b) Envelope of the filtered signal and (c) envelope spectrum for healthy bearing ($f_c = 8300$ Hz, $B_w = 3300$ Hz)*

which were listed in Tables 4.10 and 4.12, operating condition and fault information. Figures 4.41 and 4.42 show the schematic diagram and the photo of the test stand from which the test data are collected, respectively.

### *Case study 1: outer race fault*

In this case, the shaft frequency $f_r$ is 29.95 Hz (1797 rpm). The motor load is 0 hp. The sampling rate is 12 kHz. The fault is located on the outer race, so the fault frequency BPFO = 104 Hz. The diameter and depth of the fault are 0.18 and 0.28 mm, respectively. The spectrum of this signal is shown in Figure 4.57. The RFB with the highest energy should be used for envelope analysis, and the bandwidth of this frequency band should be about 30 times the BPFO [73,74,78–83], $f_{BPFO}$. In this case, the frequency $f_{BPFO}$ is equal to 104 Hz. Thus, we select the frequency band from 3 to 4 kHz for envelope analysis. Figure 4.58(c) shows the SES of the band-pass-filtered signal.

As shown in Figure 4.59, a fast kurtogram was computed using seven levels, classic kurtosis, and filter bank options. Resulted kurtograms for the healthy and damaged bearing are shown in Figures 4.59 and 4.60. The optimal dyad (frequency/frequency bandwidth) for signal filtering is chosen based on the previous kurtogram. In general, the optimal dyad is chosen by avoiding maxima that are close to border conditions or

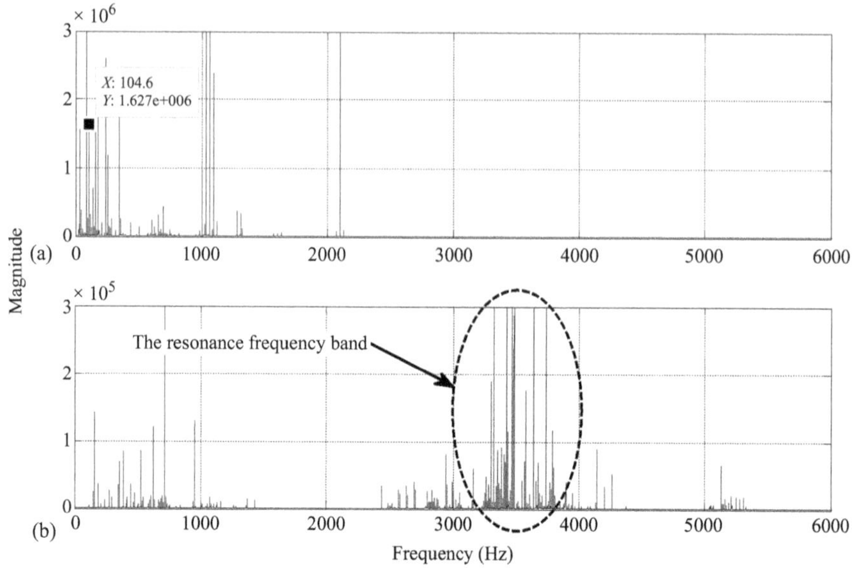


*Figure 4.57 The spectrum of the real vibration signal generated by (a) healthy bearing, and (b) bearing with an ORF*

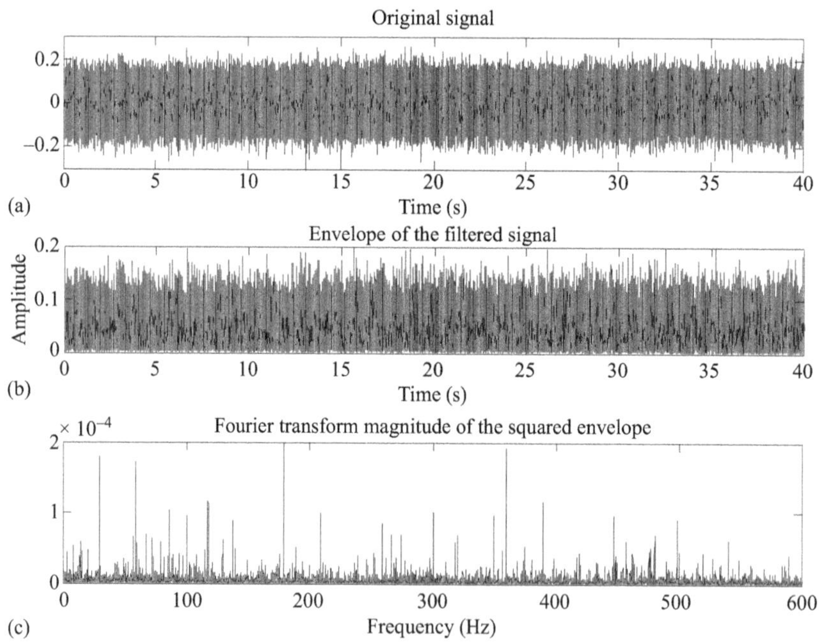


*Figure 4.58 FK results: (a) trend of the raw vibration signal, (b) envelope of the filtered signal, and (c) SESK for healthy bearing ($f_c = 1500$ Hz, $B_w = 3000$ Hz)*

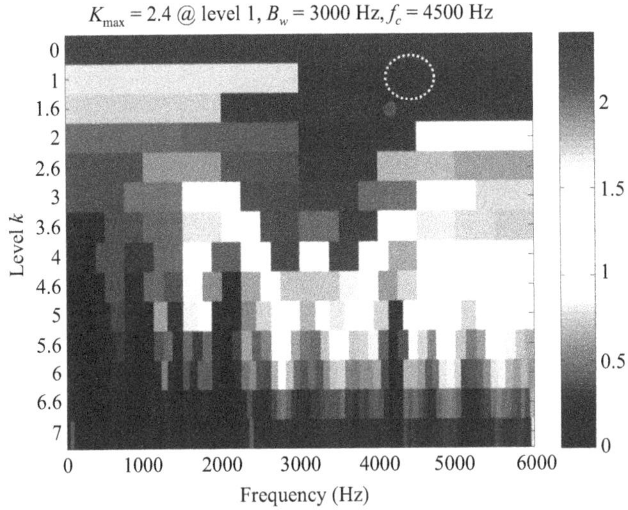


*Figure 4.59* *Fast kurtogram of vibration signal from the motor with an outer race damaged bearing. The optimal filtering band is highlighted by a dashed white circle ($f_c = 3000$ Hz, $B_w = 4500$ Hz)*

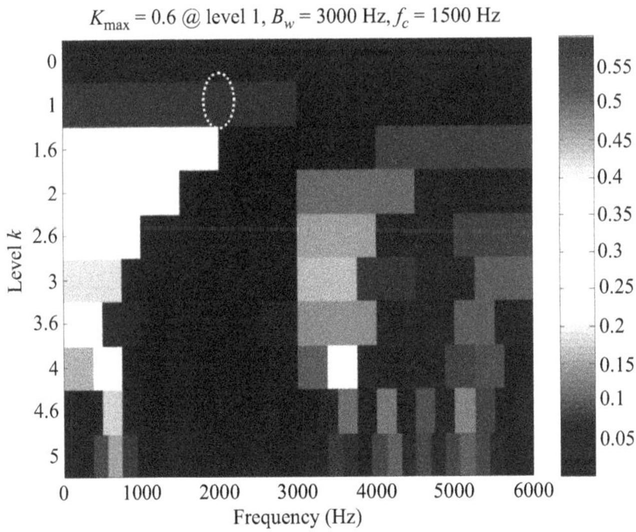


*Figure 4.60* *FK of vibration signal from the motor with healthy bearing. The optimal filtering band is highlighted by a dashed white circle ($f_c = 1500$ Hz, $B_w = 3000$ Hz)*

too far from the real vibration mode of the machine. In the case of damaged bearing, it was used $f_c = 1500$ Hz and $B_w = 3000$ Hz. The resulting envelopes of the filtered signal are illustrated in Figures 4.58 and 4.61, respectively. Figures 4.58(b) and 4.61(b) give the envelope of the filtered signal, and Figures 4.58(c) and 4.61(c) show the envelope demodulation spectrum of the filtered signal. It can be seen the displayed envelope spectrum using SK indicates the ORF frequency BPFO is approximately 104 Hz witch according to the experiment information given in Table 4.10.

The raw signal of faulty bearing and its SES is shown in Figure 4.61. With the fast kurtogram-based method, the fault characteristic frequency $f_{BPFO}$ is located at 104.556 Hz (=(1750/60) × 3.5848), and its associated harmonics at 209.11, 313.67, 418.22, 522.78, 627.34 Hz, and so on, can be easily detected.

Similarly, the faulty vibration signal (ORF) under different fault diameters and their respective SESK are given in Figure 4.62, where the ORF is diagnosed at 104.12 Hz.

*Case study 2: inner race fault*

The time signals in Figure 4.63(a)–(c) show a series of impulse responses at BPFI. The SESK in Figure 4.63(f) still has harmonics of BPFI surrounded by sidebands

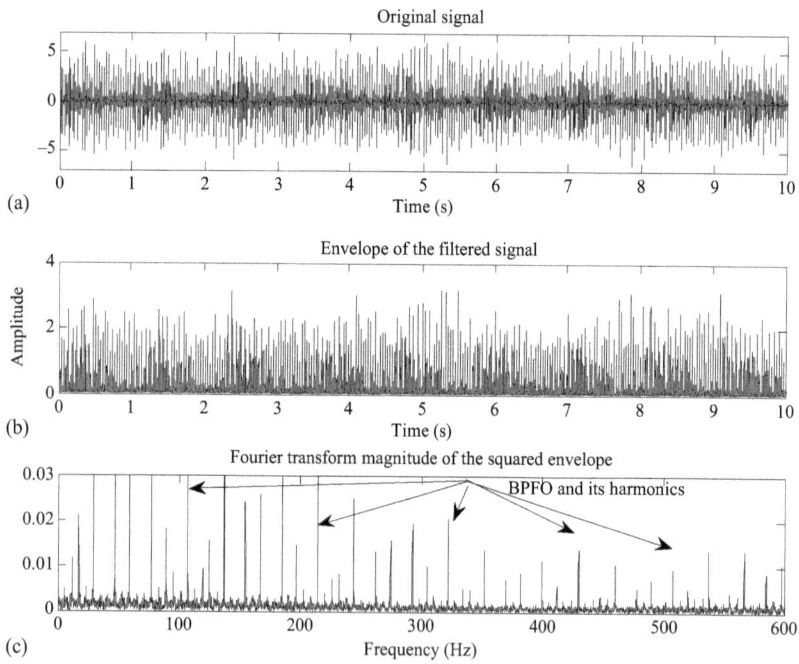


*Figure 4.61 FK results: (a) trend of raw vibration signal, (b) envelope of the filtered signal, and (c) SESK for faulty bearing ($f_c$ = 3000 Hz, $B_w$ = 4500 Hz). Outer race characteristic frequency and its harmonics are pointed by arrows. ORF is diagnosed at 104.12 Hz*

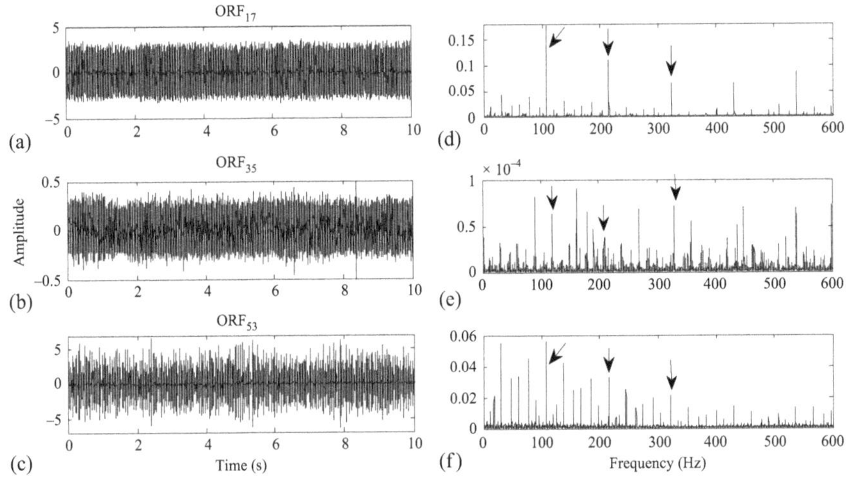


*Figure 4.62 (a), (b) and (c): faulty vibration signal (ORF) under different fault diameters and their respective SESK in (d), (e) and (f). ORF is diagnosed at 104.12 Hz*

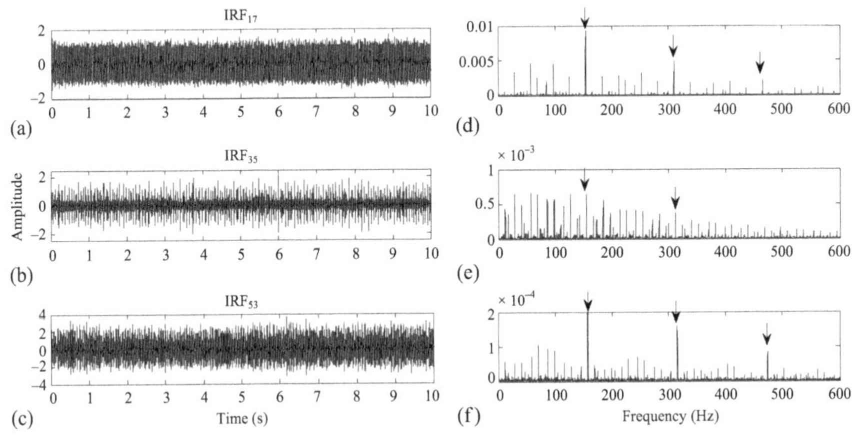


*Figure 4.63 (a), (b) and (c): faulty vibration signal (IRF) under different fault diameters and their respective SESK in (d), (e) and (f). IRF is diagnosed at 159.20 Hz*

spaced at shaft speed, though it will be seen that the spread of sidebands is greater than in Figure 4.63(e) and (d), indicating a more impulsive modulation. It is suspected that this could be a result of mechanical looseness, causing impulsive modulation of random amplitude at intervals of one revolution, but not necessarily phase-locked to the rotation. The smallest and largest faults (17 and 53 mm) in this category were all diagnosable using the SESK algorithm. The strongest ball-fault harmonics are at 2 and 4 times BPFI.

### *Case study 3: ball fault*

In Figure 4.64, the BF cases are certainly the most difficult to diagnose, with only a few giving the classic envelope spectrum symptoms of harmonics of possibly with dominant even harmonics surrounded by modulation sidebands at cage speed, and with corresponding low harmonics. The only data sets diagnosable from SESK of the raw signal (Figure 4.64) are from the 17 and 53 mm fault cases.

### *Statistical significance*

The sensitivity and robustness of the SESK method need to be explored by running a series of experiments. An ROC curve will make the results more convincing. The results presented in Figure 4.65 show the performance of the proposed impact detection algorithm, which included SESK.

The *tpr* and *fpr* are mathematically expressed in (4.63) and (4.64), respectively.

For each case (healthy, ORF, IRF, and BF) and each load and speed combinations ((0 hp, 1797 rpm), (1 hp, 1772 rpm), (2 hp, 1750 rpm), and (3 hp, 1730 rpm)), and under different BD severities (as shown in Table 4.12), a series of 70 independent Monte-Carlo experiments are conducted. For each experiment, the probability of false alarm and the probability of detection are obtained by counting detection results out of 3360 independent Monte-Carlo experiments by the SESK method. The resultant ROC curve is shown in Figure 4.65. Thus, when applied to experimental data from real bearings, the SESK method successfully identified more than 96.9% of the bearing data available with less than 1.1% error.

The total area under the ROC curve (AUC) is a single index for measuring the statistical diagnosis algorithm performance. Figure 4.65 depicts three different ROC

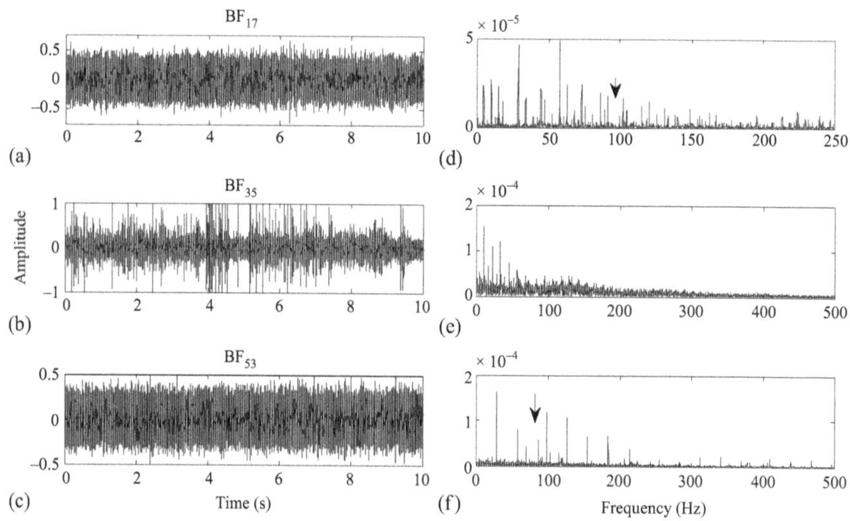


*Figure 4.64 (a), (b) and (c): faulty vibration signal (BF) under different fault diameters and their respective SESK in (d), (e) and (f). BF is diagnosed at 139 Hz*

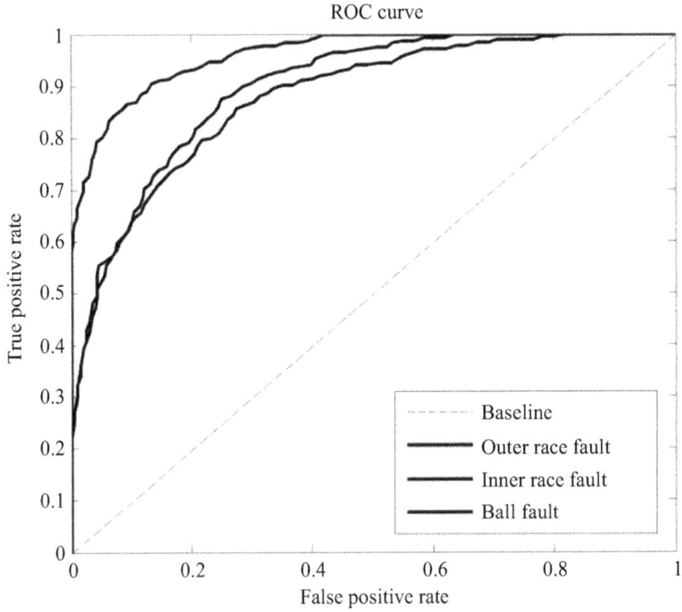


*Figure 4.65 ROC curves for SESK detection performance evaluation*

curves. Considering the AUC, the diagnosis test using ORF results is better than both with IRF and BF, and the curve is closer to the perfect discrimination. Test using IRF has good validity and test using BF has moderate one. Also, the majority of detections proved to be positive, which indicates the presence of impacts. However, a significant amount of missed detections is evident and is a consequence of weak impacts compared to the bearing damage. In Figure 4.65, we can say about the performances of the three bearing cases and statistical results; the AUC values are ORF (0.879), IRF (0.815), and BF (0.768). The relative ORF curve turns out to be superior.

## 4.5 Conclusions and perspectives

This chapter describes condition monitoring and fault diagnosis based on an HOS analysis. The crucial purpose of fault diagnosis is to increase the consistency of electromechanical system reliability and availability. In this chapter, we have revealed that, by using HOS analyses, it is possible to build a reliable induction machine diagnosis tool. Several HOS methods were proposed and investigated to extract novel features or signatures, which are useful in condition monitoring, from a single time-series. Besides, the applicability of the HOS analysis for condition monitoring is evaluated with real experimental data.

In this chapter, various HOS-based algorithms and their challenging problems are discussed. Every electromechanical system signal has its nonlinear characteristics behavior and there is probably no unified diagnostic method that applies to all machines since each diagnostic method has its advantages and shortcomings. Therefore, to obtain effective signatures from a specific system, one can use various techniques and select the most suitable diagnostic tools among them.

The bispectrum can detect and quantify QPC phenomena. Furthermore, it filters out the additive Gaussian noise. However, fault detection using the bispectrum was not completely acceptable in our experiment. Since the value of the bispectrum is determined by both the degree of QPC and the complex amplitude of the interacting frequency components, the value of the bispectrum is very sensitive to the amplitudes of the interacting spectral components.

Concerning future work, we propose the following perspective.

In this chapter, we were focused on third-order spectral moments (e.g. bispectrum or bicoherence), since the dominant frequencies satisfied the frequency selection rule, $f_3 = f_1 + f_2$, which is an index of quadratic nonlinearities. Theoretically, it is straightforward to extend the HOS analysis of this chapter to fourth-order spectral moments (e.g. trispectrum and tricoherence) which are defined in 3D frequency space and are described by a frequency selection rule $f_4 = f_1 + f_2 + f_3$. Such a step would enable the identification of cubic nonlinearities. However, because of the need to work in 3D space, the analytical and computational complexity is greatly increased and must be dealt with.

Although this chapter focused on extracting nonlinear signatures from a time-series from a single fault, a logical extension is to consider the case of multiple faults.

## Appendix A

We take (4.37) and study whether there are nonzero products between the pulses $\delta(\cdot)$; for each of the three factors of the triple product (of (4.37)), we will see the positions of $\delta(\cdot)$. If the three factors have $\delta(\cdot)$ in the same position then the product is nonzero.

$$\begin{aligned}
\hat{B}(f_1,f_2) &\approx I_a(f_1)\,I_a(f_2)\,I_a^*(f_3 = f_1 + f_2)\\
&\approx \frac{1}{8}\left(\begin{aligned} i_f\delta(f_1 - f_s)e^{j\varphi} + \sum_{k_1} i_{l,k_1}\delta(f_1 - f_{l,k_1})e^{j\varphi_{l,k_1}}\\ + \sum_{k_1} i_{r,k_1}\delta(f_1 - f_{r,k_1})e^{j\varphi_{r,k_1}}\end{aligned}\right)\\
&\times\left(\begin{aligned} i_f\delta(f_2 - f_s)e^{j\varphi} + \sum_{k_2} i_{l,k_2}\delta(f_2 - f_{l,k_2})e^{j\varphi_{l,k_2}}\\ + \sum_{k_2} i_{r,k_2}\delta(f_2 - f_{r,k_2})e^{j\varphi_{r,k_2}}\end{aligned}\right)\\
&\times\left(\begin{aligned} i_f\delta(f_3 - f_s)e^{-j\varphi} + \sum_{k_3} i_{l,k_3}\delta(f_3 - f_{l,k_3})e^{-j\varphi_{l,k_3}}\\ + \sum_{k_3} i_{r,k_3}\delta(f_3 - f_{r,k_3})e^{-j\varphi_{r,k_3}}\end{aligned}\right)
\end{aligned}$$

So, $f_3 = f_1 + f_2 = f_s + f_x; f_{l,k} = f_s(1 - 2ks)$; and $f_{r,k} = f_s(1 + 2ks)$.
The first factors are

$$\begin{cases} \frac{i_f}{2}\underbrace{\delta(0)}_{=1}\mathrm{e}^{j\varphi}; \frac{1}{2}\sum_k i_{l,k}\underbrace{\delta(2ksf_s)}_{=0}\mathrm{e}^{j\varphi_{l,k}}; \frac{1}{2}\sum_k i_{r,k}\underbrace{\delta(-2ksf_s)}_{=0}\mathrm{e}^{j\varphi_{r,k}} \\ \frac{i_f}{2}\delta(f_x - f_s)\mathrm{e}^{j\varphi}; \frac{1}{2}\sum_k i_{l,k}\delta(f_x - f_s + 2ksf_s)\mathrm{e}^{j\varphi_{l,k}}; \frac{1}{2}\sum_k i_{r,k}\delta(f_x - f_s - 2ksf_s)\mathrm{e}^{j\varphi_{r,k}} \\ \frac{i_f}{2}\delta(f_x)\mathrm{e}^{-j\varphi}; \frac{1}{2}\sum_k i_{l,k}\delta(f_x + 2ksf_s)\mathrm{e}^{-j\varphi_{l,k}}; \frac{1}{2}\sum_k i_{r,k}\delta(f_x - 2ksf_s)\mathrm{e}^{-j\varphi_{r,k}} \end{cases}$$

Note that there are nonzero terms if and only if $2ks = 1$. It gives:

$$\begin{cases} \frac{i_f}{2}\mathrm{e}^{j\varphi}; 0; 0 \\ \frac{i_f}{2}\delta(f_x - f_s)\mathrm{e}^{j\varphi}; \frac{1}{2}\sum_k i_{l,k}\delta(f_x)\mathrm{e}^{j\varphi_{l,k}}; \frac{1}{2}\sum_k i_{r,k}\delta(f_x - 2f_s)\mathrm{e}^{j\varphi_{r,k}} \\ \frac{i_f}{2}\delta(f_x)\mathrm{e}^{-j\varphi}; \frac{1}{2}\sum_k i_{l,k}\delta(f_x + f_s)\mathrm{e}^{-j\varphi_{l,k}}; \frac{1}{2}\sum_k i_{r,k}\delta(f_x - f_s)\mathrm{e}^{-j\varphi_{r,k}} \end{cases}$$

As a result, we have only the impulses $\delta(x)$ and $\delta(x + f_s)$. Hence $\hat{B}(f_1, f_2)$ is given by

$$\begin{aligned} \hat{B}(f_1 = f_s, f_2 = f_x) &= \frac{i_f^2}{8}\delta(f_x - f_s)\delta(f_x - f_s)\mathrm{e}^{2j\varphi}\sum_k i_{r,k}\mathrm{e}^{-j\varphi_{r,k}} + \delta(f_x)\delta(f_x)\sum_k i_{l,k}\mathrm{e}^{j\varphi_{l,k}} \\ &= \frac{i_f^2}{8}\left[\underbrace{\delta(f_x - f_s)}_{(f_s, f_s)}\mathrm{e}^{2j\varphi}\sum_k i_{r,k}\mathrm{e}^{-j\varphi_{r,k}} + \underbrace{\delta(f_x)}_{(f_s, 0)}\sum_k i_{l,k}\mathrm{e}^{j\varphi_{l,k}}\right] \end{aligned}$$

## Appendix B

We set $f_1 = f_2 = f$, the diagonal component of the estimated bispectrum noted $\hat{D}(f)$ is given by

$$\begin{aligned} \hat{B}(f_1, f_2)\big|_{f_1 = f_2 = f} &= \hat{D}(f) = E\{X^2(f)X^*(2f)\} \\ &= \frac{1}{8}\left(i_f\delta(f - f_s)\mathrm{e}^{j\varphi} + \sum_k i_{l,k}\delta(f - f_{l,k})\mathrm{e}^{j\varphi_{l,k}} + \sum_k i_{r,k}\delta(f - f_{r,k})\mathrm{e}^{j\varphi_{l,k}}\right) \\ &\times \left(i_f\delta(f - f_s)\mathrm{e}^{j\varphi} + \sum_k i_{l,k}\delta(f - f_{l,k})\mathrm{e}^{j\varphi_{l,k}} + \sum_k i_{r,k}\delta(f - f_{r,k})\mathrm{e}^{j\varphi_{r,k}}\right) \\ &\times \left(i_f\delta(2f - f_s)\mathrm{e}^{-j\varphi} + \sum_k i_{l,k}\delta(2f - f_{l,k})\mathrm{e}^{-j\varphi_{l,k}} + \sum_k i_{r,k}\delta(2f - f_{r,k})\mathrm{e}^{-j\varphi_{l,k}}\right) \end{aligned}$$

$$\hat{D}(f) = \frac{1}{8}\left(i_f\delta(f-f_s)e^{j\varphi} + \sum_k i_{l,k}\delta(f-f_s+2ksf_s)e^{j\varphi_{l,k}} + \sum_k i_{r,k}\delta(f-f_s-2ksf_s)e^{j\varphi_{r,k}}\right)$$
$$\times\left(i_f\delta(f-f_s)e^{j\varphi} + \sum_k i_{l,k}\delta(f-f_s+2ksf_s)e^{j\varphi_{l,k}} + \sum_k i_{r,k}\delta(f-f_s-2ksf_s)e^{j\varphi_{r,k}}\right)$$
$$\times\left(i_f\delta(2f-f_s)e^{-j\varphi} + \sum_k i_{l,k}\delta(2f-f_s+2ksf_s)e^{-j\varphi_{l,k}} + \sum_k i_{r,k}\delta(2f-f_s-2ksf_s)e^{-j\varphi_{r,k}}\right)$$

$$\hat{D}(f) = \frac{1}{8}\left[\begin{array}{r} i_f^3\delta(f-f_s)(f-f_s)(2f-f_s)e^{j\varphi} + \sum_k i_{l,k}^3\delta(f-f_{l,k})\delta(f-f_{l,k})\delta(2f-f_{l,k})e^{j\varphi_{l,k}} \\ + \sum_k i_{r,k}^3\delta(f-f_{r,k})\delta(f-f_{r,k})\delta(2f-f_{r,k})e^{-j\varphi_{r,k}} \end{array}\right]$$
$$= \frac{i_f^3}{8}\delta(f-f_s)e^{j\varphi} + \sum_k i_{l,k}^3\delta(f-f_{l,k})e^{j\varphi_{l,k}} + \sum_k i_{r,k}^3\delta(f-f_{r,k})e^{-j\varphi_{r,k}}$$

## References


[1] Nikias C.L. and Petropulu A., Higher-order Spectra Analysis: A Nonlinear Signal Processing Framework, 1993; Englewood Cliffs, NJ: Prentice-Hall.

[2] Mendel J.M., Tutorial on higher order statistics (spectra) in signal processing and system theory: theoretical results and some applications, *Proceedings of the IEEE 1991*, 1991, 287–305.

[3] Lyons R.G., Understanding Digital Signal Processing, 2012; Upper Saddle River, NJ: Prentice Hall.

[4] Courtney C.R.P., Neild S.A., Wilcox P.D., and Drinkwater B.W., Application of the bispectrum for detection of small nonlinearities excited sinusoidally, *Journal of Sound and Vibration*, 2010; **JSV-329**: 4279–4293.

[5] Picinbono B., Polyspectra of Ordered Signals, *IEEE Transactions on Information Theory*, 1999; **IT-45**: 2239–2252.

[6] Hinich M.J. and Wolinsky M., Normalizing bispectra, *Journal of Statistical Planning and Inference*, 2005; **JSPI-130**: 405–411.

[7] Vaseghi S.V., Advanced Digital Signal Processing and Noise Reduction, 2000; Second Edition, John Wiley & Sons Ltd.

[8] Priestley M.B., Nonlinear and Nonstationary Time Series Analysis, 1988; Academic Press, New York.

[9] Barnett A.G. and Wolff R.C., A time-domain test for some types of nonlinearity, *IEEE Transactions on Signal Processing*, 2005; **SP-53**: 26–33.

[10] Nichols J.M., Olson C.C., Michalowicz J.V., and Bucholtz F., The bispectrum and bicoherence for quadratically nonlinear systems subject to non-Gaussian inputs, *IEEE Transactions on Signal Processing*, 2009; **SP-57**: 3879–3890.

[11] Saidi, L., The deterministic bispectrum of coupled harmonic random signals and its application to rotor faults diagnosis considering noise immunity, *Applied Acoustics*, 2017; **AA-122**: 72–87.

[12] Peng Z.K., Zhang W., Yang M.B.T., Meng G., and Chu F.L., The parametric characteristic of bispectrum for nonlinear systems subjected to Gaussian input, *Mechanical Systems and Signal Processing*, 2013; **MSSP-36**: 456–470.

[13] Chua K.C., Chandranb V., Acharyaa U.R., and Lima C.M., Application of higher order statistics/spectra in biomedical signals—A review, *Medical Engineering & Physics*, 2010; **MEP-32**: 679–696.

[14] Kim Y.C. and Powers E.J., Digital bispectral analysis and its applications to nonlinear wave interactions, *IEEE Transactions on Plasma Science*, 1979; **PS-7**: 120–131.

[15] Zhang G.C., Ge M., Tong H., Xu Y., and Du R., Bispectral analysis for on-line monitoring of stamping operation, *Engineering Applications of Artificial Intelligence*, 2002; **EAAI-15**: 97–104.

[16] Saidi L., Fnaiech F., Capolino G.A., and Henao H., Stator current bispectrum patterns for induction machines multiple-faults detection, *IEEE Industrial Electronics Conference 2012*, 2012, 5132–5137.

[17] Gu F., Shao Y., Hu N., Naid A., and Ball A.D., Electrical motor current signal analysis using a modified bispectrum for fault diagnosis of downstream mechanical equipment, *Mechanical Systems and Signal Processing*, 2011; **MSSP-25**: 360–372.

[18] Messina A.R. and Vittal V., Assessment of nonlinear interaction between nonlinearity coupled modes using higher order spectra, *IEEE Transactions on Power Systems*, 2005; **PS-20**: 375–383.

[19] Saidi L., Henao H., Fnaiech F., Capolino G.A., and Cirrincione G., Application of higher order spectra analysis for rotor broken bar detection in induction machines, *IEEE International Symposium on Diagnostics for Electric Machines, Power Electronics and Drives, SDEMPED 2011*, 2011, 31–38.

[20] Saidi L., Fnaiech F., Capolino G.A., and Henao H., Diagnosis of broken bars fault in induction machines using higher order spectral analysis, ISA Transactions, 2013; **ISAT-52**: 140–148.

[21] Treetrong J., Sinha J.K., Gub F., and Ball A., Bispectrum of stator phase current for fault detection of induction motor, ISA Transactions, 2013; **ISTA-48**: 378–382.

[22] Nichols J.M. and Murphy K.D., Modeling and detection of delamination in a composite beam: A polyspectral approach, *Mechanical Systems and Signal Processing*, 2010; **MSSP-24**: 365–378.

[23] Park H., Jang B., Powers E.J., Grady W.M., and Arapostathis A., Machine condition monitoring utilizing a novel bispectral change detection, *IEEE Power Engineering Society General Meeting 2007*, 2007, 624–628.

[24] Wang X., Chen Y., and Ding M., Testing for statistical significance in bispectra: a surrogate data approach and application to neuroscience, *IEEE Transactions on Biomedical Engineering*, 2007; **BE-54**: 1974–1982.

[25] Momoh J. and Mili L., Dynamical models in fault tolerant operation and control of energy processing systems, *Operation and Control of Electric Energy Processing Systems*, 2010; Hoboken, NJ: Wiley-IEEE Press: 15–45.

[26] Razik H., de Rossiter Correa M.B., and da Silva E.R.C., A novel monitoring of load level and broken bar fault severity applied to squirrel-cage induction motors using a genetic algorithm, *IEEE Transactions on Industrial Electronics*, 2009; **IE-56**: 4615–4626.

[27] Nandi S., Toliyat H.A., and Li X., Condition monitoring and fault diagnosis of electrical motors—a review. *IEEE Transactions on Energy Conversion*, 2005; **EC-20**: 719–729.

[28] Bellini A., Filippetti F., Tassoni C., and Capolino G.A., Advances in diagnostic techniques for induction machines, *IEEE Transactions on Industrial Electronics*, 2008; **IE-55**: 4109–4126.

[29] Filippetti F., Franceschini Tassoni G.C., and Vas P., AI techniques in induction machines diagnosis including the speed ripple effect, *IEEE Transactions on Industry Applications*, 1998; **IA-34**: 98–108.

[30] Bellini A., Filippetti F., Franceschini G., Tassoni C., and Kliman G.B., Quantitative evaluation of induction motor broken bars by means of electrical signature analysis, *IEEE Transactions on Industry Applications*, 2001; **IA-37**: 1248–1255.

[31] Tavner P.J., Review of condition monitoring of rotating electrical machines, *IET Electric Power Applications*, 2008; **EPA-2**: 215–247.

[32] Thorsen O.V. and Dalva M., A survey of faults on induction motors in offshore oil industry, petrochemical industry, gas terminals, and oil refineries, *IEEE Transactions on Industry Applications*, 1995; **IA-31**: 1186–1196.

[33] Benbouzid M.E.H., A review of induction motors signature analysis as a medium for faults detection, *IEEE Transactions on Industrial Electronics*, 2000; **IE-47**: 984–993.

[34] Thomson W.T. and Fenger M., Current signature analysis to detect induction motor faults, *IEEE Industry Applications Magazine*, 2001; **IAM-7**: 26–34.

[35] Benbouzid M.E.H. and Kliman G.B., What stator current processing-based technique to use for induction motor rotor faults diagnosis? *IEEE Transactions on Energy Conversion*, 2003; **EC-18**: 238–244.

[36] Benbouzid M.E.H., Vieira M., and Theys C., Induction motors' faults detection and localization using stator current advanced signal processing techniques, *IEEE Transactions on Power Electronics*, 1999; **PE-14**: 14–22.

[37] Ebrahimi B.M., Takbash A.M., and Faiz J., Losses calculation in line-start and inverter-fed induction motors under broken bar fault, *IEEE Transactions on Instrumentation and Measurement*, 2013; **IM-62**: 140–152.

[38] Faiz J., Ebrahimi B.M., Akin B., and Toliyat H.A., Dynamic analysis of mixed eccentricity signatures at various operating points and scrutiny of related indices for induction motors, *IET Electric Power Applications*, 2010; **EPA-4**: 1–16.

[39] Faiz J., Ghorbanian V., and Ebrahimi B.M., EMD-based analysis of industrial induction motors with broken rotor bars for identification of operating point at different supply modes, *IEEE Transactions on Industrial Informatics*, 2014; **TII-10**: 957–966.

[40] Antonino-Daviu J.A., Riera-Guasp M., Pons-Llinares J., Roger-Folch J., Perez R.B., and Charlton-Perez C., Toward condition monitoring of damper windings in synchronous motors via EMD analysis, *IEEE Transactions on Energy Conversion*, 2012; **EC**: 432–439.

[41] Nemec M., Ambrozic V., Nedeljkovic D., Fiser R., and Drobnic K., Detection of broken bars in induction motor using voltage pattern analysis, *IEEE International Symposium on Diagnostics for Electric Machines, Power Electronics and Drives, SDEMPED 2009*, 2009, 1–6.

[42] Saidi L., Benbouzid M., Diallo D., Amirat Y., Elbouchikhi E., and Wang T., PMSG-based tidal current turbine biofouling diagnosis using stator current bispectrum analysis, *In IECON 2019—45th Annual Conference of the IEEE Industrial Electronics Society 2019*, 2019, 6998–7003.

[43] Jafarian M.J. and Nazarzadeh J., Spectral analysis for diagnosis of bearing defects in induction machine drives, *IET Electric Power Applications*, 2018; **EPA-13**: 340–348.

[44] Panagiotou P.A., Arvanitakis I., Lophitis N., Antonino-Daviu J.A., and Gyftakis K.N., On the broken rotor bar diagnosis using time-frequency analysis: "Is one spectral representation enough for the characterisation of monitored signals?," *IET Electric Power Applications*, 2019; **EPA-13**: 932–942.

[45] Faiz J., Gorbanian V., and Joksimoviæ G., Fault Diagnosis of Induction Motors, 2017; IET.

[46] Pineda-Sanchez M., Riera-Guasp M., Antonino-Daviu J.A., Roger-Folch J., Perez-Cruz J., and Puche-Panadero R., Instantaneous frequency of the left sideband harmonic during the start-up transient: a new method for diagnosis of broken bars, *IEEE Transactions on Industrial Electronics*, 2009; **IE-56**: 4557–4570.

[47] Antonio-Daviu J., Aviyente S., Strangas E.G., and Riera-Guasp M., Scale invariant feature extraction algorithm for the automatic diagnosis of rotor asymmetries in induction motors, *IEEE Transactions on Industrial Informatics*, 2013; **IA-9**: 100–108.

[48] Randall R.B. and Antoni J., Rolling element bearing diagnostics–a tutorial, *Mechanical Systems and Signal Processing*, 2011; **MSSP-25**: 485–520.

[49] Tran V.T., Yang B.S., Gua F., and Ball A., Thermal image enhancement using bi-dimensional empirical mode decomposition in combination with relevance vector machine for rotating machinery fault diagnosis, *Mechanical Systems and Signal Processing*, 2013; **MSSP-38**: 601–614.

[50] Gryllias K.C. and Antoniadis I.A., A support vector machine approach based on physical model training for rolling element bearing fault detection in industrial environments, *Engineering Applications of Artificial Intelligence*, 2012; **EAAI-25**: 326–344.

[51] Widodo A., Yang B.S., Gu D.S., and Choi B.K., Intelligent fault diagnosis system of induction motor based on transient current signal, *Mechatronics*, 2009; **M-19**: 680–689.

[52] Widodo A., Yang B.S., and Han T., Combination of independent component analysis and support vector machines for intelligent faults diagnosis of induction motors, *Expert Systems with Applications*, 2007; **ESA-32**: 299–312.

[53] Frosini L. and Bassi E., Stator current and motor efficiency as indicators for different types of bearings faults in induction motors, *IEEE Transactions on Industrial Electronics*, 2010; **IE-57**: 244–251.

[54] Urbanek J., Barszcz T., and Antoni J., Integrated modulation intensity distribution as a practical tool for condition monitoring, *Applied Acoustics*, 2014; **AA-77**: 184–194.

[55] Sawalhi N., Randall R., and Endo H., The enhancement of fault detection and diagnosis in rolling element bearings using minimum entropy deconvolution combined with spectral kurtosis, *Mechanical Systems and Signal Processing*, 2007; **MSSP-21**: 2616–2633.

[56] Barszcz T. and JabŁoński A., A novel method for the optimal band selection for vibration signal demodulation and comparison with the kurtogram, *Mechanical Systems and Signal Processing*, 2011; **MSSP-25**: 431–451.

[57] Randall R.B., Vibration-based Condition Monitoring: Industrial, Aerospace and Automotive Applications, 2011; Chichester: John Wiley & Sons Ltd.

[58] Saidi L., Ben Ali J., and Fnaiech F., Bi-spectrum based-EMD applied to the non-stationary vibration signals for bearing faults diagnosis, *ISA Transactions*, 2014; **ISAT-53**: 1650–1660.

[59] Saidi L. and Fnaiech F., Bearing defects decision making using higher order spectra features and support vector machines, *The 14th International Conference on Sciences and Techniques of Automatic Control & Computer Engineering (STA)*, 2013, 419–424.

[60] Ben Salem S., Bacha K., and Chaari A.K., Support vector machine based decision for mechanical fault condition monitoring in induction motor using an advanced Hilbert-Park transform, *ISA Transactions*, 2012; **ISAT-51**: 566–572.

[61] Stack J.R., Habetler T.G., and Harley R.G., Fault-signature modeling and detection of inner-race bearing faults, *IEEE Transactions on Industry Applications*, 2006; **IA-42**: 61–68.

[62] Arturo G.-P., Troncoso R.J.R., Yepez E.C., and Osornio-Rios R.A., The application of high-resolution spectral analysis for identifying multiple combined faults in induction motors, *IEEE Transactions on Industry Applications*, 2011; **IA-58**: 2002–2010.

[63] Yuan S. and Chu F., Fault diagnosis based on support vector machines with parameter optimisation by artificial immunisation algorithm, *Mechanical Systems and Signal Processing*, 2007; **MSSP-21**: 1318–1330.

[64] Yan R.R., Gao R.X., and Chen X., Wavelets for fault diagnosis of rotary machines: a review with applications, *Signal Processing*, 2014; **SP-96**: 1–15.

[65] Zhua D., Gaoa Q., Suna D., Lua Y., and Pengb S., A detection method for bearing faults using null space pursuit and S transform, *Signal Processing*, 2014; **SP-96**: 80–89.

[66] Baydar N. and Ball A., A comparative study of acoustic and vibration signals in detection of gear failures using Wigner-Ville distribution, *Mechanical Systems and Signal Processing*, 2001; **MSSP-15**: 1091–1107.

[67] Feldman M., Hilbert transform in vibration analysis, *Mechanical Systems and Signal Processing*, 2001; **MSSP-25**: 735–802.

[68] Saidi L., Ben Ali J., and Fnaiech F., Application of higher order spectral features and support vector machines for bearing faults classification, *ISA Transactions*, 2015; **ISAT-54**: 193–206.

[69] Ben Ali J., Chebel-Morello B., Said L., and Fnaiech F., Linear feature selection and classification using PNN and SFAM neural networks for a nearly online diagnosis of bearing naturally progressing degradations, *Engineering Applications of Artificial Intelligence*, 2015; **EEAI-42**: 67–81.

[70] Antoni J., Fast computation of the kurtogram for the detection of transient faults, *Mechanical Systems and Signal Processing*, 2007; **MSSP-21**: 108–124.

[71] Vapnik V.N., The Nature of Statistical Learning Theory, 1999; New York, NY: Springer.

[72] Ben Ali J., Fnaiech N., Saidi L., Chebel-Morello B., and Fnaiech F., Application of empirical mode decomposition and artificial neural network for automatic bearing fault diagnosis based on vibration signals, *Applied Acoustics*, 2015; **AP-89**: 16–27.

[73] Dion J.L., Stephan C., Chevallier G., and Festjens H., Tracking and removing modulated sinusoidal components: a solution based on the kurtosis and the extended Kalman filter, *Mechanical Systems and Signal Processing*, 2013; **MSSP-38**: 428–439.

[74] Wei G., Tse P.W., and Djordjevich A., Faulty bearing signal recovery from large noise using a hybrid method based on spectral kurtosis and ensemble empirical mode decomposition, *Measurement*, 2012; **M-45**: 1308–1322.

[75] Antoni J., The spectral kurtosis: a useful tool for characterising nonstationary signals, *Mechanical Systems and Signal Processing*, 2006; **MSSP-20**: 282–307.

[76] Antoni J. and Randall R., The spectral kurtosis: application to the vibratory surveillance and diagnostics of rotating machines, *Mechanical Systems and Signal Processing*, 2006; **MSSP-20**: 308–331.

[77] Lei Y., Lin J., He Z., and Zuo M.J., A review on empirical mode decomposition in fault diagnosis of rotating machinery, *Mechanical Systems and Signal Processing*, 2013; **MSSP-35**: 108–126.

[78] Borghesani P., Pennacchi P., and Chatterton S., The relationship between kurtosis- and envelope-based indexes for the diagnostic of rolling element bearings, *Mechanical Systems and Signal Processing*, 2014; **MSSP-43**: 25–43.

[79] Fasana A., Marchesiello S., Pirra M., Garibaldi L., and Torri A., Spectral kurtosis against SVM for best frequency selection in bearing diagnostics, *Mécanique & Industries*, 2010; **MI-11**: 489–494.

[80] Amirat Y., Choqueuse V., and Benbouzid M., EEMD-based wind turbine bearing failure detection using the generator stator current homopolar component, *Mechanical Systems and Signal Processing*, 2013; **MSSP-41**: 667–678.

[81] Zhang Y. and Randall R.B., Rolling element bearing fault diagnosis based on the combination of genetic algorithms and fast kurtogram, *Mechanical Systems and Signal Processing*, 2009; **MSSP-23**: 1509–1517.

[82] Guo Y., Liu T.W., Na J., and Fung R.F., Envelope order tracking for fault detection in rolling element bearings, *Journal of Sound and Vibration*, 2012; **JSV-331**; 5644–5654.

[83] Antoni J., The infogram: entropic evidence of the signature of repetitive transients, *Mechanical Systems and Signal Processing*, 2016; **MSSP-74**: 73–94.

[84] Saidi L., Ali J.B., Bechhoefer E., and Benbouzid M., Wind turbine high-speed shaft bearings health prognosis through a spectral kurtosis-derived indices and SVR, *Applied Acoustics*, 2017; **AP-120**: 1–8.

[85] Saidi L., Ali J.B., Benbouzid M., and Bechhoefer E., The use of SESK as a trend parameter for localized bearing fault diagnosis in induction machines, *ISA Transactions*, 2016; **ISAT-63**: 436–447.

[86] Ben Ali J., Saidi L., Harrath S., Bechhoefer E., and Benbouzid M., Online automatic diagnosis of wind turbine bearings progressive degradations under real experimental conditions based on unsupervised machine learning, *Applied Acoustics*, 2018; **AA-132**: 167–181.

[87] Saidi L., Ben Ali J., Benbouzid M., and Bechhofer E., An integrated wind turbine failures prognostic approach implementing Kalman smoother with confidence bounds, *Applied Acoustics*, 2018; **AA-138**: 199–208.

[88] Ben Ali J., Chebel-Morello B., Said L., Malinowski S., and Fnaiech F., Accurate bearing remaining useful life prediction based on Weibull distribution and artificial neural network, *Mechanical Systems and Signal Processing*, 2015; **MSSP-56,57**: 150–172.

[89] Loparo K.A., Bearing vibration data set, Case Western Reserve University. http://csegroups.case.edu/bearingdatacenter/pages/12k-drive-end-bearing-fault-data.

[90] Fawcett T., An introduction to ROC analysis, *Pattern Recognition Letters*, 2006; **PRL-27**: 861–874.

[91] Fawcett T., Using rule sets to maximize ROC performance, *Proc. Int. Conf. Data Mining*, 2001, 131–138.

[92] Dwyer R.F., Detection of non-Gaussian signals by frequency domain kurtosis estimation, *International Conference on Acoustics, Speech, and Signal Process ICASSP*, 1983, 607–610.

[93] Immovilli F., Cocconcelli M., Bellini A., and Rubini R. Detection of generalized-roughness bearing fault by spectral-kurtosis energy of vibration or current signals, *IEEE Transactions on Industrial Electronics*, 2009; **IE-56**: 4710–4717.

# Chapter 5
# Fault detection and diagnosis based on principal component analysis

*Tianzhen Wang*[1]

## 5.1 Introduction

With the rapid development of technology and productivity, modern industrial systems become increasingly complex. The reliability, maintainability and safety of these systems are more and more concerned by research. Therefore, the research of fault detection methods is very important. There are many fault detection and diagnosis (FDD) methods [1–7], according to different systems, which can be classified: FDD based on signal processing [2,8], FDD based on knowledge [2], FDD based on analytical model [3,9] and FDD based on reasoning [3]. Because there are mass variables, great changes of the variables in amplitude, higher response speed and more complex correlation relationship among many variables in the complex system [2,10]. The quantitative model could not be established without the equation of motion or the assistance from expert system in particular. Support vector data description (SVDD) is widely used to optimize system monitoring in the modern industries [11]. Because there are large amounts data generated and collected from the distribution control systems, some data-driven FDD methods are proposed for monitoring [12,13]. Meanwhile many statistical process control (SPC) methods are very effective [14–16], which play important role in FDD and improving the manufacturing process. Principal component analysis (PCA) is one of the most popular SPC methods [17–20] which is the core of fault detection technology based on multivariate SPC. Based on the original data space, it can construct a new set of latent variables to reduce the dimension of the original data space, and then extract the main change information from the new mapping space to extract the statistical characteristics. The basic idea of PCA is to find a group of new variables to replace the original variables, and the new variables are the linear combination of the original variables. From the point of view of optimization, the number of new variables is less than the original variables, and carries the useful information of the original variables to the maximum extent, and the new variables are not related to each other.

According to above problems in complex systems, there are several main limitations for PCA to use in FDD [21–23]: (1) The principal components (PCs) are

[1]Logistics Engineering College, Shanghai Maritime University, Shanghai, China

obtained based on the eigenvalues and eigenvectors of covariance matrix of multivariate data, but the representativeness of the PCs only depends on the magnitude of the eigenvalues, and the magnitude of the eigenvalues is correlated with the absolute value of multivariable, the error or absolute value that depends on its dimension. For example, the unit of length can be meters or centimeter, but the PCs are not the same. The number of PCs and the information contained in PCs dependent on the correlation among multivariate, because these processes may not be always correlative, it may be difficult to select significant PCs by use of classical PCA. (2) FDD based on PCA is generally used for time-invariant processes by the statistic confidence limit which is obtained by the Hotelling's $T^2$ or squared prediction error (SPE) statistics. However, when the processes are changed, the traditional statistic confidence limit, which is obtained by the Hotelling's $T^2$ or SPE statistics [24], is not available for fault detection of time-varying processes, and the performance of FDD will be degraded. In recent years, the improvements of FDD based on PCA are focused on time-varying processes monitoring. Such as, the fast-moving window PCA is proposed to use in time-varying industrial process monitoring [25]. And the recursive robust PCA model is proposed to continuously update the PCA to adapt to the time-varying process [26]. These FDD methods based on improved PCA are more effective for slow time-varying stability process. But the diagnostic performance will be degraded by the above FDD methods under nonsteady conditions [3,27–30]. (3) Most of the variables do not follow Gaussian distribution under nonsteady conditions. For the above methods, we assume the process data are Gaussian for statistic construction and determination of statistic confidence limit. When the process data does not follow Gaussian distribution, the rate of missed alarms and false alarms will be increased. So, it is better for data following Gaussian distribution when the improved PCA is used for FDD. There are many methods based on PCA are proposed to improve the FDD performance when the data are not following Gaussian processes, such as the Box-Cox transformation [31], Adaptive-PCA [32], PCA-support vector machines (SVM) model [15,18], Gaussian mixture model [33] and independent component analysis (ICA) [7,34]. Nevertheless, the computational complexity of these methods is high, which are not able to detect time-varying processes in real time. Especially, there are some singularity problems existing in Box-Cox method Gaussian mixture models (GMMs) when the data dimension is high. The indeterminacy of the ICA method will make real-time monitoring performance extremely decline. Dynamic PCA method [35] and canonical variate analysis (CVA) method [12] are widely used in FDD of dynamic systems, but non-Gaussian data are the bottleneck problem for them. The FDD based on multi-way PCA method [36] is proposed to use in batch process, but the position or quantity is very difficult to detected in every batch. And it is difficult for the PCA-SVM method to select kernel parameters which are key points of FDD performance.

In this chapter, PCA, relative PCA (RPCA) [37,38] and normalization PCA (NPCA) [27] are introduced with application in fault detection and fault diagnosis. There are some theories and applications about PCA in Section 5.2, such as the basic principles of PCA, geometrical interpretation of PCA, Hotelling's $T^2$ statistic and SPE statistic for fault detection's control limit. Then a fault detection method based on PCA is introduced for Tennessee Eastman (TE) process [39]. What's more, the fault diagnosis method based on PCA is introduced with its application for inverter [15]. There are some theories and application about RPCA in Section 5.3, such as the definition of Relative Transform, basic principles of RPCA and geometrical interpretation of RPCA. Then the fault detection method based on RPCA is introduced with its application [37,40]. In addition, in order to improve the control limit of PCA with Hotelling's $T^2$ [30], the dynamic data window control limit algorithm based on RPCA is introduced with its application. As follows, the fault diagnosis method based on RPCA is introduced with its application. There are some theories and application about NPCA in Section 5.4, such as the definition of longitudinal standardization (LS) and basic principles of NPCA. Next a fault detection method based on NPCA is presented with its application in wind power generation. Then another fault detection method based on NPCA is presented with its application in DC motor. In order to increase the control limit of PCA with Hotelling's $T^2$, a fault detection method based on NPCA-adaptive confidence limit (ACL) is presented with application in DC motor. At last, conclusions and future works was introduced in Section 5.5.

## 5.2 PCA and its application

### 5.2.1 PCA method

The PCA method is one of the most important methods in statistic method. PCA could be used to compress data with multidimensions. A small number of PCs can delegate the original system [21].

Considering the variables of system:

$$\boldsymbol{X}(K) \equiv [x_1(K), x_2(K), \ldots, x_n(K)]^T \in R^{n\times 1}$$

Each random variable $x_i(k) \in R^1,\ i = 1, 2, \ldots, n$, is a one-dimensional random variable, and obeys $\boldsymbol{x}_i(K) \sim N\left[x_{0,i}(K), \gamma_i\right],\ \gamma_i > 0.\ x_i(k) \in R^1,\ i = 1, 2, \ldots, n$, is the corresponding once realization.

The random data matrix composed by the system's random variables $\boldsymbol{x}(k)$ from $k$th to $(k + N - 1)$th is

$$\boldsymbol{X} :\equiv \boldsymbol{X}(k, k + N - 1) = [\boldsymbol{x}(k), \boldsymbol{x}(k + 1), \ldots, \boldsymbol{x}(k + N - 1)]$$

The corresponding once realization is the data matrix as follows: $[x(K), x(k + 1), \ldots, x(k + N - 1)]$. **(Note:** Random matrix is not distinguished from its realization when it was used on no confusions condition.)

$\boldsymbol{X}_i$ denotes the $i$th row vector of system matrix $\boldsymbol{X}$:

$$\mathbf{X}_i = [x_i(K), x_i\,(k + 1), \ldots, x_i\,(k + N - 1)] \tag{5.1}$$

Matrix $\boldsymbol{X}$ or $\boldsymbol{X}(k, k + N - 1)$ is expressed as follows:

$$\mathbf{X}(k, k + N - 1) \equiv \begin{bmatrix} \mathbf{X}_1 \\ \mathbf{X}_2 \\ \vdots \\ \mathbf{X}_n \end{bmatrix} \tag{5.2}$$

Considering covariance matrix $\Sigma_X$ of system matrix $\boldsymbol{X}$, if $E\{X\} = X_0$, then $\Sigma_X$ is

$$\Sigma_X = E\{[\mathbf{X}(k, k+N-1) - \boldsymbol{X}_0][\mathbf{X}(k, k+N-1) - \boldsymbol{X}_0]^T\} \tag{5.3}$$

Because

$$|\lambda \boldsymbol{I} - \Sigma_X| = 0 \tag{5.4}$$

and

$$[\lambda_i \boldsymbol{I} - \Sigma_X]\boldsymbol{e}_i = 0, \quad i = 1, \ldots, n \tag{5.5}$$

Eigenvalues $\lambda_i$ and corresponding eigenvector $e_i$ are computed from covariance matrix $\Sigma_X$, and suppose $\lambda_1 \geq \lambda_2 \geq \cdots \geq \lambda_n$.

The corresponding eigenvector $e_i$ is defined as follows:

$$\boldsymbol{E} = \begin{bmatrix} (\boldsymbol{e}_1)^T \\ (\boldsymbol{e}_2)^T \\ \vdots \\ (\boldsymbol{e}_n)^T \end{bmatrix}$$

So,

$$\boldsymbol{V} = \begin{bmatrix} V_1 \\ V_2 \\ \vdots \\ V_n \end{bmatrix} = \boldsymbol{EX} = \begin{bmatrix} \boldsymbol{e}_{11}X_1 & + & \boldsymbol{e}_{21}X_2 & + & \cdots & + & \boldsymbol{e}_{n1}X_n \\ \boldsymbol{e}_{12}X_1 & + & & + & \cdots & + & \\ \vdots & & & & \vdots & & \vdots \\ & + & & + & \cdots & + & \end{bmatrix}$$

$$\mathbf{V} = \begin{bmatrix} \mathbf{V}_1 \\ \mathbf{V}_2 \\ \vdots \\ \mathbf{V}_n \end{bmatrix} = \mathbf{EX} \tag{5.6}$$

$$= \begin{bmatrix} e_{11}\mathbf{X}_1 + e_{21}\mathbf{X}_2 + \cdots + e_{n1}\mathbf{X}_n \\ e_{12}\mathbf{X}_1 + e_{22}\mathbf{X}_2 + \cdots + e_{n2}\mathbf{X}_n \\ \vdots \\ e_{1n}\mathbf{X}_1 + e_{2n}\mathbf{X}_2 + \cdots + e_{nn}\mathbf{X}_n \end{bmatrix}$$

and satisfy with the following.

**Property 5.1** *Similar character configuration*

$$E\{V\} = \boldsymbol{EX}_0$$

$$\boldsymbol{\Sigma}_V = \boldsymbol{E\Sigma}_X\boldsymbol{E}^T$$

$$Var(V_i) = (\boldsymbol{e}_i)^T \Sigma_X \boldsymbol{e}_i = \lambda_i, \quad i = 1, 2, \ldots, n \tag{5.7}$$

$$\text{Cov}(\mathbf{V}_i, \mathbf{V}_j) = (\mathbf{e}_i)^T \Sigma_X \mathbf{e}_j = 0, \quad i \neq j \tag{5.8}$$

So, $m(m < n)$ PCs $v_1, v_2, \ldots, v_m$ are selected to explain original $n$ variables.

Mostly, the number of PCs is decided on the basis of the cumulative contribution rate $P$ as follows:

$$P\% = \frac{\Sigma_{i=1}^{m}\lambda_i}{\Sigma_{i=1}^{n}\lambda_i} \times 100\%$$

where $m$ is the number of PCs, and $P$ is determined by the user according to the requirement of system monitoring.

### 5.2.2 *The geometrical interpretation of PCA*

Suppose vector $\mathbf{x}(i)$ of matrix $\mathbf{X}$ obey $n$-dimensions normal distribution $N_n[E\{\mathbf{x}\}, \mathbf{\Sigma_X}]$. $E\{\mathbf{x}\}$ is the center of the ellipse sphere.

$$[\mathbf{x} - E\{\mathbf{x}\}]^T(\Sigma_\mathbf{X})^{-1}[\mathbf{x} - E\{\mathbf{x}\}] = c^2 \tag{5.9}$$

Every axis of the ellipse sphere is as follows:

$$\pm c\sqrt{\lambda_i}\mathbf{e}_i, \quad i = 1, 2, \ldots, n$$

Here, $(\lambda_i, \mathbf{e}_i)$ is the $i$th eigenvalue and eigenvector of covariance $\Sigma_\mathbf{X}$.

Suppose $E\{\mathbf{x}\} = 0$. So, (5.9) can be predigested to (5.10).

$$(\mathbf{x})^T(\Sigma_\mathbf{X})^{-1}\mathbf{x} = c^2 \tag{5.10}$$

Equation (5.10) is rewritten as follows:

$$(\mathbf{x})^T(\mathbf{E}^{-1}\mathbf{E})^T(\Sigma_\mathbf{X})^{-1}\mathbf{E}^{-1}\mathbf{E}\mathbf{x} = c^2 \tag{5.11}$$

$$(\mathbf{E}\mathbf{x})^T(\mathbf{E}\Sigma_\mathbf{X}\mathbf{E}^{-1})^{-1}(\mathbf{E}\mathbf{x}) = c^2 \tag{5.12}$$

$$(\mathbf{v})^T(\Sigma_\mathbf{V})^{-1}\mathbf{v} = c^2 \tag{5.13}$$

Because of (5.14)

$$\Sigma_\mathbf{X} = \sum_{i=1}^{n} \lambda_i \mathbf{e}_i(\mathbf{e}_i)^T \tag{5.14}$$

So, it can be gained:

$$(\Sigma_\mathbf{X})^{-1} = \sum_{i=1}^{n} \frac{1}{\lambda_i}\mathbf{e}_i(\mathbf{e}_i)^T \tag{5.15}$$

So,

$$(x)^T\left(\sum_{i=1}^{n}\frac{1}{\lambda_i}e_i(e_i)^T\right)(x) = \frac{1}{\lambda_1}\left[(e_1)^T x]^2 + \frac{1}{\lambda_2}[(e_2)^T x]^2 + \cdots + \frac{1}{\lambda_n}[(e_n)^T x\right]^2 \tag{5.16}$$

Here, $(e_1)^T x, (e_2)^T x, \ldots, (e_n)^T x$ are PCs of $x$, and (5.81) can be rewritten as follows:

$$c^2 = \frac{1}{\lambda_1}(v_1)^2 + \frac{1}{\lambda_2}(v_2)^2 + \cdots + \frac{1}{\lambda_n}(v_n)^2 \tag{5.17}$$

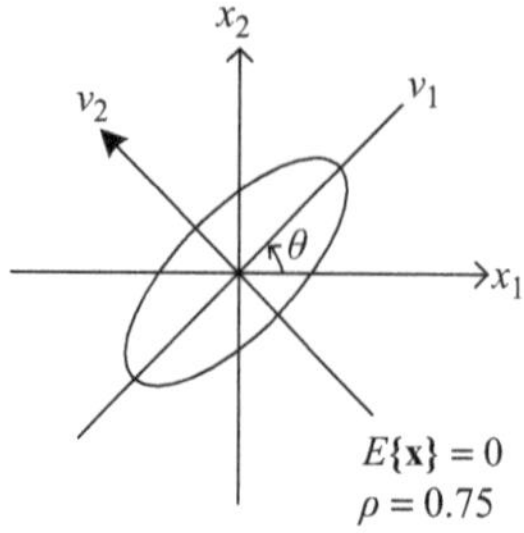


*Figure 5.1* *The constant density ellipse $(x)^T(\Sigma_X)^{-1}\mathbf{x} = c^2$ and the PC $v_1$, $v_2$ for a bivariate normal random vector* $\mathbf{x}$ *having mean 0*

where $\lambda_1$ is the maximum eigenvalue, the principal axis goes along the direction $e_1$, the rest may be deduced by analogy.

When $E\{\mathbf{x}\} \neq 0$, it is the mean-centered PC $v_i = (e_i)^T(\mathbf{x} - E\{\mathbf{x}\})$, which follows the direction of $e_i$.

As shown in Figure 5.1, there are a constant density ellipse and the PCs for a bivariate normal random vector with $E\{\mathbf{x}\} = 0$ and $\rho = 0.75$. And the PCs are obtained by rotating the original coordinate axes through an angle $\theta$ until they coincide with the axes of the constant density ellipse, which is useful for $n > 2$ dimensions as well.

### 5.2.3 Hotelling's $T^2$ statistic, SPE statistic and Q–Q plots

Retaining properly $m$ PCs, the decomposition of the different used matrices becomes (5.18) and (5.19):

$$\mathbf{E} = \left[\hat{\mathbf{E}}_{n\times m} \quad \tilde{\mathbf{E}}_{n\times(n-m)}\right] \tag{5.18}$$

$$\mathbf{V} = \left[\hat{\mathbf{V}}_{N\times m} \quad \tilde{\mathbf{V}}_{N\times(n-m)}\right] \tag{5.19}$$

The data matrix $\mathbf{X}$ can be decomposed as (5.20):

$$\mathbf{X} = \mathbf{X}\hat{\mathbf{E}}\hat{\mathbf{E}}^T + \mathbf{X}\tilde{\mathbf{E}}\tilde{\mathbf{E}}^T = \mathbf{X}\mathbf{C} + \mathbf{X}\tilde{\mathbf{C}} = \hat{\mathbf{X}} + \tilde{\mathbf{X}} \tag{5.20}$$

Matrix $\hat{\mathbf{X}}$ is the modeled variation of $\mathbf{X}$ by projection onto the PC subspace, and Matrix $\tilde{\mathbf{X}}$ is the non-modeled variation of $\mathbf{X}$ by projection onto the PC residual subspace, which has two factors such as projection matrices $\mathbf{C} = \hat{\mathbf{E}}\hat{\mathbf{E}}^T$ and $\tilde{\mathbf{C}} = (\mathbf{I} - \mathbf{C}) = \tilde{\mathbf{E}}\tilde{\mathbf{E}}^T$ provided linear combinations with large and low variances.

Hotelling's $T^2$ is used to detect variation after PCA, which is defined as

$$T^2 = \mathbf{x}^T\hat{\mathbf{E}}\hat{\mathbf{\Lambda}}^{-1}\hat{\mathbf{E}}^T\mathbf{x} \tag{5.21}$$

where $\hat{\mathbf{\Lambda}} = \text{diag}\{\lambda_1, \lambda_2, \ldots, \lambda_m\}$.

The control limit $T_\alpha$ is used for fault detection by Hotelling's $T^2$, which is determined by (5.22).

$$T_\alpha = \frac{(n^2 - 1)m}{n(n-1)} F_\alpha(m, n-m) \tag{5.22}$$

where $F_\alpha(m, n-m)$ is the critical value of the Fisher-Snedecor distribution with $m$ and $(n-m)$ degrees of freedom, and $\alpha$ is the level of significance.

The SPE or the $Q$ statistics are used to measure the lack of fit of the data for PCA, thus it is mostly used to test the variation in the residual subspace. The $Q$ statistic is defined as (5.23).

$$Q = (\tilde{\mathbf{C}}\mathbf{x})^T(\tilde{\mathbf{C}}\mathbf{x}) = \|\tilde{\mathbf{C}}\mathbf{x}\|^2 \tag{5.23}$$

The control limit $Q_\alpha$ is used for fault detection by SPE, which is determined by (5.24):

$$Q_\alpha = \theta_1\left[\frac{h_0 C_\alpha\sqrt{2\theta_2}}{\theta_1} + \frac{\theta_2 h_0(h_0-1)}{\theta_1^2} + 1\right]^{\frac{1}{h_0}} \tag{5.24}$$

where $\theta_i = \sum_{j=m+1}^{n} \lambda_j^i,\ i = 1,2,3,\ h_0 = 1 - \frac{2\theta_1\theta_3}{3\theta_2^2}$, $\lambda_j$ is the eigenvalue of $\Sigma_\mathbf{X}$, and $C_\alpha$ is the $\alpha$-quantile of normal distribution $N(0, 1)$.

The $Q$–$Q$ (quantile-quantile) plots is an exploratory graphical method used to check the validity of a distributional assumption for a data set. Here, the theoretically expected value is computed for each data point based on normal distribution. If the data follow normal distribution, then the points of the $Q$–$Q$ plots will be close to a straight line. Such as, the intervals set is selected for the quantiles. A point $(x, y)$ is corresponding to one of the quantiles of the second distribution ($y$-coordinate) plotted against the same quantile of the first distribution ($x$-coordinate).

To simplify notion, let $x_1, x_2, \ldots, x_n$ represent $n$ observations on any single characteristic $X_i$. Let $x_{(1)} \leq x_{(2)} \leq \cdots \leq x_{(n)}$ represent these observations after they are ordered according to magnitude. The $x_{(j)}$'s are the sample quantiles. When the $x_{(j)}$'s are distinct, exactly $j$ observations are less than or equal to $x_{(j)}$. The proportion $j/n$ of the sample at or to the left of $x_{(j)}$ is often approximated by $(j-\frac{1}{2})/n$ for analytical convenience.

The quantiles $q_{(j)}$ are defined for a standard normal distribution by relation (5.25).

$$P[Z \leq q_{(j)}] = \int_{-\infty}^{q(j)} \frac{1}{\sqrt{2\pi}} e^{-z^2/2} dz = p_{(j)} = \frac{j-\frac{1}{2}}{n} \tag{5.25}$$

Here $p_{(j)}$ is the probability to get a value which is less than or equal to $q_{(j)}$ from a standard normal population in a single drawing.

As shown in the pairs of quantiles $(q_{(j)}, x_{(j)})$ with the same associated cumulative probability $(j-\frac{1}{2})/n$, if the data arise from a normal population, the pairs $(q_{(j)}, x_{(j)})$ will be approximately linearly related, since $\sigma q_{(j)} + \mu$ is nearly the expected sample quantile.

In order to construct a $Q$–$Q$ plots, an example is in the following, where the observation values are obtained from the sample of $n = 10$ as shown in Table 5.1. There is $Q$–$Q$ plots for the foregoing data shown in Figure 5.2, which is a plot of the ordered data $x_{(j)}$ against the normal quantiles $q_{(j)}$. The pairs of points $(q_{(j)}, x_{(j)})$ are very close to a straight line, so these data are following normal distribution with.

*Table 5.1 The test data for Q–Q plots*

| Ordered observations $x_{(j)}$ | Probability levels $\left(j - \frac{1}{2}\right)/n$ | Standard normal quantiles $q_{(j)}$ |
|---|---|---|
| −1.00 | 0.05 | −1.645 |
| −0.10 | 0.15 | −1.036 |
| 1.16 | 0.25 | −0.674 |
| 0.41 | 0.35 | −0.385 |
| 0.62 | 0.45 | −0.125 |
| 0.80 | 0.55 | 0.125 |
| 1.26 | 0.65 | 0.385 |
| 1.54 | 0.75 | 0.674 |
| 1.71 | 0.85 | 1.036 |
| 2.30 | 0.95 | 1.645 |

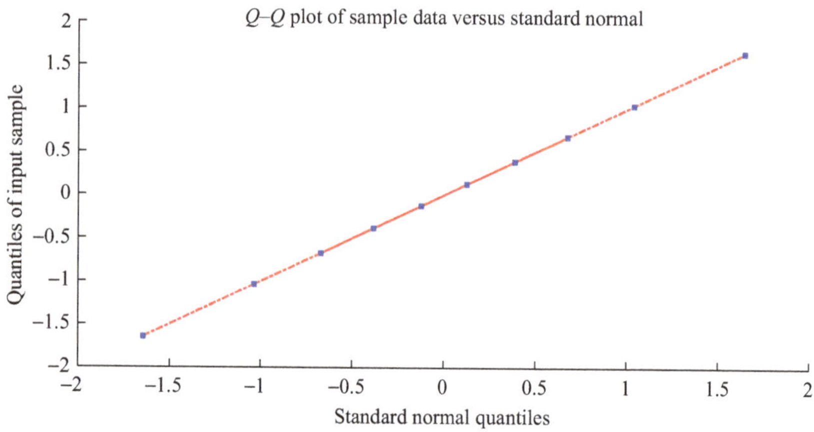


*Figure 5.2 A Q–Q plot for the data in an example*

### 5.2.4 Fault detection based on PCA for TE process

In this section, FDD based on PCA is used for TE process data. TE process is a benchmark process to compare FDD methods, which is developed by Downs and Vogel. There are five major unit operations in the TE process simulator: a recycle compressor, a vapor–liquid separator, a product condenser, a reactor and a product stripper. Two products are produced by two simultaneous gas–liquid exothermic reactions, and a by-product is generated by two additional exothermic reactions. The TE process has 22 continuous process measurements, 12 manipulated variables and 19 compositions, and all the process measurements include Gaussian noise. Almost all state variables will be affected when a fault happens in the TE process. The control scheme of the TE process and the simulation code of the open loop can be downloaded from the website as follows: http://brahms.scs.uiuc.edu. A normal process data set (500 samples) has been collected to build the FDD based on PCA monitoring model. Then the set of 21 faults are simulated by programming and the faults process data are sampled from the TE process for testing.

#### 5.2.4.1 Case study on Fault 4

There is a step change of the reactor cooling water inlet temperature in Fault 4 which will affect the reactor cooling water flow rate. If Fault 4 happens, a sudden temperature increasing will occur in the reactor, which is compensated though control loops. The other 50 testing variables remain stable after Fault 4 happens; that is, the mean of the variables is less than 2% between Fault 4 and the normal operating condition, and mostly the same to the standard deviation of the variables. FDD performance of Fault 4 is shown in Figure 5.3 by PCA.

#### 5.2.4.2 Case study on Fault 11

Fault 11 is another fault in the reactor cooling water inlet temperature in TE process, which is a random variation. Fault 11 will induce large oscillations in the reactor cooling water flow rate, which makes the reactor temperature fluctuate. But the other variables are able to keep around the set value, as same as the behave in the normal operating conditions. FDD performance of Fault 11 is shown in Figure 5.4 by PCA.

### 5.2.5 *Fault diagnosis based on PCA for multilevel inverter*

With the technological development of renewable energy in recent years, inverter, the intermediate link of electrical energy conversion, developed rapidly. Cascaded multilevel inverter (CMI) has been widely applied to mesohigh voltage and high-power industrial occasions due to some good properties, such as low harmonic content of output signals and convenient modularization, etc. While as the levels of CMI

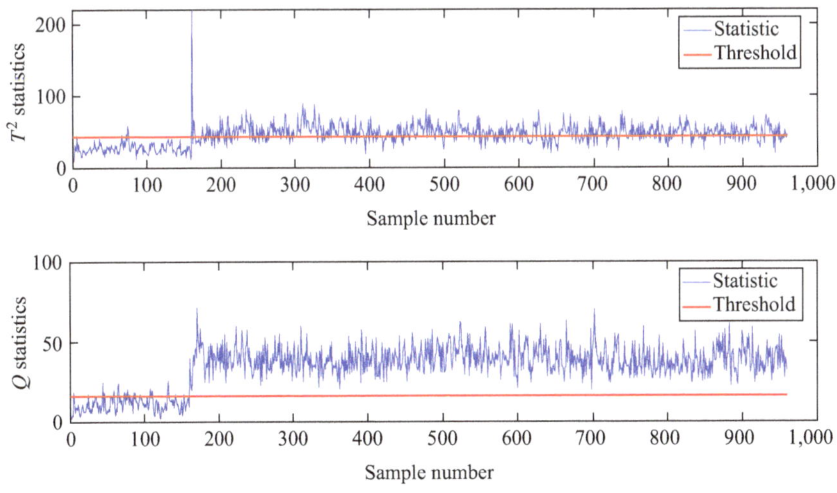


*Figure 5.3 FDD performance of Fault 4 by PCA*

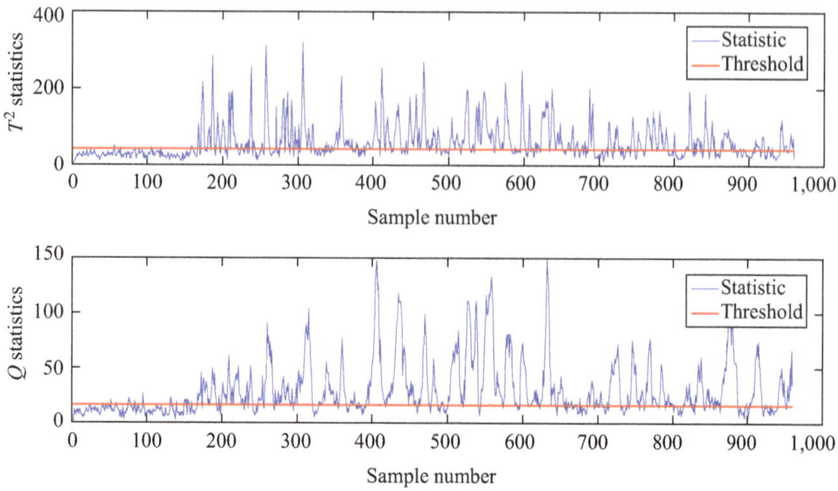


*Figure 5.4 Monitoring performances of Fault 11 based on PCA*

increase, the number of its power electronic devices will be more, whose losses and faults cannot usually be avoided. To improve the system's reliability, the fault diagnosis of CMI is particularly critical in actual production and life. Additionally, as the premise of fault diagnosis, whether fault feature extraction and the subsequent feature representation are good directly affects the performance of the fault classifier. To improve the efficiency of feature extraction and performance of fault diagnosis, it is of great necessity to find a high-efficient feature representation method and the fault diagnostic strategy based on this method. In this context, a fault diagnosis strategy based on PCA is proposed. First, the output voltage signal is selected as input signal. Then, fast Fourier transformation (FFT) is used to transform time signal to frequency signal. Next PCA is used to compress the frequency signal's dimension to extract main features. Finally, the FDD results are obtained by different classifier models.

#### 5.2.5.1 Time–frequency transform based on FFT

Figure 5.5 shows the output voltage from the cascaded H-bridge multilevel inverter switch (CHMLIS) under no-load condition. From Figure 5.5, it is difficult to find the difference between health and fault. Because there are no obvious features among different faults and health condition in time signal the FFT technique is used to transform time signal to frequency signal, which has a good identity feature to classify normal and abnormal features.

As shown in Figure 5.6, the frequency signal of the output voltage is more obvious than the time signal for the faults and normal data after FFT preprocessing. But if a fault of the switching devices occurs simultaneously at H1 (S1 & S2) or at H2 (S3 & S4), the frequency signal of the output voltage does not change, just like the phase's changing. So, it is difficult to make a distinction between the different faults if only using the spectrum. But if the phase information of the voltage data is added

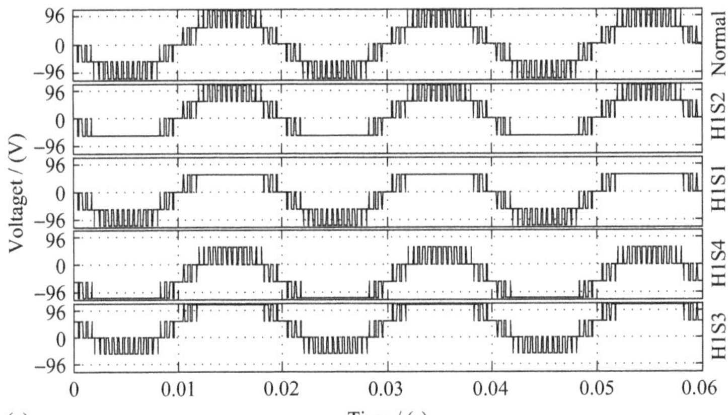

(a)

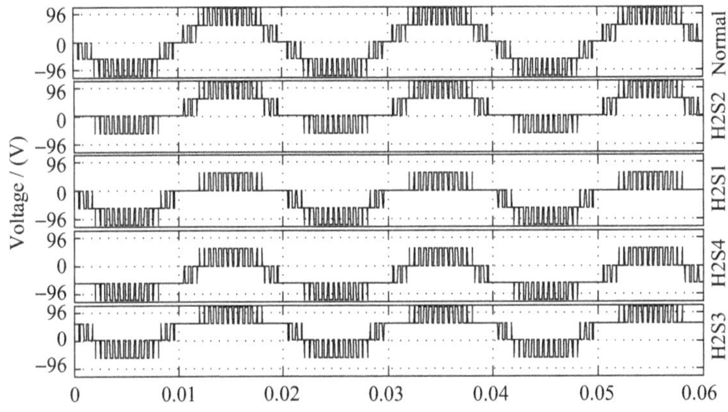

(b)

*Figure 5.5 Output when cascaded H-bridge switch S1–S4 open circuit. (a) S1 open circuit. (b) S2 open circuit [15]*

after FFT, the discrimination of these two faults is easier to distinguish. However, the dimension of data after preprocessing will be increased in the next step. So only the real part of the DC component is used to classify the features among the different faults, which is as shown in Figure 5.7.

#### 5.2.5.2 FDD based on PCA

In order to further compress the data and reduce the amount of subsequent calculation after FFT preprocessing. PCA is used to extract data features and get a new lower-dimensional set of variables, and the new features data are uncorrelated or orthogonal with each other. Here, the cumulative contribution rate $p$ is 85%. The PCs are selected to replace the original frequency signal, which are input for next step. At last, some classification algorithms are used to classify the faults after feature extraction by PCA, such as BP, SVM and mRVM.

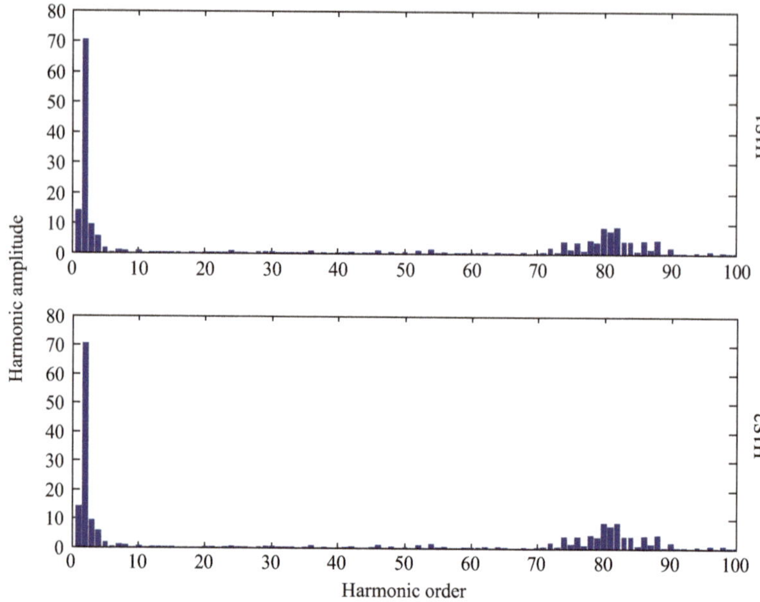


*Figure 5.6 Harmonic amplitude for H1S1 and H1S2 by FFT [15]*

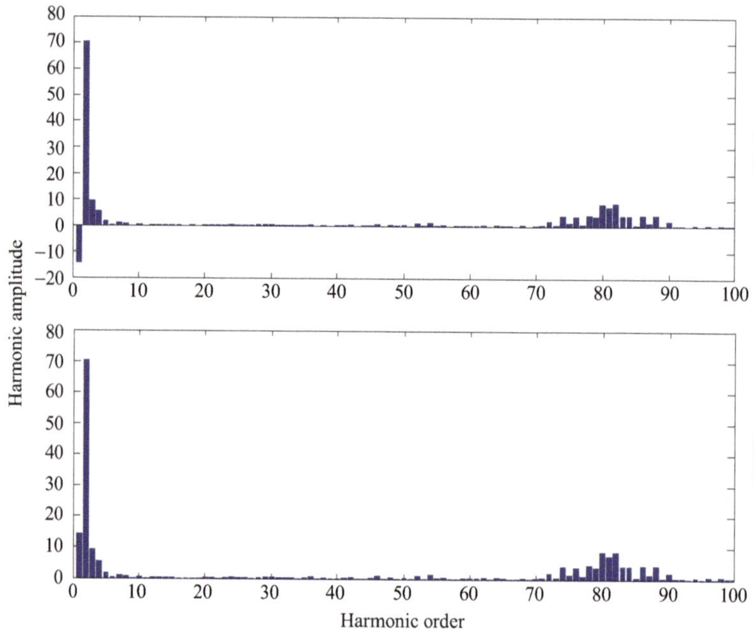


*Figure 5.7 Harmonic amplitude for H1S1 and H1S2 after special handling [15]*

#### 5.2.5.3 Experimental tests

There is a single-phase cascaded five-level inverter fault diagnosis experimental built based on dSPACE1104 control system as shown in Figure 5.8 with the experimental circuit structure, wherein the integrated power modules TLP250 is consisted in the drive circuit. The carrier phase-shifted sinusoidal pulse width modulation (CPS-SPWM) is used to control the single-phase CHMLIS. And there are the main parameters of the testing system in Table 5.2. There are some $N$-channel power IGBT IRGP35B60PD selected as the power switch transistors in the testing system, where the built-in reverse diode is included. There are two switching power supplies S-480-48 selected as switch powers, including the input 115 V–230 V/AC and the output +48 V/DC. The oscilloscope is used to show the output voltage of the testing system.

The CHMLIS experimental setup covers driving circuit, dead-zone circuit and H-bridge main circuit. The short-circuit faults are generally changed into open-circuit (OC) faults by implanting rapidly fuses into the inverter circuit. So, the OC faults of CHMLIS are sampled in the CHMLIS experimental setup.

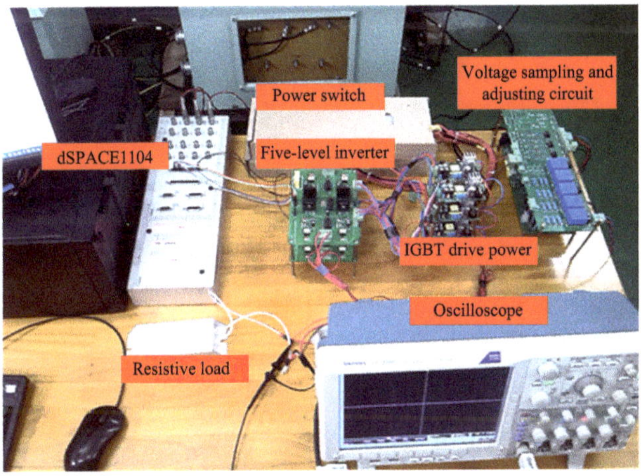


*Figure 5.8 The CHMLIS experimental setup [15]*

*Table 5.2 Main parameters of the testing system [15]*

| Symbol | Quantity | Value |
|---|---|---|
| $U_{dc}$ | DC-link voltage | 48 V |
| $R_{load}$ | Resistance load | 1 k Ω |
| $f_s$ | Switching frequency | 1 kHz |
| $f_{esample}$ | Experimental sample frequency | 25 kHz |
| $f_{ssample}$ | Simulation sample frequency | 40 kHz |

For the FFT-PCA-Classifier strategy, several classifiers such as mRVM, BP Neural Network and SVM are applied for CHMLIS fault diagnosis. After FFT transformation, 50 groups of frequency signal are selected, where there are nine kings of faults in each group. The sampling time is 0.02 s, the sampling frequency 25 kHz. The harmonics of output voltage are calculated by the FFT. Because after FFT, the waveform is symmetrical in one cycle, it is enough to use the first half cycle to be as output voltage's feature for the next step. And the feature is unique and stable, so only the inverter output voltage is utilized to diagnose different IGBT simple fault in the testing system. However, the features of the inverter output voltage are not small enough to be classified after FFT. So, it is necessary to compress the dimension of the features. Hence, the dimension of features is reduced by PCA after FFT. The optimal parameter is trained through many times testing as shown in Table 5.3.

Here, for FFT-PCA-BP method, the *input* is the size of BP's input layer, *hide* is the size of BP's hidden layer, *output* means the size of BP's output layer, $f(x)$ is the activation function in hidden layer of BP, and *lr* is the learning rate. After training many times, the parameters of the FFT-PCA-BP method are set, which has better performance. For FFT-PCA-SVM method, $\boldsymbol{K}(x, y)$ is the kernel function of SVM, $c$ is the penalty factor of SVM and $\sigma$ is the kernel parameter for SVM. The setting of kernel function is up to the accuracy of classification. $P$ is the predetermined limit for PCA, which is calculated by $P\% = \frac{\sum_{i=1}^{m} \lambda_i}{\sum_{i=1}^{n} \lambda_i} \times 100\%$. The value of $p$ is 0.85, which can be used to select the number of PCs. So, the first PCs selected are used to delegate the original date.

Here, PCA is useful to extract main features which improve the efficiency of the whole method by greatly reducing classifier's training time. At the same time, some noise is filtered out from the original data to main features as shown in Table 5.4, which improves the diagnosis performance.

The main objectives for the FFT-PCA-Classifier method are as follows: (1) FFT is used to transform time domain signal into frequency domain signal to get more distinct features; (2) PCA is used to extract main features, compress dimension and remove noise, which helps to increase the fault diagnosis's efficiency and accuracy; (3) different classifiers are used for fault recognition.

*Table 5.3 Parameter configuration for different methods [15]*

| Different classifier | PCs | Parameter configuration |
|---|---|---|
| FFT-PCA-BP | 2 | $input = 2, hide = 8, output = 9, f(x) = \frac{1}{1+e^{-x}}, lr = 0.01$ |
| FFT-PCA-SVM | 2 | $\boldsymbol{K}(x,y) = \exp\left(-\frac{\Vert x-y\Vert^2}{2\sigma^2}\right), P = 0.85, c = 5, \sigma = 0.02$ |
| FFT-PCA-mRVM | 2 | $\boldsymbol{K}(x,y) = \exp\left(-\frac{\Vert x-y\Vert^2}{2\sigma^2}\right), CL = 0.85, \sigma = 0.5$ |

*Table 5.4 Results of fault diagnosis method based on different classifiers [15]*

| Testing samples (groups) | Average testing time (s) | | | Average diagnosis accuracy (%) | | |
|---|---|---|---|---|---|---|
| | BP | SVM | mRVM | BP | SVM | mRVM |
| 5 | 0.076 | 0.068 | 0.047 | 85.63 | 100 | 100 |
| 10 | 0.103 | 0.071 | 0.052 | 86.21 | 97.2 | 100 |
| 20 | 0.122 | 0.083 | 0.056 | 78.31 | 96.5 | 100 |

## 5.3 RPCA and its application

When the system tallies with Rotundity Scatter distribution, it is difficult to get representative PCs. This section proposes the concept of relative principal component (RPC), and puts forward RPCA method.

### 5.3.1 RPCA method

There are two parts of RPCA: first is Relative Transform, and another is computing RPCs.

#### 5.3.1.1 Relative Transform

Consider (5.26) as the data matrix made up of system variable:

$$\mathbf{X}(n,N) = \begin{bmatrix} x_1(1) & x_1(2) & \dots & x_1(N) \\ x_2(1) & x_2(2) & \dots & x_2(N) \\ \vdots & \vdots & \ddots & \vdots \\ x_n(1) & x_n(2) & \dots & x_n(N) \end{bmatrix} \tag{5.26}$$

**Definition 5.3.1** *Relative Transform*

Define:

$$\begin{aligned} \mathbf{X}^R &= M \cdot \mathbf{X}^* \\ &= \begin{bmatrix} M_1 & 0 & \cdots & 0 \\ 0 & M_2 & \cdots & 0 \\ \vdots & \vdots & \ddots & \vdots \\ 0 & 0 & \dots & M_n \end{bmatrix} \begin{bmatrix} x_1^*(1) & x_1^*(2) & \dots & x_1^*(N) \\ x_2^*(1) & x_2^*(2) & \dots & x_2^*(N) \\ \vdots & \vdots & \ddots & \vdots \\ x_n^*(1) & x_n^*(2) & \dots & x_n^*(N) \end{bmatrix} \\ &= \begin{bmatrix} x_1^R(1) & x_1^R(2) & \dots & x_1^R(N) \\ x_2^R(1) & x_2^R(2) & \dots & x_2^R(N) \\ \vdots & \vdots & \ddots & \vdots \\ x_n^R(1) & x_n^R(2) & \dots & x_n^R(N) \end{bmatrix} \end{aligned} \tag{5.27}$$

$$\mathbf{X}_i^R = M_i \mathbf{X}_i^* \tag{5.28}$$

Here:

$$\mathbf{X}_i^* = \frac{\mathbf{X}_i - E(\mathbf{X}_i)}{m_i} \tag{5.29}$$

So (5.27) is Relative Transform, $M$ is the operator of Relative Transform and $\boldsymbol{X}^R$ is relative matrix of system matrix $\boldsymbol{X}$. $m_i$ is the corresponding standardization gene, such as $m_i = max_{1 \le k \le N} |x_i(K)|$, or $m_i = (Var(X_i))^{1/2}$. $M_i$ is the proportion coefficient. Equation (5.29) is the standardization process.

**Property 5.3.1** *Relative Transform does not change the relativity of data matrix.*

**Proof:**

$$\rho(\mathbf{X}_i^R, \mathbf{X}_j^R) = \frac{Cov(\mathbf{X}_i^R, \mathbf{X}_j^R)}{\sqrt{\sigma^2(\mathbf{X}_i^R)\sigma^2(\mathbf{X}_j^R)}} \tag{5.30}$$

From (5.28) and (5.30), we can educe

$$Cov(\mathbf{X}_i^R, \mathbf{X}_j^R) = \frac{M_i}{m_i}\frac{M_j}{m_j} Cov(\mathbf{X}_i - E(\mathbf{X}_i), \mathbf{X}_j - E(\mathbf{X}_j)) \tag{5.31}$$

$$\sqrt{\sigma^2(X_i^R)\sigma^2(X_j^R)} = \sqrt{\left(\frac{M_i}{m_i}\right)^2 \sigma^2(X_i - E(X_i))\left(\frac{M_j}{m_j}\right)^2 \sigma^2(X_j - E(X_j))} \tag{5.32}$$

Because of

$$Cov(\mathbf{X}_i - E(\mathbf{X}_i), X_j - E(\mathbf{X}_j)) = Cov(\mathbf{X}_i, \mathbf{X}_j) \tag{5.33}$$

$$\sigma^2(\mathbf{X}_i - E(\mathbf{X}_i))\sigma^2(\mathbf{X}_j - E(\mathbf{X}_j)) = \sigma^2(\mathbf{X}_i)\sigma^2(\mathbf{X}_j) \tag{5.34}$$

So we can prove:

$$\rho(\mathbf{X}_i^R, \mathbf{X}_j^R) = \rho(M_i\mathbf{X}_i^*, M_j\mathbf{X}_j^*) = \frac{Cov(\mathbf{X}_i, \mathbf{X}_j)}{\sqrt{\sigma^2(\mathbf{X}_i)\sigma^2(\mathbf{X}_j)}} \tag{5.35}$$

**Definition 5.3.2** *Rotundity Scatter*
If the eigenvalues $\lambda_1, \lambda_2, \ldots, \lambda_n$ are approximately equal to each other in (5.5), system matrix $\boldsymbol{X}$ is defined as Rotundity Scatter.

The system matrix $X$ with Rotundity Scatter is satisfied with the property as follows.

**Property 5.3.2** If system matrix $\boldsymbol{X}$ tallies with Rotundity Scatter, the vectors $[X(1), X(2), \ldots, X(n)]$ will constitute a hypersphere of $n$ dimensions.

The rule of modeling Relative Transform $M$:

1. System matrix $\boldsymbol{X}$ should do its best to fall short of Rotundity Scatter by Relative Transform.
2. RPCs selected should more exactly delegate the system by Relative Transform; that is, the information contained in first $m$ RPCs is much more than in first $m$ PCs.

3. The energy of new system though RT should be equal to the energy of original system, or should be $K$ times of the energy of original system; that is, $\|\boldsymbol{X}^R\|_2 = K\|\boldsymbol{X}\|_2$, $K$ is a certain proportion constant.

#### 5.3.1.2 Computing RPCs

All RPCs $v_1^R, v_2^R, \ldots, v_m^R$ can be gained by the following steps:

1. Computing the covariance matrix $\Sigma_{X^R}$ of $\boldsymbol{X}^R$ from (5.27):

$$\Sigma_{X^R} = E\{[\boldsymbol{X}^R - E(\boldsymbol{X}^R)][X^R - \boldsymbol{E}(\boldsymbol{X}^R)]^T\} \tag{5.36}$$

2. Calculating eigenvalues $\lambda_i^R$ and its corresponding eigenvector $e_i^R$ respectively by

$$\left|\lambda^R I - \Sigma_{X^R}\right| = 0 \tag{5.37}$$

and

$$\left[\lambda_i^R I - \Sigma_{X^R}\right]e_i^R = 0 \quad i = 1, \ldots, n \tag{5.38}$$

where $e_i^R = \left[e_i^R(1), e_i^R(2), \ldots, e_i^R(n)\right]^T$, $\lambda_1 \geq \lambda_2 \geq \cdots \geq \lambda_n$.

3. Obtaining the RPCs
Given the following transformation:

$$\begin{Bmatrix} v_1^R \\ v_2^R \\ \vdots \\ v_n^R \end{Bmatrix} = \begin{bmatrix} e_1^R(1) & e_1^R(2) & \ldots & e_1^R(n) \\ e_2^R(1) & e_2^R(2) & \ldots & e_2^R(n) \\ \vdots & \vdots & \ddots & \vdots \\ e_n^R(1) & e_n^R(2) & \ldots & e_n^R(n) \end{bmatrix} \begin{Bmatrix} X_1^R \\ X_2^R \\ \cdot \\ X_n^R \end{Bmatrix} \tag{5.39}$$

or

$$v^R = e^R X^R$$

then select $m(m < n)$ vectors $v_1^R, v_2^R, \ldots, v_m^R$ as RPCs. Similar to PCA, the effect of RPC $v_i^R$ is

$$P_i^R\% = \frac{\lambda_i^R}{\sum_{i=1}^n \lambda_i^R} \times 100\% \tag{5.40}$$

In general, the RPCA of system is as follows:

1. Get the contribution of each variable to system through computing $P_i^R\%$.
2. According to the requirement of system, $m(m < n)$ RPCs $v_1^R, v_2^R, \ldots, v_m^R$ are chosen to delegate the original system matrix $\boldsymbol{X}^R$, and analyze the character of $\sum_{X^R}$; consequently, the RPCA model is found to process FDD in the matrix data with Rotundity Scatter.

### 5.3.2 The geometrical interpretation of RPCA

The RPCs can provide more information than PCs. Suppose vector $\boldsymbol{x}^R(i)$ of relative matrix $\boldsymbol{X}^R$ obey $n$ dimensions normal distribution $N_n[E\{\boldsymbol{x}^R\}, \Sigma_{X^R}]$. $E\{\boldsymbol{x}^R\}$ is the center of the ellipse sphere.

$$[\boldsymbol{x}^R - E\{\boldsymbol{x}^R\}]^T(\Sigma_{X^R})^{-1}[\boldsymbol{x}^R - E\{\boldsymbol{x}^R\}] = c^2 \tag{5.41}$$

Every axis of the ellipse sphere is as follows:

$$\pm c\sqrt{\lambda_i^R}\boldsymbol{e}_i^R, \quad i = 1, 2, \ldots, n$$

Here, $(\lambda_i^R, \boldsymbol{e}_i^R)$ is the $i$th eigenvalue and eigenvector of relative covariance $\Sigma_{X^R}$.

Suppose $E\{x^R\} = 0$. So, (5.41) can be predigested as follows:

$$(\boldsymbol{x}^R)^T(\Sigma_{X^R})^{-1}\boldsymbol{x}^R = c^2 \tag{5.42}$$

It can be rewritten as follows:

$$(\boldsymbol{x}^R)^T(\boldsymbol{E}^T\boldsymbol{E})^T(\Sigma_{X^R})^{-1}\boldsymbol{E}^{-1}\boldsymbol{E}\boldsymbol{x}^R = c^2 \tag{5.43}$$

$$(\boldsymbol{E}\boldsymbol{x}^R)^T(\boldsymbol{E}\Sigma_{X^R}\boldsymbol{E}^{-1})^{-1}(\boldsymbol{E}\boldsymbol{x}^R) = c^2 \tag{5.44}$$

$$(\boldsymbol{v}^R)^T(\Sigma_{V^R})^{-1}\boldsymbol{v}^R = c^2 \tag{5.45}$$

Because of (5.46):

$$\Sigma_{X^R} = \sum_{i=1}^{n}\lambda_i^R\boldsymbol{e}_i^R(\boldsymbol{e}_i^R)^T \tag{5.46}$$

So, it can be gained:

$$(\Sigma_{X^R})^{-1} = \sum_{i=1}^{n}\frac{1}{\lambda_i^R}\boldsymbol{e}_i^R(\boldsymbol{e}_i^R)^T \tag{5.47}$$

So:

$$(x^R)^T\left(\sum_{i=1}^{n}\frac{1}{\lambda_i^R}e_i^R(e_i^R)^T\right)(x^R) = \frac{1}{\lambda_1^R}[(e_1^R)^T x^R]^2 + \frac{1}{\lambda_2^R}[(e_2^R)^T x^R]^2 + \cdots + \frac{1}{\lambda_n^R}[(e_n^R)^T x^R]^2 \tag{5.48}$$

Here, $(e_1^R)^T x^R, (e_2^R)^T x^R, \ldots, (e_n^R)^T x^R$ is RPCs of $x^R$, so, (5.79) can be rewritten as follows:

$$c^2 = \frac{1}{\lambda_1^R}(v_1^R)^2 + \frac{1}{\lambda_2^R}(v_2^R)^2 + \cdots + \frac{1}{\lambda_n^R}(v_n^R)^2 \tag{5.49}$$

$\lambda_1^R$ is the maximum eigenvalue; the principal axis goes along the direction $e_1^R$, and the rest may be deduced by analogy.

The RT can change the scatter of data, and alter the Rotundity Scatter. That is, it can make the axis of ellipse sphere be distinct, and ultimately enhance the effect of RPCs.

In this section, a simulation example is used to illustrate what the performance of RPCA is better than PCA's when system matrix $X$ is Rotundity Scatter. Parameter setting and simulation result are as shown in Table 5.5.

Figure 5.9(a) shows the shape of original multivariate sequence matrix $\boldsymbol{X}$ of system, which is Rotundity Scatter due to $\lambda_1 \cong \lambda_2$ obtained by PCA; this ellipse is just like a round, and $V_1$ cannot "replace" the original matrix $\boldsymbol{X}$ of system. $\boldsymbol{X}$ with Rotundity Scatter is changed through Relative Transform, as shown in Figure 5.9(b) $\lambda_1^R \gg \lambda_2^R$ is obtained by RPCA, and $V_1$ can "replace" the original matrix $\boldsymbol{X}$ of system.

The result of simulation shows: if the system data is Rotundity Scatter, PCs can be gained by PCA. RPCA can change the Rotundity Scatter through RT, and

*Table 5.5 Parameter setting and simulation result [21]*

| | | | |
|---|---|---|---|
| System matrix $\boldsymbol{X}$ | Rotundity Scatter | $\lambda_1$ | 0.0765 |
| The number of multivariate sequence $N$ | 62 | $\lambda_2$ | 0.0504 |
| The number of variable $n$ | 2 | $\lambda_1^R$ | 0.6221 |
| $E\{\boldsymbol{X}^R\}$ | (0.5, 0.5) | $\lambda_2^R$ | 0.0573 |
| The proportion coefficient $\mu_1$ | 1 | $P_1\%$ | 60.72% |
| The proportion coefficient $\mu_2$ | 3 | $P_1^R\%$ | 91.57% |

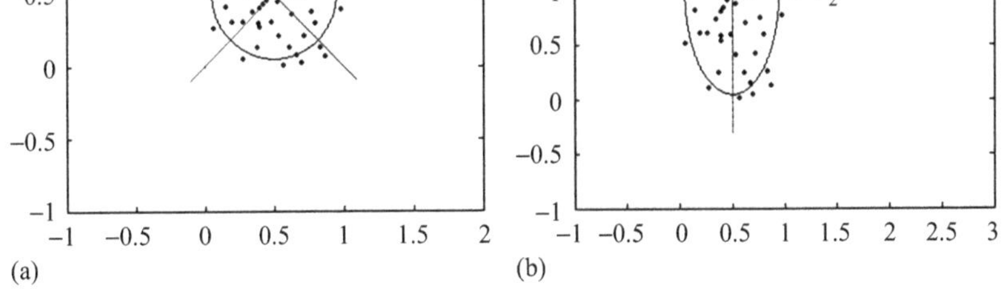


*Figure 5.9 The constant density ellipse of original multivariate sequence matrix $\boldsymbol{X}$ with Rotundity Scatter. (a) $\lambda_1 \equiv \lambda_2$, (b) $\lambda_1^R \gg \lambda_2^R$ [21]*

consequently, several RPCs can be gained to replace the original system so as to process FDD.

### 5.3.3 Fault detection based on RPCA for assembly

The RPCA model is found: collect normal history data, do RT and gain the RPCs. If the data from the real-time system do not tally with the RPCA model, maybe there are faults in the system. And after analyzing the effect of real-time data to the RPCA model, the particular fault can be found out.

If the data from real-time system is $\boldsymbol{X}_{\text{new}}$, then $\boldsymbol{X}_{\text{new}}^R$ can be got. Finally, the RPCs $\boldsymbol{v}_{\text{new}}^R$ can be gained, and Hotelling's $T^2$ is used to test. If the process of the system runs in gear, the $\boldsymbol{v}_{\text{new}}^R$ should satisfy with (5.50):

$$T^2 < UCL \tag{5.50}$$

Here:

$$T^2 = \sum_{i=1}^{m} \frac{(v_{\text{new}_i}^R)^2}{S^2_{v_{\text{new}_i}^R}} \tag{5.51}$$

$S^2_{v_{\text{new}_i}^R}$ is the estimate variance $v_{\text{new}_i}^R$.

$$UCL = \frac{m(n^2 - m)}{n(n - m)} F_{\alpha}(m, n - m) \tag{5.52}$$

UCL is the upper control limit of Hotelling's $T^2$, and if $T^2 > UCL$, there are some abnormal circs in the process of system.

There is a simulation example to introduce the application of RPCA in FDD. The data are from the assembly of a driveshaft for an automobile, which requires the circle welding of tube yokes to a tube. In order to make the machine produce welds of good quality, we must control the inputs to the automated welding machines to keep in certain operating limits. In order to monitor the process, there are four critical variables as shown in Table 5.6 measured form the process engineer as follows: $X_1$, the voltage; $X_2$, the current; $X_3$, the feed speed; and $X_4$, the gas flow.

Some random noise is added to the points after the 20th point as system fault. The eigenvalues are almost equal to each other, so it is difficult to select PCs to delegate the original system. However, the Rotundity Scatter can be changed by RPCA, a few RPCs can be selected to delegate the system. The faults cannot be detected by PCA when $\alpha = 0.01$ and $\alpha = 0.05$ in Figure 5.10(a). It is distinct to detect faults by RPCA in Figure 5.10(b), as shown in Table 5.7.

When the data matrix is Rotundity Scatter, it is difficult to select PCs to detect fault. Or the PCs do not delegate the original system without thinking over the dimension of different variable. But it is better by RPCA, which resolves successfully above problems as follows:

1. Dimension problems are avoided by RPCA; that is, if the numerical value of a system variable is bigger, the selection of system PCs is more influenced. So, the RPCs have stronger representativeness than PCs.

*Table 5.6 Welder data [21]*

| Case | Voltage ($X_1$) | Current ($X_2$) | Feed speed ($X_3$) | Gas flow ($X_4$) |
|---|---|---|---|---|
| 1 | 23.0000 | 276 | 289.6000 | 51.0000 |
| 2 | 22.0000 | 281 | 289.0000 | 51.7000 |
| 3 | 22.8000 | 270 | 288.2000 | 51.3000 |
| 4 | 22.1000 | 278 | 288.0000 | 52.3000 |
| 5 | 22.5000 | 275 | 288.0000 | 53.0000 |
| 6 | 22.2000 | 273 | 288.0000 | 51.0000 |
| 7 | 22.0000 | 275 | 290.0000 | 53.0000 |
| 8 | 22.1000 | 268 | 289.0000 | 54.0000 |
| 9 | 22.5000 | 277 | 289.0000 | 52.0000 |
| 10 | 22.5000 | 278 | 289.0000 | 52.0000 |
| 11 | 22.3000 | 269 | 287.0000 | 54.0000 |
| 12 | 21.8000 | 274 | 287.6000 | 52.0000 |
| 13 | 22.3000 | 270 | 288.4000 | 51.0000 |
| 14 | 22.2000 | 273 | 290.2000 | 51.3000 |
| 15 | 22.1000 | 274 | 286.0000 | 51.0000 |
| 16 | 22.1000 | 277 | 287.0000 | 52.0000 |
| 17 | 21.8000 | 277 | 287.0000 | 51.0000 |
| 18 | 22.6000 | 276 | 290.0000 | 51.0000 |
| 19 | 22.3000 | 278 | 287.0000 | 51.7000 |
| 20 | 23.0000 | 266 | 289.1000 | 51.0000 |
| 21 | 22.9000 | 271 | 288.3000 | 51.0000 |
| 22 | 21.3000 | 274 | 289.0000 | 52.0000 |
| 23 | 21.8000 | 280 | 290.0000 | 52.0000 |
| 24 | 22.0000 | 268 | 288.3000 | 51.0000 |
| 25 | 22.8000 | 269 | 288.7000 | 52.0000 |
| 26 | 22.0000 | 264 | 290.0000 | 51.0000 |
| 27 | 22.5000 | 273 | 288.6000 | 52.0000 |
| 28 | 22.2000 | 269 | 288.2000 | 52.0000 |
| 29 | 22.6000 | 273 | 286.0000 | 52.0000 |
| 30 | 21.7000 | 283 | 290.0000 | 52.7000 |
| 31 | 21.9000 | 273 | 288.7000 | 55.3000 |
| 32 | 22.3000 | 264 | 287.0000 | 52.0000 |
| 33 | 22.2000 | 263 | 288.0000 | 52.0000 |
| 34 | 22.3000 | 266 | 288.6000 | 51.7000 |
| 35 | 22.0000 | 263 | 288.0000 | 51.7000 |
| 36 | 22.8000 | 272 | 289.0000 | 52.3000 |
| 37 | 22.0000 | 277 | 287.7000 | 53.3000 |
| 38 | 22.7000 | 272 | 289.0000 | 52.0000 |
| 39 | 22.6000 | 274 | 287.2000 | 52.7000 |
| 40 | 22.7000 | 270 | 290.2000 | 51.0000 |

2. RPCs can still be obtained by RPCA when the system matrix $\boldsymbol{X}$ is following Rotundity Scatter. The RPCs are more representative than PCs to detect fault and diagnose.

This section presents the geometrical interpretation of RPCA, and the results of simulation and experiment demonstrate the effectiveness as follows.

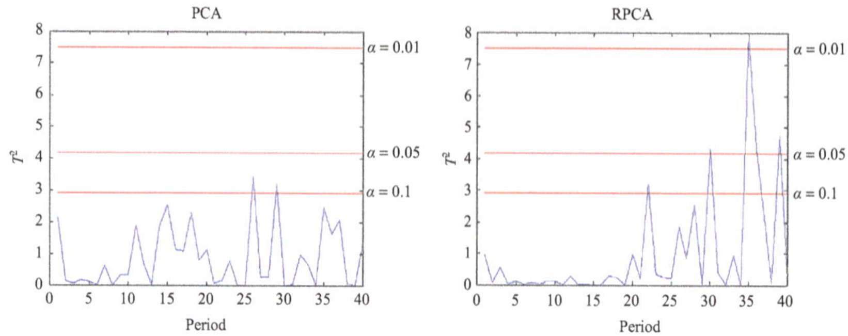


*Figure 5.10 The $T^2$ chart of (a) PCA and (b) RPCA [21]*

*Table 5.7 The contrast of parameters of PCA and RPCA [21]*

| **The parameters of PCA** | | | | | | **The parameters of RPCA** | | | |
|---|---|---|---|---|---|---|---|---|---|
| $\lambda_1$ | 1.3966 | $P_1\%$ | 34.91% | $M_1$ | 10 | $\lambda_1^R$ | 100.0294 | $P_1^R\%$ | 95.95% |
| $\lambda_2$ | 1.0757 | $P_2\%$ | 26.89% | $M_2$ | 5 | $\lambda_2^R$ | 3.9727 | $P_2^R\%$ | 3.81% |
| $\lambda_3$ | 0.8144 | $P_3\%$ | 20.36% | $M_3$ | 1 | $\lambda_3^R$ | 0.2479 | $P_3^R\%$ | 0.24% |
| $\lambda_4$ | 0.7134 | $P_4\%$ | 17.83% | $M_4$ | 0.01 | $\lambda_4^R$ | 9.7879e−005 | $P_4^R\%$ | 0.0001% |

### 5.3.4 Dynamic data window control limit based on RPCA

Dynamic data window fault detection method based on RPCA needs to be modeled by the data during normal operation. The flowchart of system fault detection is as shown in Figure 5.11. The fault detection method is mainly divided into two parts: (1) According to the historical data of the system, RPCA modeling is used to construct the control limit as the basis for fault detection. (2) Through online real-time sampling data, through RPCA analysis, the $T^2$ statistic of the PC is obtained; and then according to the comparison between the $T^2$ statistic and the control limit, it is determined whether the system is working normally.

The dynamic data window control limit based on RPCA is

$$T_{\mathrm{ucl}}(k) = \omega^* T_{\mathrm{ucl1}} + (1 + \omega) * T_{\mathrm{ucl2}}(k) \tag{5.53}$$

where $\omega$ is the weight, and $0 < \omega < 1, 0 < k \leq a$.

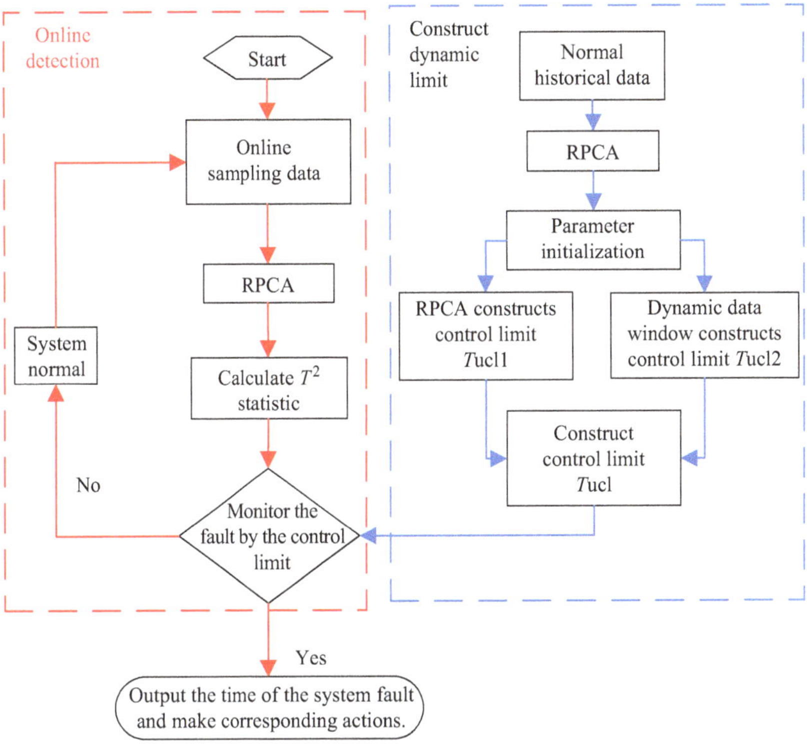


*Figure 5.11 Flowchart of fault monitoring [38]*

The steps of the dynamic data window algorithm based on RPCA are described in detail as follows:

Step 1: Historical data sampling

In the historical data set, historical data $\boldsymbol{X}$ is collected according to the certain period length, and the period length is $a$ (the length of sampling time).

Because the randomness and instability of wind speed changes are significant, there are a large number of unsteady factors in the wind turbine. On the basis of the mathematical formula of wind turbine $P = \frac{1}{2}\rho\pi R^2 C_p(\lambda)V^3$, a model of power feedback control strategy is established as shown in Figure 5.11, where $\rho$ is the air density, $R$ is the radius of the wind wheel, $\lambda$ is the tip speed ratio, $C_p(\lambda)$ is the wind energy utilization factor when the tip speed ratio is $\lambda$, and $V$ is the wind speed. The wind speed is from 6 to 11 m/s during normal operation, and some of the main parameters are shown in Table 5.8 about the wind turbine. This section assumes that the sampling time length is fixed, so that a series of periodically sampled data is obtained from the wind turbine system, and when a new set of observations is available, the RPCA is periodically updated. The fault detection performance of the dynamic data window based on RPCA is as follows.

*Table 5.8 The main parameters of the wind generator power feedback model [38]*

| | | | |
|---|---|---|---|
| Mechanical output power (W) | 39,900 | Inductance (mH) | 0.635 |
| Generator power (V/A) | 39,900/0.9 | Pole pairs (P) | 36 |
| Maximum wind energy utilization factor | 0.48 | Optimal tip speed ratio | 8.1 |
| Wind speed range (m/s) | 6–11 | Stator resistance (Ω) | 0.05 |
| Friction coefficient (N m s) | 0.001889 | Wind turbine radius $R$(m) | 15 |
| Rotor flux (Wb) | 0.192 | Pitch angle (°) | 0 |

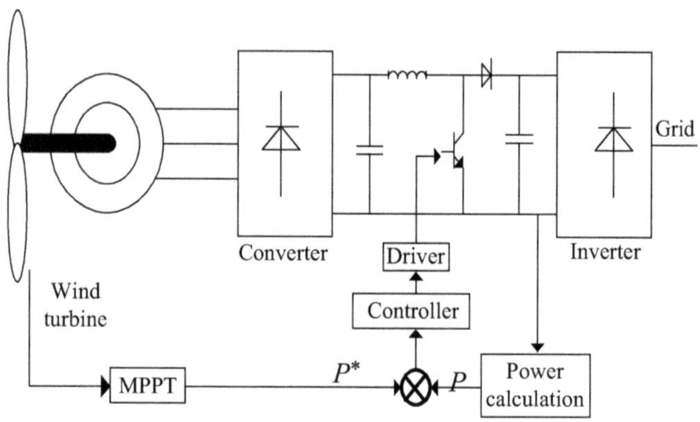


*Figure 5.12 The model of wind generator [38]*

The normal wind speed is from 6 to 11 m/s. There are several normal wind turbine parameters (rotating speed, power, voltage, three-phase rotor current, etc.) obtained with white Gaussian noise in the simulation model as shown in Figure 5.12. In this section, the fault detection method based on RPCA is built by using the data of three-phase voltage, three-phase current and rotating speed in total seven dimensions, which is used to construct the control limit of fault detection. The size of the period is 951. The warning limits and action limits are 95% and 99%, respectively.

Mostly, when the wind speed exceeds 11 m/s, the mechanical stress of the wind turbine reaches the limit. Because the wind turbine will be damaged when it works in this speed for long term, the fault wind speed is about 14 m/s in the simulation model, as a set of fault data. At other times, the wind speed of the wind turbine is random, but they are all within the normal wind speed range.

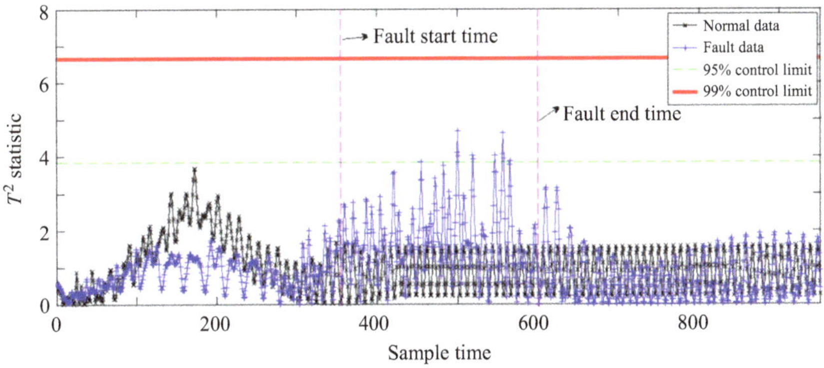


*Figure 5.13 Fault detection results of wind power generation system based on PCA [38]*

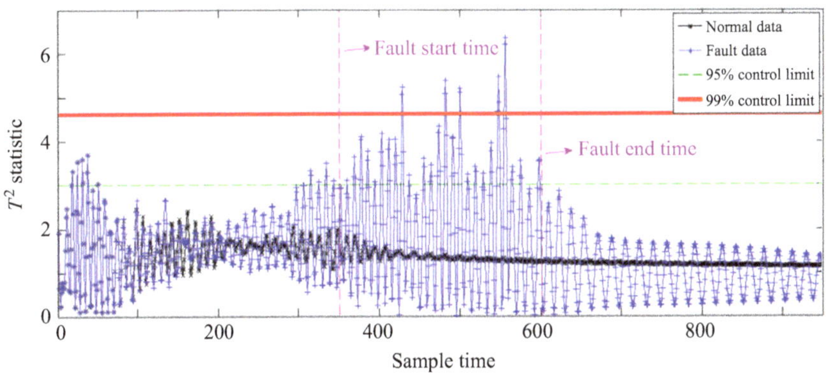


*Figure 5.14 Fault detection results of wind power generation system based on recursive PCA [38]*

The traditional PCA algorithm, recursive PCA algorithm, PCA-based dynamic window algorithm and RPCA-based dynamic window algorithm are used to monitor the working process of the wind turbine.

In general, an effective fault detection system should keep the false alarm rate and missed detection rate as small as possible. Comparing the four methods, it can be seen from Figures 5.13–5.15 that when the traditional PCA algorithm is used in periodic nonsteady conditions, the fault cannot be detected effectively, there is a serious missed detection phenomenon and the detection sensitivity is low. Although the recursive PCA algorithm does not need to establish a model based on normal historical data, it can be applied to nonperiodic situations, but there are a large number of missed detections in the detection process, and the reliability of system detection is relatively poor. When the dynamic data window is used directly in PCA, the missed detection rate

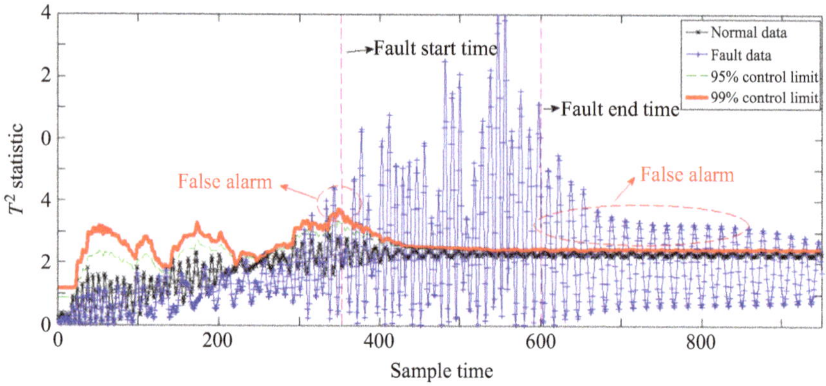


*Figure 5.15 Dynamic data window fault detection results of wind power generation system based on PCA [38]*

*Table 5.9 Four kinds of fault detection methods performance comparison [38]*

| Multivariate statistical fault detection method | | | | False alarm rate (%) | Missed detection rate (%) |
|---|---|---|---|---|---|
| Traditional PCA | | | | 0 | 26.28 |
| Recursive PCA | | | | 0 | 25.76 |
| Dynamic data window based on PCA | | | | 40.20 | 9.57 |
| Dynamic data window based on RPCA | $\omega = 0.1$ | $L_1$ | 4 | 2.11 | 2.53 |
| | | $L_2$ | 19 | 1.79 | 5.05 |
| | | $L_3$ | 49 | 1.82 | 5.16 |
| | Average | | | 1.91 | 4.25 |
| | $L = 4$ | $\omega_1$ | 0.2 | 0.74 | 9.47 |
| | | $\omega_2$ | 0.1 | 2.11 | 2.53 |
| | | $\omega_3$ | 0.05 | 3.37 | 3.26 |
| | Average | | | 2.07 | 5.09 |

is greatly reduced, but serious false alarm appears. The dynamic data window method based on RPCA proposed in this section can not only detect fault data effectively, but also has relatively high sensitivity. Compared with other methods, the dynamic data window based on RPCA method has the lower rate of missed detection and false alarm rate. As can be seen from Table 5.9, this method makes the fault detection system maintain a low false alarm rate and missed detection rate, and greatly improves the effectiveness of system monitoring.

In view of the fact that traditional multivariate statistical methods are applied to nonsteady conditions, false alarms or missed detections may occur, and fault detection reliability is poor. This section proposed a dynamic data window method based on RPCA, which can effectively improve the fault detection process and improve

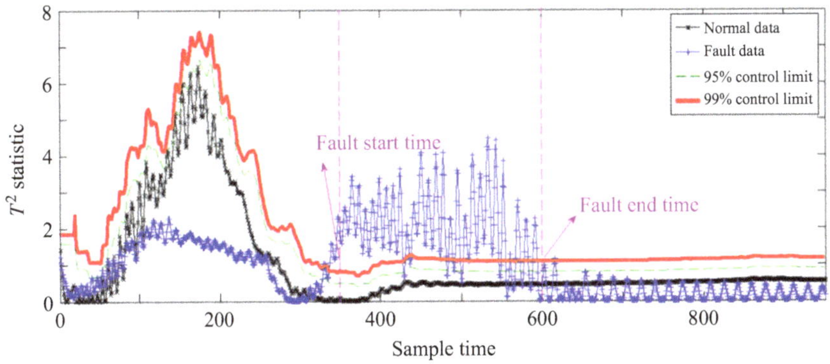


*Figure 5.16 Dynamic data window fault detection results of wind power generation system based on RPCA [38]*

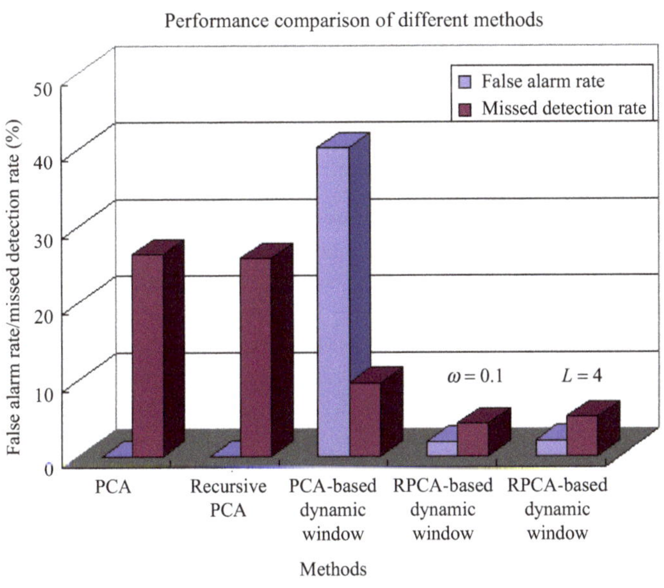


*Figure 5.17 When $\omega = 1$ and $L = 4$ the performance comparison of different methods [38]*

the effectiveness of system monitoring as shown in Figure 5.16. These are mainly embodied in the following aspects: (1) it can effectively detect faults and reduce the occurrence of false alarms; (2) the sensitivity of monitoring can be adjusted by adjusting the weight $\omega$ to achieve a kind of balance between missed detection and false alarms as shown in Figure 5.17; and (3) it is applicable to process monitoring under nonsteady conditions.

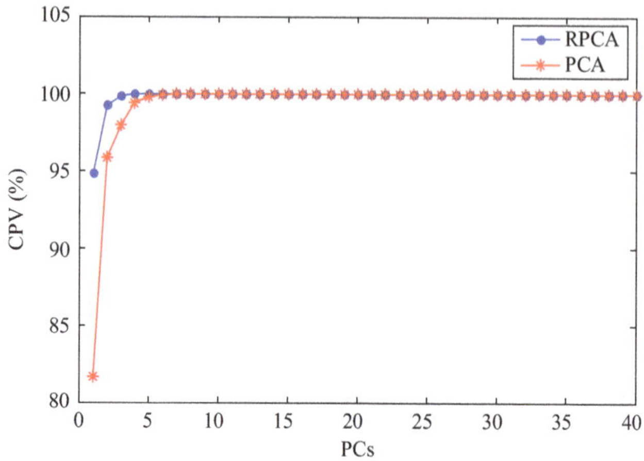


*Figure 5.18 CPV with different PCs [40]*

### 5.3.5 Fault diagnosis based on RPCA for multilevel inverter

FFT-RPCA-Classifier method can be used in multilevel inverter's fault diagnosis. First, FFT is used to transform time-domain signal into frequency-domain signal. Second, RPCA is used to extract main features, compress dimension and remove noise. Finally, different classifiers are used for fault recognition, training the RPCA-Classifier when the diagnosis results achieve the expected goals, or go to the second step. This kind of fault diagnosis method can be used in the system of Section 5.2.5.

The same test data are from Section 5.2.5. As shown in Figure 5.18, the first PC of PCA contains 82% main features, but the first PC of RPCA contains almost 94% main features when $M = \{M_1 = 20, M_2 = M_3 = M_4 = 10, M_i = 1|_i = 5, 6, \ldots, 40\}$. So, the first PC of RPCA can extract more main features from the original data than the first PC of PCA.

The first PC of PCA and RPCA are shown in Figure 5.19, where there are 40 sets of samples contained in each fault. "1, 2, 3, ..." mean the category labels of different faults. It is difficult to distinguish Faults 7 and 8 by PCA when PC = 1 as shown in Figure 5.19(a). However, RPCA can do this much easier as shown in Figure 5.19(b). So, the feature of first PC is more representative by RPCA, which is more useful for fault diagnosis.

According to Table 5.10, the average computing time of FFT-BP is the longest among all the methods, and the average diagnostic accuracy rate is about 70%. Although the average diagnostic accuracy rate of FFT-SVM is similar as FFT-RPCA-SVM, its average computing time is much longer than that of FFT-RPCA-SVM's. The performance of FFT-PCA-SVM is better than FFT-SVM in diagnostic accuracy and average computing time. The diagnostic accuracy of FFT-PCA-SVM comes down

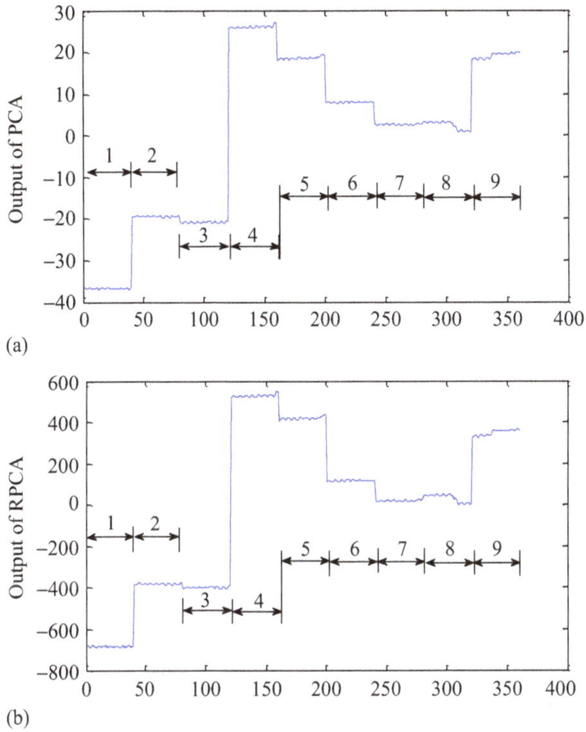


*Figure 5.19 The output of PCA and RPCA when PC = 1. (a) The output of PCA. (b) The output of RPCA [40]*

with the decrease in PCs. But the diagnostic accuracy of FFT-RPCA-SVM is basically unchanged with the decrease of PCs. Because RPCA is used to extract the key features of signal, the first PCs can represent the whole data. FFT-RPCA-SVM is useful to improve the diagnostic accuracy, and reduce the computing time comparing with other three methods from the experimental results.

In the FFT-RPCA-Classifier fault diagnosis method, RPCA is used to further compress data, extract main features and de-noise. So, the Classifiers easily increase fault detection's accuracy after RPCA, and the generalization ability of FFT-RPCA-Classifier is strong.

## 5.4 NPCA and its application

In order to solve the data problems such as higher data dimension, non-Gaussian distribution, more complex correlation among variables, the signal mutations and so

*Table 5.10 Results based on four methods [40]*

| Different methods | | | Test samples | | | |
|---|---|---|---|---|---|---|
| | | | 45 groups | 72 groups | 108 groups | 153 groups |
| Computing time (ms) | FFT-BP | | 1,617 | 1,541 | 710 | 1,677 |
| | FFT-SVM | | 27.5 | 28.1 | 42.4 | 58.3 |
| | FFT-PCA-SVM | PC = 1 | 9.8 | 10.7 | 12.1 | 13.1 |
| | FFT-RPCA-SVM | PC = 1 | 9.9 | 11.1 | 12.0 | 13.7 |
| Diagnostic accuracy (%) | FFT-BP | | 70.29 | 69.76 | 73.44 | 68.76 |
| | FFT-SVM | | 95.97 | 94.02 | 91.93 | 90.58 |
| | FFT-PCA-SVM | PC = 4 | 100 | 97.21 | 96.52 | 95.47 |
| | | PC = 1 | 91.11 | 89.81 | 88.88 | 88.88 |
| | FFT-RPCA-SVM | PC = 4 | 100 | 100 | 100 | 100 |
| | | PC = 1 | 100 | 100 | 100 | 100 |

on for systems' monitoring, this section introduces an NPCA method to transform the non-Gaussian normal data into Gaussian data through an LS.

### *5.4.1 NPCA method*

There are many errors existing in observations during practical industrial processes in error theory. So, the true value ($\psi$) and the random error ($\varsigma$) are consisted in the observed value ($x$), which means $x = \psi + \varsigma$. At same time, there are many random errors caused by many uncertainty factors, which are mostly following Gaussian distribution. In the section, the periodic data should satisfy the following constraint conditions for the NPCA method.

**Constraint condition 5.4.1:** *The test errors are random errors, which follow Gaussian distribution.*

Suppose $\boldsymbol{X}$ is a cyclical multivariables matrix under periodic nonsteady conditions.

$$\boldsymbol{X} = \left[\boldsymbol{X}^1, \boldsymbol{X}^2, \ldots, \boldsymbol{X}^j, \ldots\right] \tag{5.54}$$

where $\boldsymbol{X}^j$ has $n$ variables and $N$ samples, which is the $j$th period of sampling data with, i.e.:

$$\boldsymbol{X}^j = \left[x_1^j, x_2^j, \ldots, x_n^j\right] \tag{5.55}$$

$$x_i^j = \left[x_i^j(l), x_i^j(2), \ldots, x_i^j(N)\right], \quad i = 1, 2, \ldots, n \tag{5.56}$$

As shown in (5.57), $x_i^j(l)$ is the observed value of the $i$th variable from the $l$th sampling point of the $j$th period.

$$x_i^j(l) = \psi_i^j(l) + \varsigma_i^j(l), \quad l = 1, 2, \ldots, N \tag{5.57}$$

where $\psi_i^j(l)$ is the true value of the $i$th variable and $\varsigma_i^j(l)$ is the error of the $i$th variable from the $l$th sampling point of the $j$th period.

On the basis of the periodicity knowledge, the true values are equal for the same sampling points in different periods, i.e.:

$$\psi_i^j(l) = \psi_i(l) \tag{5.58}$$

where $\psi_i(l)$ is the true value of the $i$th variable at the $l$th sampling point, then (5.84) is transformed into (5.85):

$$x_i^j(l) = \psi_i(l) + \varsigma_i^j(l), \quad l = 1, 2, \ldots, N \tag{5.59}$$

When $\boldsymbol{A}_i(l) = [x_i^1(l), x_i^2(l), \ldots, x_i^j(l), \ldots]$. Here $\{\boldsymbol{A}_i(l),\ l = 1, 2, \ldots, N\}$ is a series of sampling data of the $i$th variable at the $l$th sampling point in all different periods. Property 5.4.1 and Property 5.4.2 can be obtained as follows.

**Property 5.4.1** *If the system is under periodic unstable condition, then* $\{\boldsymbol{A}_i(l),\ l = 1, 2, \ldots, N\}$ *follows Gaussian distribution. That means, the data of different sampling periods at the same sampling point of any process variable is subject to Gaussian distribution.*

**Proof:**

Equation (5.60) is transformed from (5.59) under periodic nonsteady conditions:

$$\begin{aligned}
\boldsymbol{A}_i(l) &= \left[x_i^1(l), x_i^2(l), \ldots, x_i^j(l), \ldots\right] \\
&= \left[\psi_i^1(l) + \varsigma_i^1(l), \psi_i^2(l) + \varsigma_i^2(l), \ldots, \psi_i^j(l) + \varsigma_i^j(l), \ldots\right] \\
&= \left[\psi_i(l) + \varsigma_i^1(l), \psi_i(l) + \varsigma_i^2(l), \ldots, \psi_i(l) + \varsigma_i^j(l), \ldots\right] \\
&= \psi_i(l) + \left[\varsigma_i^1(l), \varsigma_i^2(l), \ldots, \varsigma_i^j(l), \ldots\right]
\end{aligned} \tag{5.60}$$

On the basis of Constraint Condition 5.4.1, the random errors $[\varsigma_i^1(l), \varsigma_i^2(l), \ldots, \varsigma_i^j(l), \ldots]$ are following Gaussian distributions, which are from $N(\mu_i(l), \chi_i^2(l))$, where $\chi_i(l)$ is the standard deviation of random fluctuation errors and $\mu_i(l)$ is the mean value of random fluctuation errors. According to the additivity of Gaussian distribution, $\{\boldsymbol{A}_i(l),\ l = 1, 2, \ldots, N\}$ is following Gaussian distributions $N(\psi_i(l) + \mu_i(l), \chi_i^2(l))$. Thus, Property 5.4.1 is proved by the above analysis. Based on the above analysis of the periodic unstable condition system, a new data standardization method is proposed as follows.

**Definition 5.4.1:** *Longitudinal standardization*
Suppose $x_i^j$ is the case of a cycle time of sampling data, the equation of LS is defined as (5.61):

$$x_i^{j*}(l) = \frac{x_i^j(l) - \bar{A}_i(l)}{S_i^J(l)}, \quad l = 1, 2, \ldots, N \tag{5.61}$$

where $\bar{A}_i(l)$ is the mean value of $\boldsymbol{A}_i(l)$, and $S_i^j(l)$ is the standard deviation of $\boldsymbol{A}_i(l)$. On the basis of Property 5.4.1, $\{A_i(l), l = 1, 2, \ldots, N\}$ is following Gaussian distributions $N(\psi_i(l) + \mu_i(l), \chi_i^2(l))$. Equations (5.62) and (5.63) are obtained according to the law of large numbers.

$$\bar{A}_i(l) = \lim_{J\to\infty} \frac{1}{J}\sum_{j=1}^{J} x_i^j(l) = \psi_i(l) + \mu_i(l) \tag{5.62}$$

$$S_i^j(l) = \lim_{J\to\infty} \sqrt{\frac{1}{J}\sum_{j=1}^{J}\left(x_i^j(l) - \bar{A}_i(l)\right)^2} = \chi_i(l) \tag{5.63}$$

When the number of cycles $J$ tends to infinity, the mean $\bar{A}_i(l)$ of $\boldsymbol{A}_i(l)$ tends to $\psi_i(l) + \mu_i(l)$, and the standard deviation $S_i(l)$ of $\boldsymbol{A}_i(l)$ tends to $\chi_i(l)$, and the LS transform does not affect the final detection results under large sample data. Then (5.61) is transformed into (5.64):

$$\begin{aligned} x_i^{*j}(l) &= \frac{x_i^j(l) - \bar{A}_i(l)}{S_i(l)} = \frac{(\psi_i(l) + \varsigma_i^j(l)) - (\psi_i(l) + \mu_i(l))}{\chi_i(l)} \\ &= \frac{\varsigma_i^j(l) - \mu_i(l)}{\chi_i(l)} \end{aligned} \tag{5.64}$$

Therefore, the mean value of $\boldsymbol{x}_i^{*j}(l)$ is 0 and the standard deviation is 1 after LS; that is, $\boldsymbol{x}_i^{*j}(l)$ is taken from the standard normal distribution $N(0, 1)$. Then the Property 5.4.2 is obtained.

**Property 5.4.2** *The transformed data is following standard normal distributions after LS under periodic nonsteady conditions when J tends to infinity.*

**Proof:**
Because $\psi_i(l)$ is a deterministic term in $x_i^j(l) = \psi_i(l) + \varsigma_i^j(l)$, (5.60) can be replaced by (5.64) when $J$ is tending to infinity. Thus, $x_i^{j*}(l)$ is following standard Gaussian distribution because the random errors $[\varsigma_i^1(l), \varsigma_i^2(l), \ldots, \varsigma_i^j(l), \ldots]$ are independent identically distributed $N(\mu_i(l), \chi_i^2(l))$ random variables. The transformed data are following Gaussian distributions after LS under periodic nonsteady conditions, which satisfies the requirement of $T^2$ chart for fault detection.

The *Q–Q* plots is usually used to evaluate if the data are following Gaussian assumption as the number of cycles $J$ does not tend to infinity. When the shape of the points scatter is close to a straight line, the normality hypothesis still holds. Normality is suspect if the shape of the points scatter is far from a straight line.

Here the *Q–Q* plots is used to verify Property 5.4.1 and Property 5.4.2 as shown in the following example. The data $\boldsymbol{X}_{\text{test}}$ has been selected with Gaussian noise, $\boldsymbol{X}_{\text{test}} = [X_{\text{test1}}, X_{\text{test2}}, X_{\text{test3}}, X_{\text{test4}}]$, there are 100 samples in one cycle and $J = 50$.

1. Square wave:

$$X_{\text{test1}} = \begin{cases} 10, & t \in [a, a+0.5). \\ 0, & t \in [a+0.5, a+1). \end{cases} \quad a \in N.$$

2. Sine wave:

$$X_{\text{test2}} = 30 * \sin(2\pi t) + 40, t \in [0, +\infty).$$

3. Sawtooth wave:

$$X_{\text{test3}} = 5 * (t - a), t \in [a, a+1), a \in N.$$

4. Step wave:

$$X_{\text{test4}} = a, a = [0, 1, \ldots, 7], t \in \left[\frac{a}{8} + l, \frac{a+1}{8} + l\right], l \in N.$$

The *Q–Q* plots of $\boldsymbol{A}_1(35), \boldsymbol{A}_2(35), \boldsymbol{A}_3(35), \boldsymbol{A}_4(35)$ are shown in Figure 5.20; the shape of the points scatter is close to a straight line, which means that $\boldsymbol{A}_1(35), \boldsymbol{A}_2(35), \boldsymbol{A}_3(35), \boldsymbol{A}_4(35)$ are following Gaussian distributions. As shown from Figures 5.21–5.24, one periodic data $\boldsymbol{X}_{\text{text}}$ is selected randomly to make *Q–Q* plots before LS and after LS. The sampling data are not close to a straight line before LS. However, the data are close to a straight line after LS. Property 5.4.1 and Property 5.4.2 are verified under periodic nonsteady conditions from the above analysis.

### 5.4.2 *Fault detection based on NPCA for wind power generation*

The flowchart of fault detection based on NPCA is as shown in Figure 5.25. Direct-drive permanent-magnet (DDPM) synchronous generation is one of the main directions of wind power generation. Because the permanent magnet material has a high demand for stability, the weight of the motor increases, and the capacity of the inverter also increases. So, the cost of the generator is higher than before. If the generator faults happen, it will cause great economic loss, making fault detection very important for DDPM. The structure of DDPM is shown in Figure 5.12. There are wind turbine, inverter, rectifier and maximum power point tracking (MPPT) composed in the system.

Because the wind speed varies greatly in different seasons, the model of DDPM synchronous generation is built on the basis of $P = \frac{1}{2}\rho\pi R^2 C_p(\lambda)V^3$, where $\rho$ is the air density, $R$ is radius of the rotor, $\lambda$ is the tip-speed ratio, $C_p(\lambda)$ is the utilization coefficient of wind power, which is related with tip velocity ratio, and $V$ is the wind speed. In order to maximize absorption wind, wind turbines always run in the

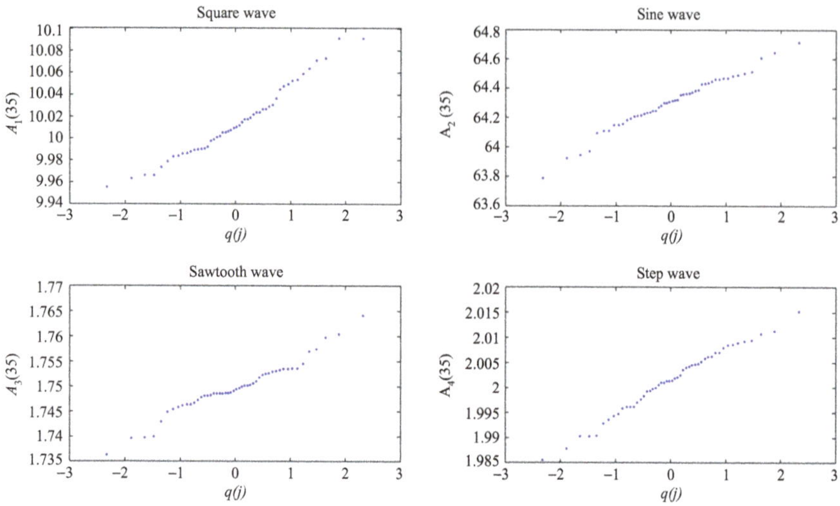


*Figure 5.20 Q–Q plots of $A_1(35)$,$A_2(35)$, $A_3(35)$, $A_4(35)$ [27]*

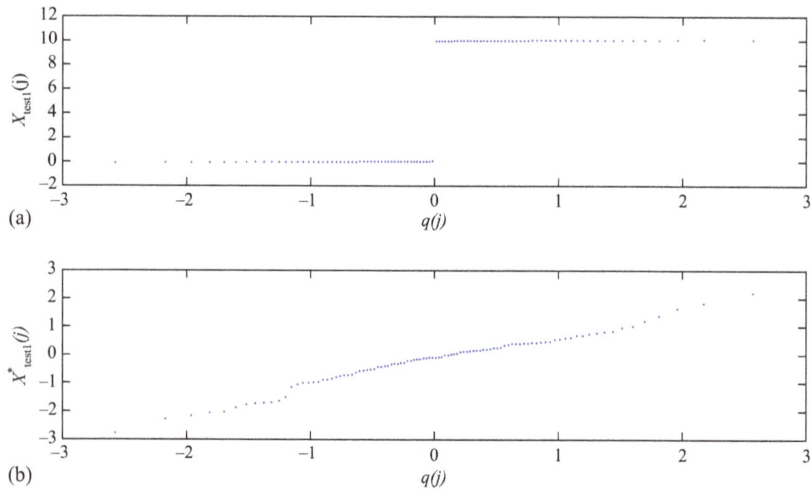


*Figure 5.21 Q–Q plots of square wave (a) before LS and (b) after LS [27]*

maximum power point and generator system output power must match wind turbines capture mechanical power strictly. Mostly, the normal wind speed range is from 6 to 11 m/s. Table 5.11 is the main parameters of the wind generator power. Then 400 groups of normal fan parameters data are obtained from the simulation model; here the fan parameters are speed, voltage, power, three-phase rotor current, and so on.

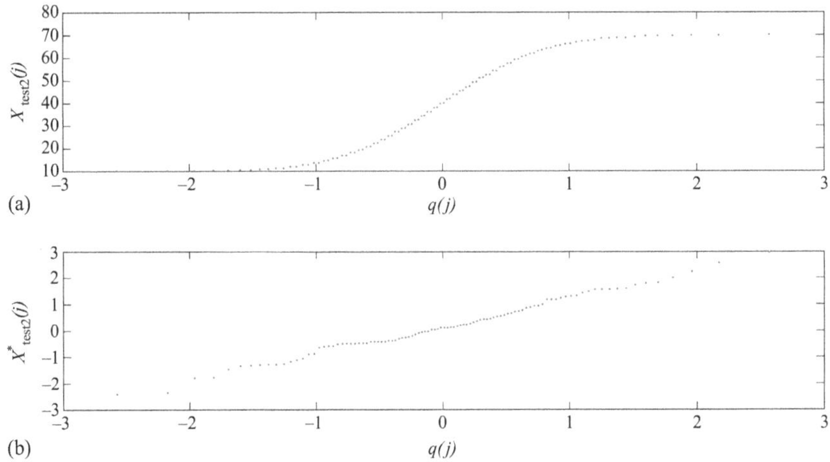

Figure 5.22 Q–Q plots of sine wave (a) before LS and (b) after LS [27]

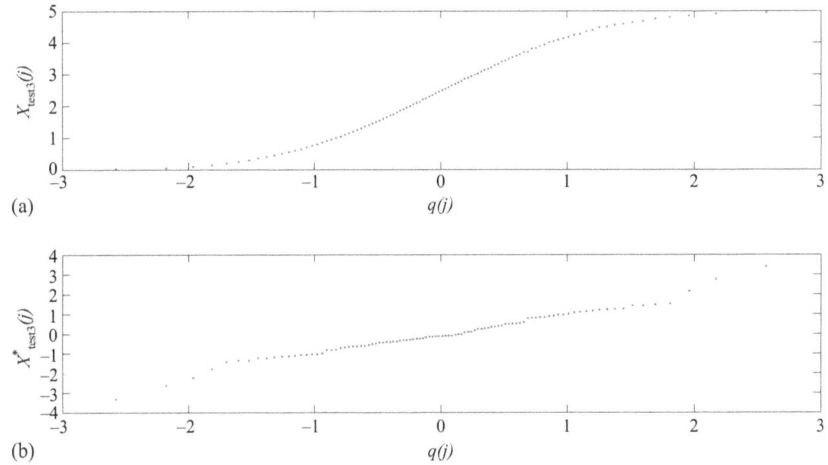

Figure 5.23 Q–Q plots of sawtooth wave (a) before LS and (b) after LS [27]

Next the mean and variance are calculated for LS, and the cycle size is 800, and the confidence of $T^2$ control limit is 95%.

The wind speed is set about 14 m/s in sampling time 200–400 as a group of fault data from the simulation model, and the wind speed of other time is in normal wind speed range. When the wind turbine woks in the speed exceeding 11 m/s for long time, the fan will be damaged. It's because the fan's withstand mechanical stress is greater than the rated maximum stress.

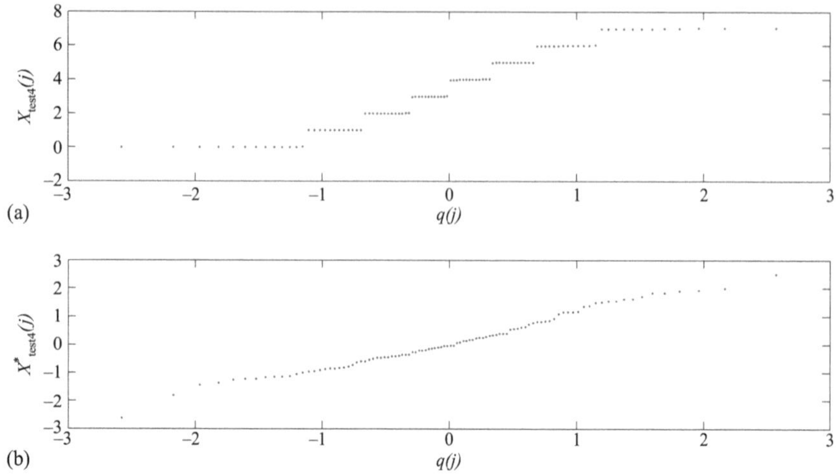


*Figure 5.24 Q–Q plots of step wave (a) before LS and (b) after LS [27]*

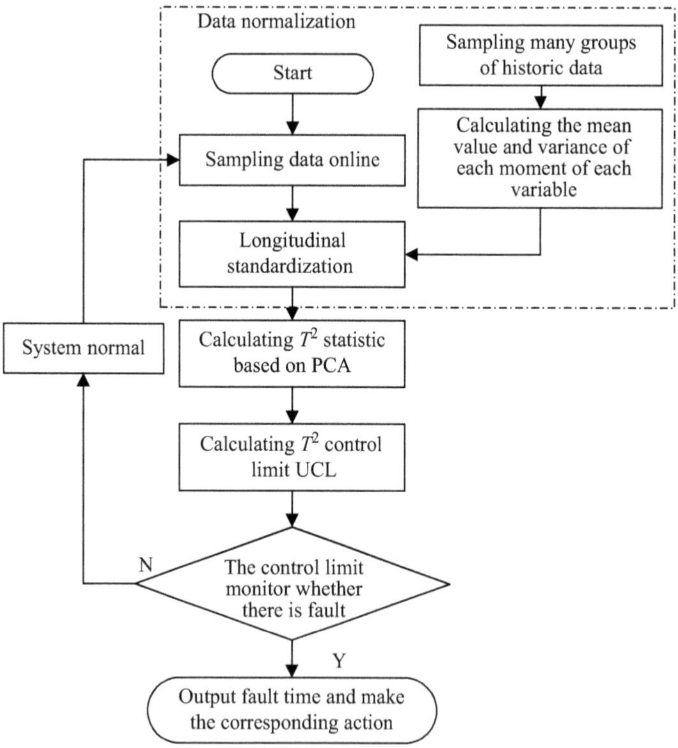


*Figure 5.25 Flowchart of fault detection based on NPCA [24]*

*Table 5.11 The main parameters of the wind generator power [24]*

| | | | |
|---|---|---|---|
| Mechanical ($W$) | 39,900 | Rotor flux ($Wb$) | 0.192 |
| Generator power (V/A) | 39,900/0.9 | Friction coefficient (N m s) | 0.001889 |
| Pitch angle (°) | 0 | The optimal tip speed ratio | 8.1 |
| Stator resistance (Ω) | 0.05 | Fan radius $R$ (m) | 15 |
| Inductance (H) | 0.000635 | Wind speed range (m/s) | 6–11 |
| Pole logarithmic (P) | 36 | The biggest wind power utilization coefficient | 0.48 |

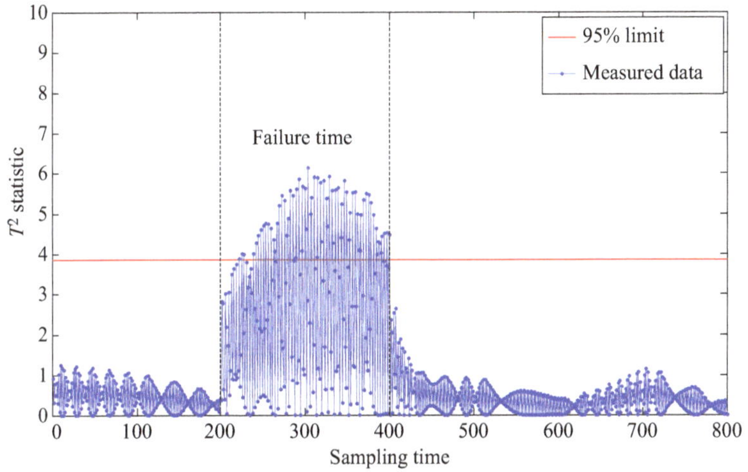


*Figure 5.26 Fault detection result of wind power generation system by PCA [24]*

The fault detection results are shown in Figures 5.26 and 5.27 for wind power generation system by PCA and NPCA, respectively. The false alarm rate and the missing alarm rate of are listed by PCA and NPCA as shown in Table 5.12. The result of RPCA is better than PCA's for fault detection, which increases the effectiveness of system monitoring.

When the data are not following normal distribution, it is not good to detect fault by the $T^2$ control limit based on PCA. According to this problem, the NPCA model is proposed to transform the non-normal distribution data into normal distribution data to reduce false alarm greatly and improve the effectiveness of monitor tremendously under the periodic nonsteady conditions. The NPCA model includes three steps: data normalization, PCA and fault detection based on $T^2$ control limit. LS is the main part of data normalization, which is used to transform the non-normal distribution data

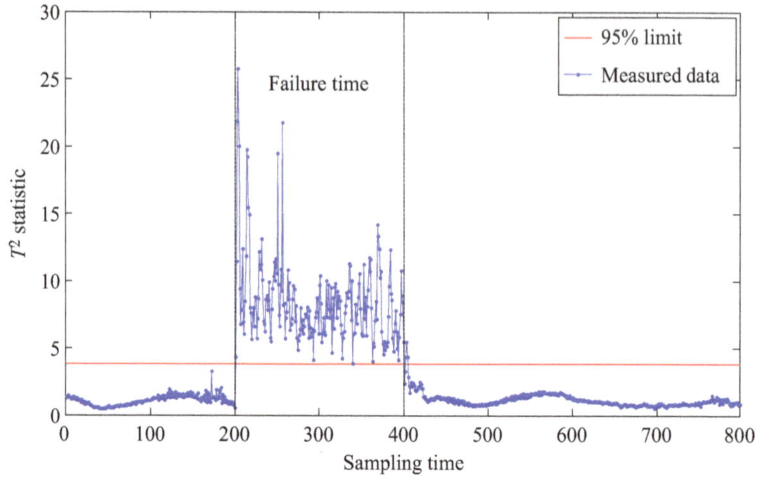


*Figure 5.27 Fault detection result of wind power generation system by NPCA [24]*

*Table 5.12 Two kinds of fault detection methods performance comparison [24]*

| Multivariate statistical fault detection method | False alarm rate (%) | Missing alarm rate (%) |
|---|---|---|
| PCA | 0 | 16.38 |
| NPCA | 0.37 | 0.13 |

into normal distribution data under the periodic nonsteady conditions to ensure fault detection valid through $T^2$ control limit.

### *5.4.3 Fault detection based on NPCA for DC motor*

In this section, the NPCA model is applied to excite DC motor for fault detections. The basic parameters are selected as follows: voltage, $U_{d0} = 60$ V; pole logarithmic, $P_0 = 1$; armature resistance, $R_a = 25\ \Omega$; armature inductance, $L_a = 0.3$ H; rotary inertia, $I = 0.0004$ kg m$^2$; rated excitation, $Ce = 0.05236$. The failure-free historical data, with sample size 100 for each cycle and $J = 400$, are obtained to calculate the mean value $\bar{A}_i^J(k)$ and the standard deviation $S_i^J(k)$ by running the simulation model of the excited DC repeatedly under the normal assumption of the error.

The data include three variables: speed $\eta_0 1$, electromagnetic torque $T_L$ and current $i_a$. The changing situations of the motor load torque are shown in Figure 5.28. The normal load torque range changes between 0.2 and 0.6 N m. The first and third periods of the test data are obtained during normal operation, while the load torque of

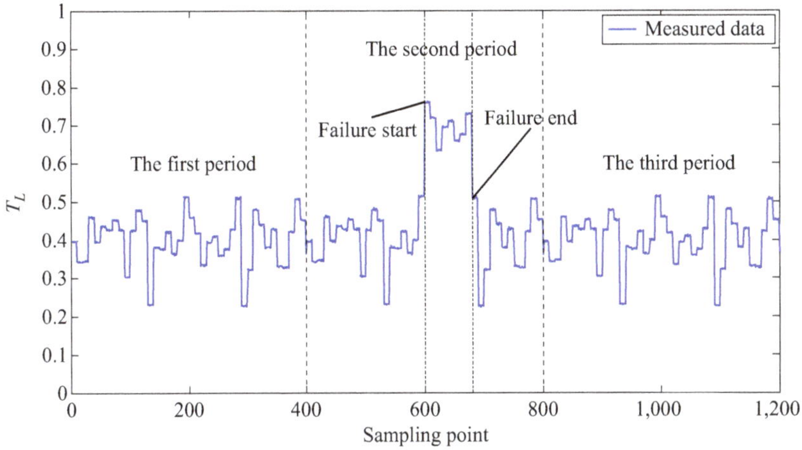


*Figure 5.28 The motor load torque in three periods [28]*

the second period suddenly increases to 0.3 N m during the sampling times 600–680, which has exceeded the normal range, so the second period of data is regarded as fault data. The number of the PCs is determined by cumulative percent variance $P \geq 90\%$. In this section, the LSPCA model is adopted to detect the three periods of the test data, which is compared with the PCA method and the dynamic data window based on RPCA (D-RPCA) method.

Fault detection results of the three periods of data are shown in Figures 5.29–5.31, which are based on PCA, D-RPCA and NPCA, respectively. As can be seen from the comparison results in Table 5.12, the following conclusion could be obtained: for the normal data, the false alarm rate is high based on PCA or D-RPCA, so their detection reliability is low. For the abnormal data, the missing alarm rate is high by the fault detection based on PCA, so its detection sensitivity is low, but the fault detection based on D-RPCA method could maintain a low false alarm rate and missing alarm rate while the fault detection based on NPCA could maintain a low false alarm rate and missing alarm rate under the normal situations and abnormal situations, which improves the efficiency of system monitoring greatly.

The NPCA model is suitable for fault detection under periodic nonsteady conditions as shown in experimental results, which is useful to reduce false alarm rate.

## 5.4.4 ACL based on NPCA

It is difficult to detect faults effectively by the Hotelling's $T^2$ confidence limit when the system is under periodic transient conditions. According to this problem, the ACL is built by the dynamic data window algorithm for periodic transient conditions.

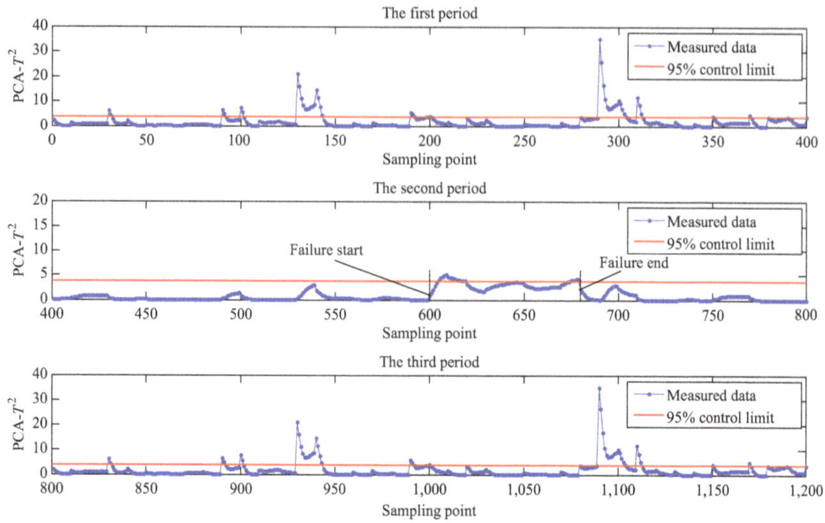


*Figure 5.29 Fault detection results by the PCA method [28]*

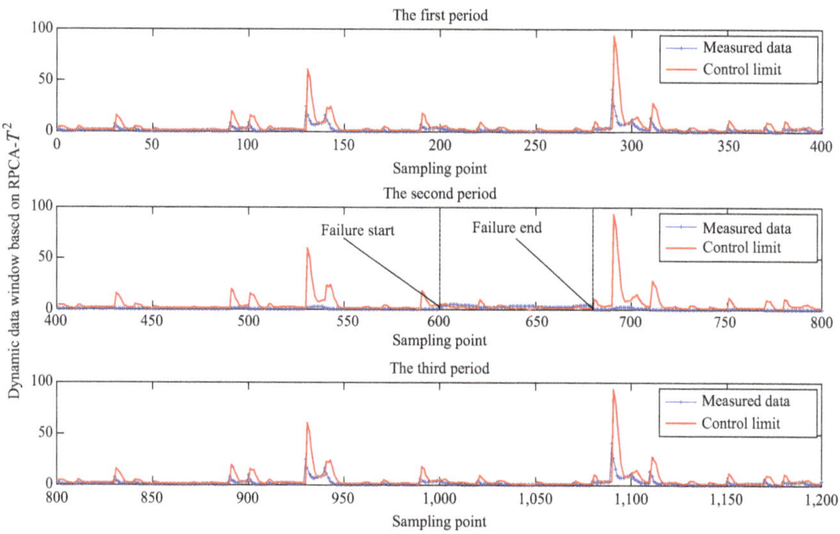


*Figure 5.30 Fault detection results by D-RPCA method [28]*

The ACL is built by (5.65) as follows:

$$T_{\mathrm{ucl}} = \omega \times T_{\mathrm{ucl1}} + (1 - \omega) \times T_{\mathrm{ucl2}} \tag{5.65}$$

where $T_{\mathrm{ucl}}$ is from the standard confidence limit by Hotelling's $T^2$, and $T_{\mathrm{ucl2}}$ is the confidence limit of every sampling point, $\omega$ $(0 < \omega < 1)$ is set by different users'

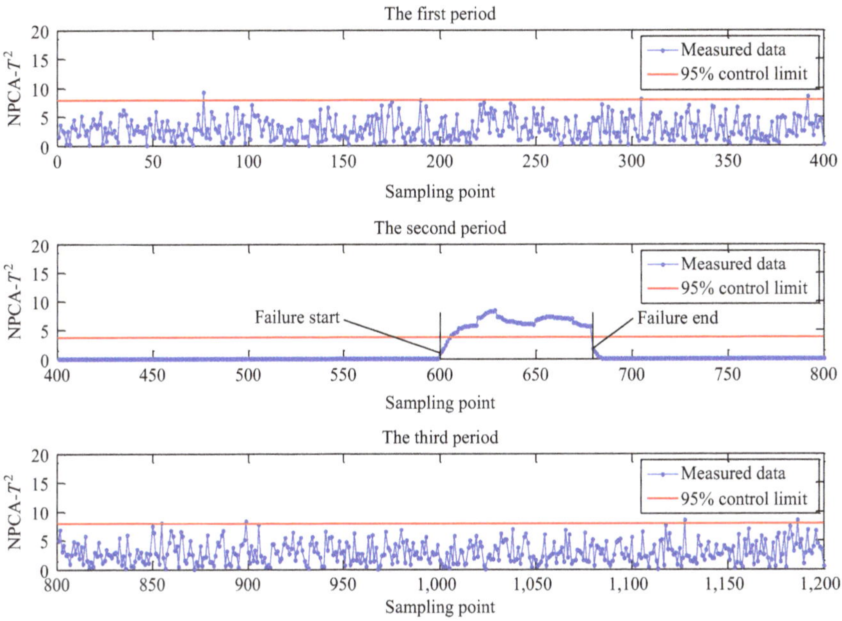


*Figure 5.31 Fault detection results by the NPCA model [28]*

*Table 5.13 Comparison results of the three detection methods [28]*

| **Detection precision** | | **Process data** | | |
|---|---|---|---|---|
| | | **The 1st period** | **The 2nd period** | **The 3rd period** |
| False alarm rate (%) | PCA | 10.75 | 0 | 7.25 |
| | D-RPCA | 7.5 | 0 | 7.75 |
| | NPCA | 0.75 | 0 | 1 |
| Missing alarm rate (%) | PCA | 0 | 15.25 | 0 |
| | D-RPCA | 0 | 4.75 | 0 |
| | NPCA | 0 | 1.75 | 0 |

requirement. The calculating procedure are as follows for the three parameters $T_{ucl1}$, $T_{ucl2}$ and $\omega$:

1. *$T_{ucl1}$, and $T_{ucl2}$ calculated by mPCA*
   In order to compress the data and extract key features from the original data, and then improve the fault detection accuracy, the mPCA method is proposed to preprocess the original data. There are the historical normal data and the real-time

test data composed in the original multi-period test data. The mPCA method is detailed as follows:

a. Multi-period data for input: $\boldsymbol{X} = \begin{pmatrix} x_1(l) & \cdots & x_n(1) \\ \vdots & \ddots & \vdots \\ x_1(N) & \cdots & x_n(N) \end{pmatrix}$ is selected stochastically as one-period historical normal data, $\boldsymbol{X}_{\text{test}} = \begin{pmatrix} x_{1\text{test}}(l) & \cdots & x_{n\text{test}}(l) \\ \vdots & \ddots & \vdots \\ x_{1\text{test}}(\text{N}) & \cdots & x_{n\text{test}}(\text{N}) \end{pmatrix}$ is selected stochastically as one period real-time test data, where is the size of variables of the multi-period data and is the length of the multi-period data.

b. LS Transform: $\boldsymbol{X}_{\text{test}}$ is transformed to $\boldsymbol{X}^*_{\text{test}} = \begin{pmatrix} x^*_{1\text{test}}(l) & \cdots & x^*_{n\text{test}}(l) \\ \vdots & \ddots & \vdots \\ x^*_{1\text{test}}(N) & \cdots & x^*_{n\text{test}}(N) \end{pmatrix}$ and $\boldsymbol{X}$ is transformed to $\boldsymbol{X}^* = \begin{pmatrix} x^*_1(l) & \cdots & x^*_n(l) \\ \vdots & \ddots & \vdots \\ x^*_1(N) & \cdots & x^*_n(N) \end{pmatrix}$, respectively, after LS transform by (5.61). $\boldsymbol{X}^*$ is following standard Gaussian distribution according to the Property 5.4.2.

c. Transform Data Combination: Let:

$$\boldsymbol{Y}^* = \begin{pmatrix} \boldsymbol{X}^* \\ \boldsymbol{X}^*_{\text{test}} \end{pmatrix} = \begin{pmatrix} x^*_1(l) & \cdots & \cdots & x^*_n(l) \\ \vdots & \vdots & \vdots & \vdots \\ x^*_1(N) & \cdots & \cdots & x^*_n(N) \\ x^*_{1\text{test}}(l) & \cdots & \cdots & x^*_{n\text{test}}(l) \\ \vdots & \vdots & \vdots & \vdots \\ x^*_{1\text{test}}(N) & \cdots & \cdots & x^*_{n\text{test}}(N) \end{pmatrix}.$$

$$= \begin{pmatrix} y^*_1(l) & \cdots & y^*_n(l) \\ \vdots & \ddots & \vdots \\ y^*_1(2N) & \cdots & y^*_n(2N) \end{pmatrix} \tag{5.66}$$

d. Feature extraction by PCA: The multivariable dimensionality of $\boldsymbol{Y}^*$ is reduced by PCA, then covariance matrix, eigenvalues and corresponding eigenvector are calculated to decide the number $m$ of PCs by the cumulative percent variance $P$.

$T_{\text{ucl1}}$ is calculated according to Hotelling's $T^2$ by (5.67).

$$T_{\text{ucl1}} = \frac{m(2N-1)}{2N-m} F_\alpha(m, 2N-m) \tag{5.67}$$

where $T_{\text{ucl1}}$ is as the standard confidence limit of Hotelling's $T^2$, the length of $\boldsymbol{Y}^*$ is $2N$, and the number of reserved PCs is $m$. $F_\alpha(m, 2N-m)$ is the critical value of $F$ distribution corresponding to the test level $\alpha$, the degree of freedom $m$ and $2N-m$. Confidence $(1-\alpha)$ can be determined by users' requirement. For example, when $\alpha = 0.05$ or $\alpha = 0.10$, the confidence is 95% or 90%, Hotelling's $T^2$ is calculated by (5.68):

$$\begin{aligned} T^2_{Y^*}(l_0) &= \boldsymbol{Y}^*(l_0)\,\boldsymbol{P}_m\boldsymbol{\Lambda}_m^{-1}\boldsymbol{P}_m^T\boldsymbol{Y}^*(l_0)^T \\ &= \left[T^2_{Y^*}(l),\ldots,T^2_{Y^*}(N),T^2_{Y^*}(N+1),\ldots,T^2_{Y^*}(2N)\right] \end{aligned} \tag{5.68}$$

where $l_0 = 1,2,3,\ldots,2N$, and $\boldsymbol{\Lambda}_m = diag(\lambda_1,\lambda_2,\ldots,\lambda_m)$ is the diagonal matrix composed of the top $m$ eigenvalues, and the number of PCs is $m$. The load vector matrix is $\boldsymbol{P}_m = [p_1,p_2,\ldots,p_m]$.

Here (5.69) is the $T^2$ statistics of one-period normal data, and (5.70) is the $T^2$ statistics of one-period real-time test data.

$$T_0^2 = [T^2(l),\ldots,T^2(N)] \tag{5.69}$$

$$T_1^2 = [T^2(N+1),\ldots,T^2(2N)] \tag{5.70}$$

The validity of the mPCA method is proved as follows.

On the basis of (5.68), $T^2$ statistics of the same $l$th sampling point can be expressed as (5.71) for the test data $\boldsymbol{X}^*_{\text{test}}$ and the normal data $\boldsymbol{X}^*$.

$$T^2_{Y^*}(l) = \boldsymbol{Y}^*(l)\boldsymbol{P}_m\boldsymbol{\Lambda}_m^{-1}\boldsymbol{P}_m^T\boldsymbol{Y}^*(l)^T = \boldsymbol{X}^*(l)\boldsymbol{P}_m\boldsymbol{\Lambda}_m^{-1}\boldsymbol{P}_m^T\boldsymbol{X}^*(l)^T \tag{5.71}$$

$$\begin{aligned} T^2_{Y^*}(N+l) &= \boldsymbol{Y}^*(N+l)\boldsymbol{P}_m\boldsymbol{\Lambda}_m^{-1}\boldsymbol{P}_m^T\boldsymbol{Y}^*(N+l)^T \\ &= \boldsymbol{X}^*_{\text{test}}(l)\boldsymbol{P}_m\boldsymbol{\Lambda}_m^{-1}\boldsymbol{P}_m^T\boldsymbol{X}^*_{\text{test}}(l)^T \end{aligned} \tag{5.72}$$

$\Delta T^2(l)$ is obtained from normal data at the same $l$th sampling point from (5.71) and (5.72), which is visible between test and normal data, because $T^2$ statistics are calculated for test and normal data in the same PC space.

$$\begin{aligned} \Delta T^2(l) &= T^2_{Y^*}(N+l) - T^2_{Y^*}(l) \\ &= \left(\boldsymbol{X}^*_{\text{test}}(l) - \boldsymbol{X}^*(l)\right)\boldsymbol{P}_m\boldsymbol{\Lambda}_m^{-1}\boldsymbol{P}_m^T(\boldsymbol{X}^*_{\text{test}}(l) - \boldsymbol{X}^*(l))^T \end{aligned} \tag{5.73}$$

As shown in (5.73), $\left(\boldsymbol{X}^*_{\text{test}}(l) - \boldsymbol{X}^*(l)\right)$ is the deviation of test data from normal data at $l$th sampling point, and the value of $\Delta T^2(l)$ is increased with the increasing of $\left(\boldsymbol{X}^*_{\text{test}}(l) - \boldsymbol{X}^*(l)\right)$. $\Delta T^2$ is a small shift if $\boldsymbol{X}_{\text{text}}(l)$ is normal, but when $\boldsymbol{X}^*_{\text{text}}(l)$ is fault, the deviation $(\boldsymbol{X}^*_{\text{text}}(l) - \boldsymbol{X}^*(l))$ will increase and the value of $\Delta T^2(l)$ is increased too. So, the faults can be detected and monitoring accuracy will be improved by detecting the statistics differences between normal data and test data at same sampling point in one period.

The confidence limit $T_{\text{ud2}}$ is built as follows:

a. The statistics $T_0^2$ of one-period normal data is converted into $\xi$ in (5.69). Let:

$$S_k = \{\xi_{k-L},\ldots,\xi_{k-1}\} \tag{5.74}$$

where $\xi$ is contained in $S_k$, the size of the dynamic window data is $L$ and the values of the circulation beginning is $k$, $k = L+1$.

b. $S_k$ is rewritten by (5.74), the value of $g_k$ and $f_k$ are calculated, then the confidence limit is obtained for the $k$th value $T_{\text{ucl2}}(k)$:

$$g_k = \delta_\xi / 2\bar{\xi} \tag{5.75}$$

$$h_k = 2\bar{\xi}^2 / \delta_\xi \tag{5.76}$$

$$T_{\text{ucl2}}(k) = g_k \cdot \chi^2(\alpha, f_k) \tag{5.77}$$

where $\bar{\xi}$ is the mean value and $\delta_\xi$ is the variance of $\xi$ on the basis of $S_k$; $\xi^2$ is the Chi-Squared distribution function.

c. Determining if the above factors lead to circulation turnoff. If $k \le N$, $k = k + 1$, the working procedure will return to the Step (a), or the circulation condition will be stopped.

$T_{\text{ucl2}}$ is obtained by the above steps.

2. *Setting ω based on the fault sections determination procedure*
A new fault sections determination procedure is proposed to set $\omega$ in this section. We set:

$$X_{\text{ucl}} = 3 \tag{5.78}$$

as the standard line. The fault sections would be located only if there exist some continuous sampling points after LS above the standard line $X_{\text{ucl}}$. That is, $(\boldsymbol{X}^*_{\text{test}}(l) > X_{\text{ucl}})$ & $(\boldsymbol{X}^*_{\text{test}}(l+1) > X_{\text{ucl}})$& $(X^*_{\text{test}}(l+2) > X_{\text{ud})}$ & …. According to the test experience, the value of continuous sampling points is from 3 to 6. If the value is more than 6, the missing alarm rate will be increased, at the same time the computation time will be increased. If the value is less than 3, the false alarm rate will be increased.

The reason is explained for the standard line $X_{\text{ucl}} = 3$, which is shown in the following statement.

If the faults begin at $l_1$th sampling point and end at $l_2$th sampling point in test data, $l_1, l_2 \in (1, N)$, $l_1 \le l_2$. Thus, the fault data $\boldsymbol{X}_{\text{fault}} = \begin{pmatrix} x_{1\text{fault}}(l_1) & \cdots & x_{n\text{fault}}(l_1) \\ \vdots & \ddots & \vdots \\ x_{1\text{fault}}(l_2) & \cdots & x_{n\text{fault}}(l_2) \end{pmatrix}$ can be obtained, and $l_3$th is a sampling point $l_1 \le l_3 \le l_2$. Then the fault data of first variable at th sampling point are $x_{1\text{fault}}(l_3)$, and the normal data as

$$x_1(l_3) = \bar{A}_1^J(l_3) + \varsigma_1(l_3) \tag{5.79}$$

Here $\bar{A}_1^J(l_3)$ is the mean value of historical normal data and $\varsigma_1(l_3)$ is the random errors sampled from the Gaussian distribution $N(0, S_1^2(l_3))$ according to (5.79). Based on the $3\sigma$ principle of Gaussian distributions, i.e. if $X \sim N(\mu, \sigma^2)$, then $P_{X \notin [\mu-3\sigma, \mu+3\sigma]} > 99.7\%$, it is obtained from (5.80):

$$P_{\varsigma_1(l_3) \in [-3S_1(l_3), 3S_1(l_3)]} > 99.7\% \tag{5.80}$$

The normal data of first variable of the $l_3$th the sampling point can therefore be transformed to $x_1^*(l_3) = \frac{x_1(l_3) - \bar{A}_1^J(l_3)}{S_1(l_3)} = \frac{\varsigma_1(l_3)}{S_1(l_3)}$ after LS, the conclusion (5.81) is obtained according to (5.80):

$$P_{x_1^*(l_3) \in [-3,3]} > 99.7\% \tag{5.81}$$

Let $e$ represent the large deviation of first variable at $l_3$th sampling point, so

$$|e| > 3S_1(l_3) \tag{5.82}$$

$$x_{1\text{fault}}(l_3) = \bar{A}_1^J(l_3) + e \tag{5.83}$$

After LS, $x_{1\text{fault}}(l_3)$ is transformed to (5.84):

$$x_{1\text{fault}}^*(l_3) = \frac{e}{S_1(l_3)} > 3 \text{ or } \frac{e}{S_1(l_3)} < -3 \tag{5.84}$$

It is clear to conclude that the normal data transformed to (5.85) according to the above analysis.

$$P_{x_i^*(l) \in [-3,3]} > 99.7\% \tag{5.85}$$

and the fault data are transformed to (5.86):

$$\left|x_{i\text{fault}}^*(l)\right| > 3 \tag{5.86}$$

$i = 1, 2, \ldots, n, l = 1, 2, \ldots, N$ after LS. Thus $X_{\text{ucl}} = 3$ is set as the standard line. In order to avoid misinformation, the fault sections are located only if there are some continuous sampling points higher than the standard line $X_{\text{ucl}}$. Because the accuracy of fault detection is not good by the determined fault sections, fault section determination procedure is only used to set $\omega$, but not used to fault detection directly. $\omega$ is set through matching one-period real-time test data with one-period fault data of historical database by fault sections as shown in Figure 5.32.

a. *The database of the optimal* $\omega$: First the missing alarm rate and false alarm rate are calculated about different $\omega$ values for $\omega = 0 : 0.01 : 1$ by ACL from one-period fault data which is chosen randomly from history database. Then the optimal $\omega$ is selected on the basis of missing alarm rate and false alarm rate of users' requirement, at the same time the optimal $\omega$ and its fault sections are saved for the next step. Next the above procedure is repeated in different fault sections. At last, the optimal $\omega$ values different from the corresponding fault sections are saved in history database.
b. *Fault section decision*: Fault section decision procedure is used to decide the fault sections of real-time test data. The most representative variable is to be selected to estimate the fault sections when there are much more variables.
c. *Setting* $\omega$: On the basis of the fault sections in the test data from (b), the optimal $\omega$ value is found on the basis of the fault sections in the $\omega$ history database. If the test fault sections do not match the corresponding fault sections in the historical database, then $\omega$ is set by Step (a).

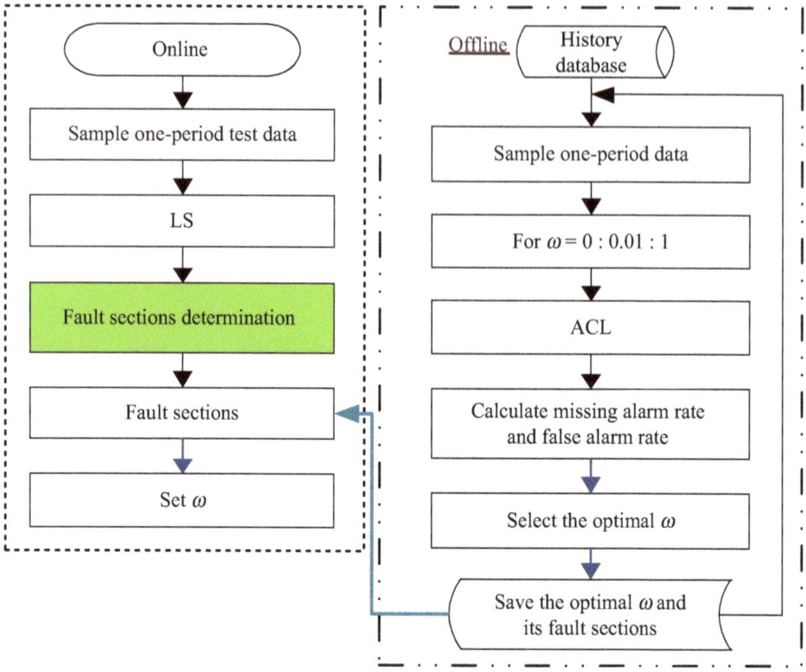


*Figure 5.32 ω setting flowchart [27]*

The fault detection procedure based on NPCA-ACL is illustrated as shown in Figure 5.33.

### 5.4.5 Fault detection based on NPCA-ACL for DC motor

In this section, the fault detection based on NPCA-ACL is used to detect faults for a DC motor of plastic bag-making system, where the DC motor works under periodic nonsteady conditions and there are four conditions in its motion process: constant speed, speed recovery, acceleration and deceleration. The real-time RT-Lab platform is used to simulate the periodic nonsteady conditions' data, which helps to get hardware in the loop (HIL) simulations. The external panel is connected to the platform for hardware in the loop simulation through the I/O interface as shown in Figure 5.34. The basic parameters of DC motor are as follows: voltage, $U = 60$; armature resistance, $R_a = 25\ \Omega$; pole pair number, $P_0 = 1$; rated excitation, $Ce = 0.05236$; rotary inertia, $I = 0.0004\ \mathrm{kg\,m^2}$; and armature inductance, $L_a = 0.3$ H. There are three variables selected as follows: armature current $i_a$, load torque $T_L$ and speed $n_0$. For the load, the variance of the load torque is equal to 0.005, and the mean of the load torque is 0.4 N m. The changeable range of the load torque is from $-0.1$ N m to $+0.1$ N m, where the whole system would malfunction if the load torque was out of this range.

The mean value $\bar{A}_i(l)$ and the standard deviation $S_i^J(l)$ are calculated form the normal historical data obtained from the DC motor under normal situation. Here

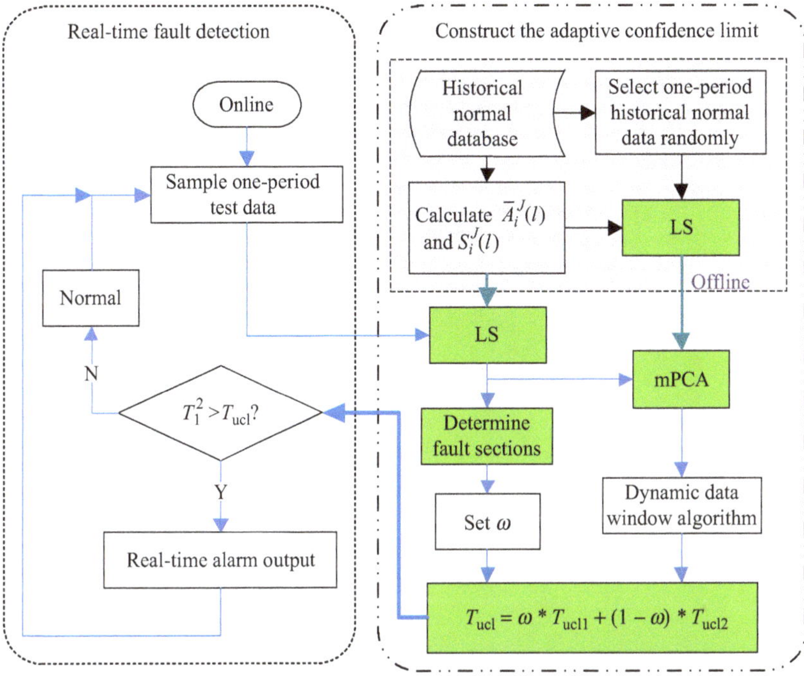


*Figure 5.33 Flowchart of real-time fault detection based on NPCA-ACL [27]*

$J = 500$ and $N = 400$. Fault data (during the sampling points 200–280) are sampled while load torque suddenly increases $\Delta T_L$ 0.30 N m, which exceeds the normal range of the load torque. All the data are with Gaussian white noise, where the SNR is about 17 dB. Cumulative percent variance $P \geq 85\%$ is set to select the number of the PCs for each method. Different fault detection methods are compared as follows.

The $Q$–$Q$ plots of load torque $T_L$ are shown in Figure 5.35 for one-period normal data and one-period test data with faults, which shows that the lines do not conform to linearity before LS. Figure 5.35 indicates that all the normal data and fault data do not follow Gaussian distribution. But the normal data follows Gaussian distribution after LS, and it is more obvious to show different from the normal data when the fault data of $T_L$ is exceeding the normal range after LS, which is useful for fault detection.

The fault detection results of PCA are shown in Figure 5.36(a) with $T^2$ statistics, where the first PC is selected to delegate the original system data. From Figure 5.36(a), we can find that the changing of $T^2$ chart is obvious when it is under periodic transient conditions. The missing alarm rate of fault detection based on PCA is close to 20.25% when the confidence limit is 99%. But the false alarm and missing alarm will be increased when the confidence limit is 90%. The fault detection results of PCA is shown in Figure 5.36(b) with SPE statistics, where the missing alarm rate is close to 19.25% and the false alarm rate is 1.5%. So, it is not good for PCA to use in fault detection under periodic transient conditions for the system. Because when the system is under periodic transient conditions, most of the

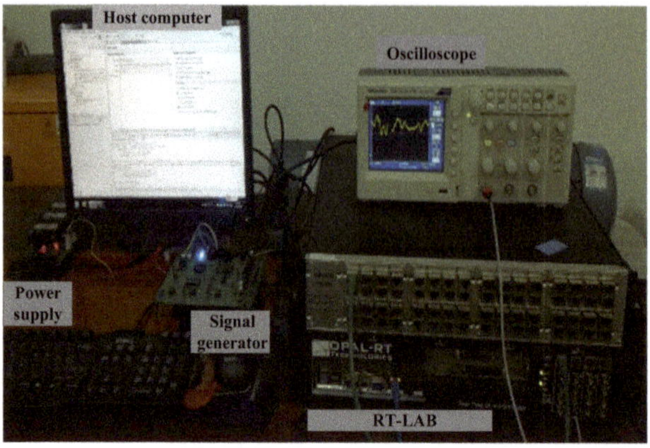


*Figure 5.34 Hardware in the loop simulation platform [27]*

variables do not follow Gaussian distribution, which are not taken in consideration the system's time-varying characteristics. It's essential that it is not good to use the same confidence limit to detect different sampling points under various working conditions.

The fault detection results of the recursive PCA are shown in Figure 5.37, where the first PC is chosen. The missing alarm rate of recursive PCA reaches 15.5% when the confidence limit is 99%, which is better than the fault detection method based on PCA. But the false alarm rate is 4.75%, which is higher than the fault detection method based on PCA's. As shown in Figure 5.37, the fault detection performance of RPCA is not very good when the system is working under periodic transient conditions. Because the fault detection method based on RPCA is more suitable for monitoring slow time-varying stable signals, it is not good to use RPCA to process non-Gaussian distribution data and changing signals.

The ACL of NPCA-ACL is shown in Figure 5.38(a), which is built by (5.65). The final fault detection result of the NPCA-ACL method is shown in Figure 5.38(b). Because the various conditions of different sampling time are considered by the ACL of the NPCA-ACL method and the parameter is used to adjust the detection sensitivity, it will satisfy the requirement of missing alarms and false alarms from the user. As shown in Table 5.13, the false alarm of the NPCA-ACL method is high close to 0.25%, which is lower than fault detection based on PCA, but is higher than fault detection based on RPCA. However, the missing alarm rate is highest of the other two methods. So, the comprehensive detection accuracy of NPCA-ACL method is still the best among the three methods.

All the above detection results are summarized in Table 5.14.

From Figures 5.36–5.38 and Table 5.14, it is obvious to show that

1. the periodic transient conditions characteristics are not handled by fault detection based on PCA;

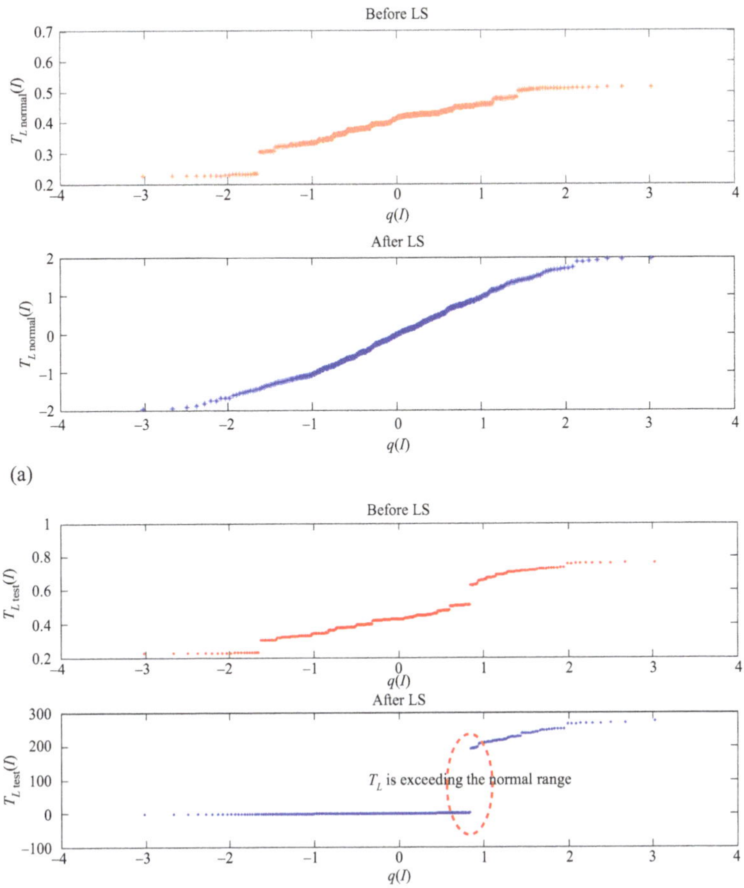


(a)

(b)

*Figure 5.35 Q–Q plots of $T_L$ in normal and test data. (a) The normal data. (b) The test data with faults [27]*

2. when the signal changes a lot under periodic transient conditions, RPCA is not effective to deal with this kind of signals.

However, there are some advantage parts in the NPCA-ACL method: (1) The dimension of data is reduced by mPCA. (2) The non-Gaussian normal data is transformed into Gaussian data by LS. (3) The adaptive control limit is built by dynamic data window method. (4) The $T^2$ of historical normal data and test data are calculated in the same PC space, which could improve the fault detection accuracy. The average computational time reflects the algorithm complexity of each methods as shown in Table 5.14. The computing time of NPCA-ACL method is less than RPCA.

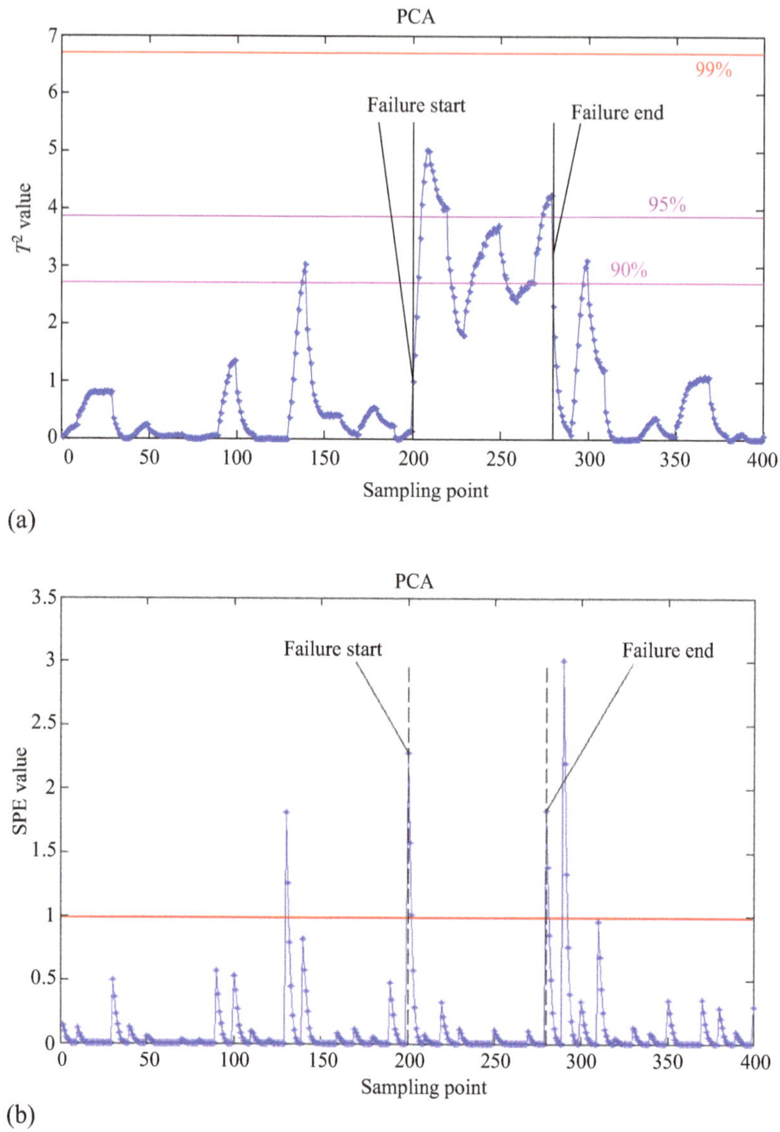


*Figure 5.36 Fault detection results by the PCA approach. (a) With $T^2$ statistics. (b) With SPE statistics [27]*

## 5.5 Conclusions and future works

PCA and its improved methods such as RPCA and NPCA are introduced with their application for fault diagnosis in this chapter. The Hotelling's $T^2$ statistic and SPE statistic are useful for fault detection based on PCA to set control limit. The selection

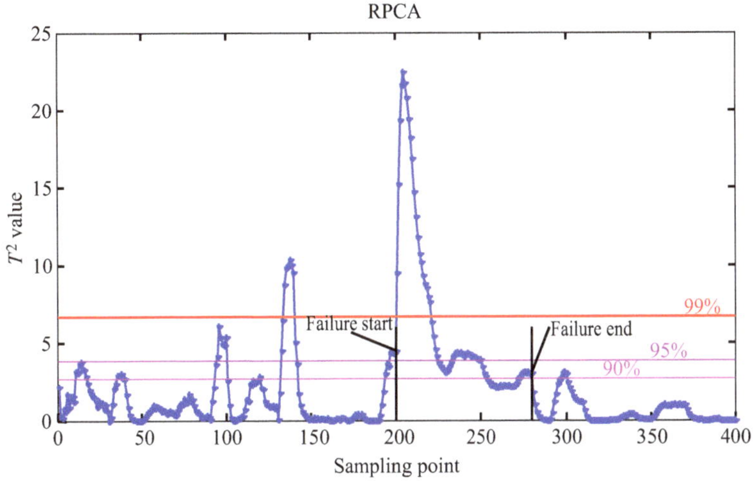


*Figure 5.37 Fault detection results by RPCA [27]*

*Table 5.14 Fault detection results of different methods [27]*

| **Methods** | | **Confidence limit (%)** | **False alarm rate (%)** | **Missing alarm rate (%)** | **Computational time (s)** |
|---|---|---|---|---|---|
| PCA | With $T^2$ | 90% | 1.5 | 6.75 | 0.1568 |
| | | 95 | 0 | 15 | 0.1568 |
| | | 99 | 0 | 20.25 | 0.1568 |
| | With SPE | | 1.5 | 19.25 | 0.2415 |
| RPCA | | 90 | 10.5 | 4.5 | 0.3127 |
| | | 95 | 4.75 | 9.5 | 0.3127 |
| | | 99 | 1.5 | 15.25 | 0.3127 |
| NPCA-ACL | | 90 | 0.50 | 0 | 0.2467 |
| | | 95 | 0.25 | 0 | 0.2467 |
| | | 99 | 0 | 0.75 | 0.2467 |

of PC and how to set control limit are important for fault detection based on PCA. In the previous research, different PCs can be selected to detect different faults, or the first few PCs can be used to detect the whole system. The control limit can be set according to the monitoring requirements, such as dynamic control limit, adaptive control limit and so on. PCA can also be very useful in fault diagnosis by reducing the dimension of the samples and extract representative features, which improves the accuracy and efficiency of diagnosis. The number of PCs is key point, and sometimes is decided by different systems and requirements.

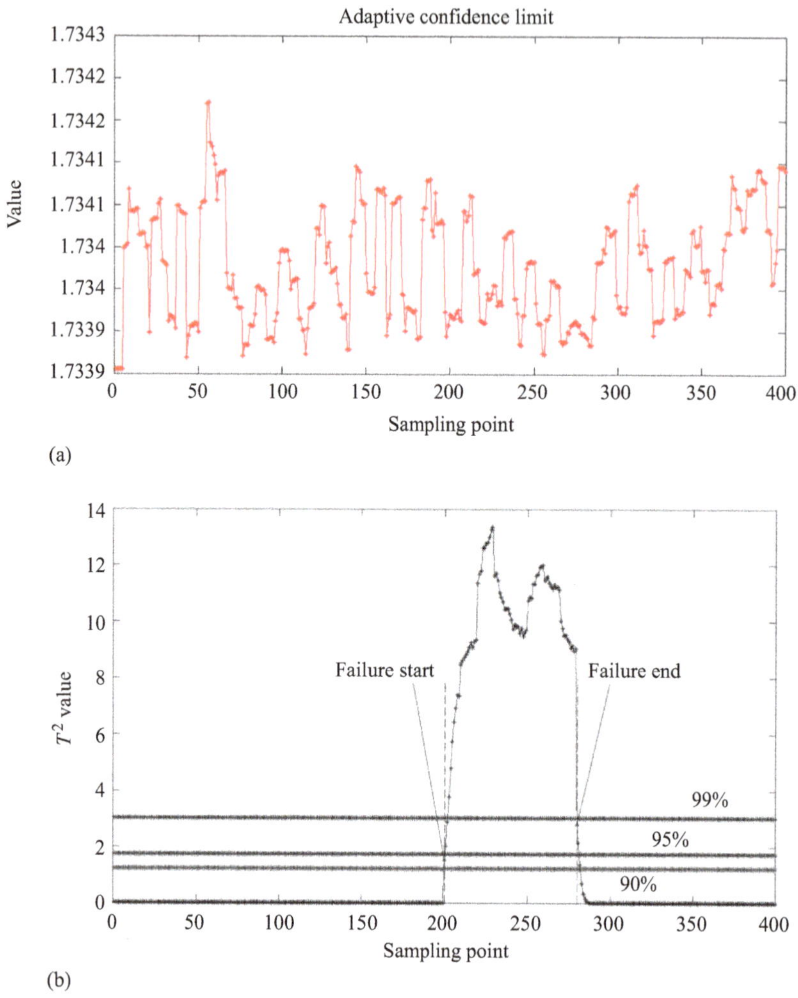


*Figure 5.38 Fault detection results by NPCA-ACL [27]. (a) ACL of 95%. (b) Detection result*

When the data are fitted Rotundity Scatter distribution, it is difficult to get representative PCs. RPCA is introduced in this chapter. The key point of RPCA is to select the operator of Relative Transform $M$. The optimal $M$ can reduce PCs; that is to say, less PCs can delicate the original data. So, RPCA is useful in fault detection and fault diagnosis, and some new control limits and feature representation methods were proposed to make the RPCA more be useful.

When PCA is used, mostly the data were required to be approximately normal. According to periodic nonsteady conditions' operating systems, NPCA method is introduced in this chapter. In this context, LS helps the normal data transformed from non-Gaussian distribution into standard Gaussian distribution, which satisfies the pre-condition that $T^2$ chart can effectively detect faults. In addition, the multi-period PCA algorithm and the fault section determination procedure is used to set up the adaptive statistic confidence limit to solve the problem of signal mutations, and also to achieve de-noising and dimensionality reduction. There are several advantages in the NPCA-ACL strategy as follows: (1) The ACL is set up to detect the signal mutation in real time. (2) De-noising and dimensionality reduction are achieved. (3) Non-Gaussian distribution data are allowed to be detected. (4) The robust performance is improved specially under Gaussian white noise. (5) It is sensitive tiny faults. (6) The monitoring sensitivity is adjusted by regulating $\omega$ to achieve the best for missing and false alarms according to user's requirement. The experiments results have clearly shown that the NPCA-ACL strategy is useful for fault detection under periodic nonsteady conditions, and can be used in detecting tiny faults.

Prospective investigations regarding PCA should be as follows: (1) Data standardization for dimensional unification: Indeed, PCA is only applicable to Gaussian data, while most actual data are non-Gaussian. As existing data standardization methods cannot transform non-Gaussian normal data into Gaussian ones, new data standardization methods need therefore to be targeted. (2) Selection of retained PCs for feature extraction: Most of the classical methods, such as cumulative percent variance, cross-validation and variance of reconstruction error, just consider normal operational data and select the first PCs with large variance; however, while PCs with larger variance of normal data cannot guarantee online capture of the largest variations in fault data. (3) Selection of statistics for fault detection: Traditional Hotelling's $T^2$ and $Q$ statistics will produce large false or missed detections, resulting in a decrease in fault detection accuracy. It is necessary to select new statistics that are sensitive to faults while being insensitive to system disturbances.

## References


[1] Odiowei P.E.P. and Cao Y. "Nonlinear dynamic process monitoring using canonical variate analysis and kernel density estimations." *IEEE Transactions on Industrial Informatics*. 2010;6(1):36–45.

[2] Yin S., Ding S.X. and Zhou D. "Diagnosis and prognosis for complicated industrial systems—Part I." *IEEE Transactions on Industrial Electronics*. 2016;63(4):2501–2505.

[3] Yin S., Ding S.X. and Zhou D. "Diagnosis and prognosis for complicated industrial systems—Part II." *IEEE Transactions on Industrial Electronics*. 3016;63(5):3201–3204.

[4] Prieto M.D., Cirrincione G., Espinosa A.G. and Ortega J.A. "Bearing fault detection by a novel condition-monitoring scheme based on statistical-time features and neural networks." *IEEE Transactions on Industrial Electronics*. 2013;30(8):3398–3407.

[5] Geiger B.C. and Kubin G. "Relative information loss in the PCA." *Information Theory Workshop (ITW)*; Lausanne, Switzerland, Sep 2012. New York: IEEE; 2012. pp. 562–566.

[6] Mukherjee A. and Sengupta A. "Estimating the probability density function of a nonstationary non-Gaussian noise." *IEEE Transactions on Industrial Electronics.* 2010;57(4):1429–1435.

[7] Tsai D., Wu S. and Chiu W. "Defect detection in solar modules using ICA basis images." *IEEE Transactions on Industrial Informatics.* 2013;9(1):122–131.

[8] Zhang S.M. and Zhao C.H. "Slow-feature-analysis-based batch process monitoring with comprehensive interpretation of operation condition deviation and dynamic anomaly." *IEEE Transactions on Industrial Electronics.* 2018;66(5):3773–3783.

[9] Sakellariou J.S. and Fassois S.D. "Vibration based fault detection and identification in an aircraft skeleton structure via a stochastic functional model based method." *Mechanical Systems and Signal Processing.* 2008;22(3):557–573.

[10] Puyati W. and Walairacht A. "Efficiency improvement for unconstrained face recognition by weightening probability values of modular PCA and wavelet PCA." *International Conference on Advanced Communication Technology*; Gangwon-Do, South Korea, Feb 2008. New York: IEEE; 2008. pp. 1449–1453.

[11] Nomikos P. and MacGregor J.F. "Monitoring batch processes using multi-way principal component analysis." *AIChE Journal.* 1994;40(8):1361–1375.

[12] Zhu X., Zhang Y. and Zhu Y. "Bearing performance degradation assessment based on the rough support vector data description." *Mechanical Systems & Signal Processing.* 2013;34(1–2):203–217.

[13] Xu H., Tang T.H., Wang T.Z., *et al.* "A PCA-mRVM fault diagnosis strategy and its application in CHMLIS." *IEEE IECON*; Dallas, TX, USA, Nov 2014. New York: IEEE; 2015. pp. 1124–1130.

[14] Rao P.S. and Ratnam C. "Health monitoring of welded structures using statistical process control." *Mechanical Systems & Signal Processing.* 2012;27(1):683–695.

[15] Wang T.Z., Xu H., Han J.G., *et al.* "Cascaded H-bridge multilevel inverter system fault diagnosis using a PCA and multi-class relevance vector machine approach." *IEEE Transactions on Power Electronics.* 2015;30(12):7006–7018.

[16] Dong J., Wang T.Z., Tang T.H., *et al.* "Application of a KPCA-KICA-HSSVM hybrid strategy in bearing fault detection." *Power Electronics & Motion Control Conference*; Hefei, China, May 2016. New York: IEEE; 2016. pp. 1–5.

[17] Harrou F., Nounou M.N. and Nounou H.N. "Enhanced monitoring using PCA-based GLR fault detection and multiscale filtering." *IEEE Symposium on Computational Intelligence in Control and Automation*; Singapore, Singapore, Apr 2013. New York: IEEE; 2013. pp. 1–8.

[18] Wang Y., Liu M., Bao Z., *et al.* "Stacked sparse autoencoder with PCA and SVM for data-based line trip fault diagnosis in power systems." *Neural Computing and Applications.* 2019;31:6719–6731.

[19] Zhang K., Peng K., Li G., *et al.* "New kernel independent and principal components analysis-based process monitoring approach with application to hot strip mill process." *IET Control Theory and Applications.* 2014;8(16): 1723–1731.

[20] Maroua S., Radhia F., Khaoula B.A., *et al.* "Decentralized fault detection and isolation using bond graph and PCA methods." *The International Journal of Advanced Manufacturing Technology*. 2018;99:517–529.

[21] Wang T.Z., Tang T.H., Wen C.L., *et al.* "Relative principal component analysis algorithm and its application in fault detection." *Journal of System Simulation.* 2007;19(13):2889–2894.

[22] Bartkowiak A. and Zimroz R. "Sparse PCA for gearbox diagnostics." *Federated Conference on Computer Science and Information Systems (FedCSIS)*; Szczecin, Poland, Sep 2011. New York: IEEE; 2011. pp. 25–31.

[23] Zhang M., Wang T.Z. and Tang T.H. "A multi-mode process monitoring method based on mode-correlation PCA for marine current turbine." *IEEE 11th International Symposium on Diagnostics for Electrical Machines, Power Electronics and Drives (SDEMPED)*; Tinos, Greece, Sep 2017. New York: IEEE; 2017. pp. 286–291.

[24] Wang T.Z., Xu M., Tang T.H. *et al.* "The normalization PCA model and its application in fault detection of wind power generation system." *IEEE International Symposium on Industrial Electronics*; Taipei, Taiwan, May 2013. New York: IEEE; 2013. pp. 1–6.

[25] Wang X., Kruger U. and Irwin G. "Process monitoring approach using fast moving window PCA." *Industrial and Engineering Chemistry Research.* 2005;44(15):5691–5702.

[26] Wang X., Kruger U. and Lennox B. "Recursive partial least squares algorithms for monitoring complex industrial process." *Control Engineering Practice.* 2003;11(6):613–632.

[27] Wang T.Z., Wu H., Ni M., *et al.* "An adaptive confidence limit for periodic non-steady conditions fault detection." *Mechanical Systems and Signal Processing.* 2016;72–73:328–345.

[28] Wang T.Z., Xu M. and Tang T.H. "The normalization PCA model and its application under the periodic non-steady conditions." *Control and Decision Conference (CCDC)*; Guiyang, China, May 2013. New York: IEEE; 2016. pp. 4313–4318.

[29] Wang T.Z., Gao D.J., Liu P. and Tang T.H. "A fault detection model based on dynamic limit under non-periodic non-steady conditions." *Shanghai Jiaotong University*. 2012;46(4):607–612.

[30] Wang T.Z., Xu M. and Tang T.H. "A fault detection method based on dynamic peak-valley limit under the non-steady conditions." *IECON 2013—39th Annual Conference of the IEEE Industrial Electronics Society*; Vienna, Austria, Nov 2013. New York: IEEE; 2014. pp. 7346–7351.

[31] Sennaroglu B. and Senvar O. "Performance comparison of Box-Cox transformation and weighted variance methods with Weibull distribution." *Journal of Aeronautics & Space Technologies*. 2015;8(2):49–55.

[32] Poekaew P. and Champrasert P. "Adaptive-PCA: an event-based data aggregation using principal component analysis for WSNs." *International Conference on Smart Sensors and Application*; Kuala Lumpur, Malaysia, May 2015. New York: IEEE; 2015. pp. 50–55.

[33] Wang J., Hu Y. and Shi H.B. "Fault detection for batch processes based on Gaussian mixture model." *Zidonghua Xuebao/Acta Automatica Sinica*. 2015;41(5):899–905.

[34] Fan J.C., Wang Y.Q. and Qin S.J. "Combined indices for ICA and their applications to multivariate process fault diagnosis." *Acta Automatica Sinica*. 2014;39(5):494–501.

[35] Yao Y. and Gao F. "Batch process monitoring and fault diagnosis based on multi-time-scale dynamic PCA models." *IFAC International Symposium on Advanced Control of Chemical Processes*; 2009.

[36] Gao X. "On-line monitoring of batch process with multiway PCA/ICA." *Principal Component Analysis*. InTech, 2012.

[37] Xu H., Zhang J., Qi J., *et al.* "RPCA-SVM fault diagnosis strategy of cascaded H-bridge multilevel inverters." *International Conference on Green Energy*; Sfax, Tunisia, Mar 2014. New York: IEEE; 2014. pp. 164–169.

[38] Wang T.Z., Liu Y., Tang T.H. and Chen Y. "Dynamic data window fault detection method based on relative principal component analysis." *Transactions of China Electrotechnical Society*. 2013;28(1):142–148.

[39] Fuente M.J., Garcia-Alvarez D., Sainz-Palmero G.I. "Fault detection and identification method based on multivariate statistical techniques." *Emerging Technologies and Factory Automation*; Mallorca, Spain, Sep 2009. New York: IEEE; 2009. pp. 1–6.

[40] Wang T., Qi J., Xu H., *et al.* "Fault diagnosis method based on FFT-RPCA-SVM for cascaded-multilevel inverter." *ISA Transactions*. 2016;60: 156–163.

# Conclusion

*Mohamed Benbouzid*[1] *and Demba Diallo*[2]

In summary, this book has identified opportunities of some advanced signal processing techniques for electromechanical systems' fault detection and diagnosis. It has provided methodologies and algorithms with several illustrative examples and practical case studies, while highlighting some prospective investigations.

Further investigations are needed for the parametric spectral estimation techniques generalization. Indeed, these techniques must be adapted to transients in electrical machines and drives, and generator operation. This generalization will necessarily involve a model with a higher number of parameters. Therefore, a higher number of faster and more accurate optimization techniques will be required.

Regarding demodulation-based fault detection and diagnosis, both empirical mode decomposition (EMD) and ensemble empirical mode decomposition (EEMD) still lack theoretical background. Further investigations are therefore expected to address this critical issue. An alternative has been recently proposed, such as the variational mode decomposition in addition to the complete EEMD that solves the original signal exact reconstruction problem, and later provides better mode separations. These demodulation techniques should be better assessed and evaluated.

Regarding the high-order spectra application, it is straightforward to extend the analysis to fourth-order spectral moments, which are defined in three-dimensional frequency space. While such a step would enable the identification of cubic nonlinearities, the need to work in three-dimensional space will significantly increase the analytical and computational complexity. This issue must be addressed.

PCA enhancement requires further investigations addressing data standardization for dimensional unification. New data standardization methods need to be targeted to address the issue of non-Gaussian data transform into Gaussian ones. In addition, the selection of retained principal components for feature extraction is another issue to be further investigated to guarantee online capture of the most substantial variations in fault data. Moreover, to increase fault detection accuracy, new statistics sensitive

[1]Institut de Recherche Dupuy de Lôme, CNRS, University of Brest, Brest, France
[2]Group of Electrical Engineering Paris, CNRS, CentraleSupelec, University of Paris-Saclay, Gif/Yvette, France

to faults and insensitive to system disturbances, apart from traditional Hotelling's $T^2$ and $Q$ statistics, should be investigated.

In terms of automatic feature extraction and analysis, machine-learning techniques are currently highly promoted. Other techniques developed in the signal processing and telecommunication communities, such as source separation, optimization techniques or non-cooperative game theory are potential candidates.

# Index

www.ingramcontent.com/pod-product-compliance
Lightning Source LLC
LaVergne TN
LVHW051050210926
848486LV00001B/57